Heritage on Stage

Heritage on Stage

The Invention of Ethnic Place in America's Little Switzerland

Steven D. Hoelscher

The University of Wisconsin Press

The University of Wisconsin Press
2537 Daniels Street
Madison, Wisconsin 53718

3 Henrietta Street
London WC2E 8LU, England

Library of Congress Cataloging-in-Publication Data
Hoelscher, Steven D.
Heritage on stage: the invention of ethnic place in America's
Little Switzerland / Steven D. Hoelscher.
348 pp.   cm.
Includes bibliographical references and index.
ISBN 0-299-15950-7 (cloth: alk. paper).
ISBN 0-299-15954-X (paper: alk. paper).
1. Swiss Americans—Wisconsin—New Glarus—History.
2. Swiss Americans—Wisconsin—New Glarus—Ethnic identity.
3. Swiss Americans—Wisconsin—New Glarus—Social life and customs.
4. New Glarus (Wis.)—Social life and customs. 5. New Glarus (Wis.)—
Ethnic relations.   I. Title.
F589.N5H64   1998
977.5'86—dc21   98-2766

To Barb, Doug, and Kristen

In the crucible of the frontier the immigrants were Americanized, liberated, and fused into a mixed race.

—*Frederick Jackson Turner,*
"The Significance of the Frontier in American History," 1893

Throughout the years, descendants of the original settlers have carefully nurtured their Swiss customs, and today the village welcomes all visitors with the warmth of Old World charm and Swiss hospitality.

—*promotional brochure, New Glarus, Wisconsin, 1993*

Contents

Contents

Illustrations

Tables

Acknowledgments

No extended work can be written in isolation and it is a great pleasure to acknowledge the considerable support and intellectual sustenance from the many people who have seen a sketchy idea develop into a book. Most immediately, the sort of research involved in a case study such as this would not be possible without the interest, involvement, and participation of the community's members. From my first visit to New Glarus I received unfailing assistance and encouragement at every corner. Although it is impossible to acknowledge everyone personally here (a full list of informants is found in the bibliography), I do want to single out my considerable indebtedness to: Doris Arn, Marilyn and Howard Christensen, Pete Etter, Hans Lenzlinger, Chuck Phillipson, Elda Schiesser, and Millard Tschudy. On the other side of the Atlantic, I want to extend thanks to Kaspar Marti and Claudia Kock of Engi, Switzerland, for not only supplying me with important historical materials, but for sharing their views of American ethnic place. Also, I am indebted to Hans Rhyner of the Swiss Friends of New Glarus in Glarus, Switzerland, for his insights on contemporary Swiss travel to the United States.

Several institutions gave important financial assistance without which it would have been very difficult to do this work: the Swiss American Historical Society of Chicago; the University of Wisconsin Department of Geography; the State Historical Society of Wisconsin's Division of Historic Preservation; Louisiana State University, for a Manship Fellowship for Research in the Humanities and a Council on Research Summer Stipend; and the Department of Geography and Anthropology at LSU for assistance with photographic and indexing help.

I wish to thank the following individuals and institutions for providing me access to the rich photographic materials that illustrate this book and for permission to reproduce them here: the New Glarus Historical Society; the Wilhelm Tell Community Guild; the State Historical Society of Wisconsin; the University of Wisconsin Archives; National Geographic Society; the Immigration and Refugee Services of America; and once again Millard Tschudy and Elda Schiesser of New Glarus, Wisconsin. Additionally, my thanks go to Phil Zarrilli and Deborah Neff for providing me access to their considerable materials on the early years of the *Wilhelm Tell* drama, and to the University of Wisconsin Cartographic Laboratory for the production of figure 1.2. Portions of this book have appeared, in different form, in the *Journal of Cultural Geography, Wisconsin Preservation,* and *Ecumene.* I am grateful to the publishers and editors of each for permission to use that material here.

This book began at the University of Wisconsin, and I would like to acknowledge my gratitude to my mentors there. The experience of studying with so many talented, encouraging, and inspiring teachers has been a gift for which I will be always grateful. Within the Department of Geography, I want to thank Bob Ostergen for his unfailing support for a project that must have seemed rather odd at first and far removed from our shared interest in immigration history; I hope that he recognizes that this book, in many ways, is an extension of his own. I owe a special debt to Yi-Fu Tuan, who, through his own scholarship and during our many long conversations over coffee, continually demonstrates the real value of higher education; this book is unimaginable without his dedicated influence. Likewise, I am indebted to Bob Sack for his suggestion at an early stage of the power of place and the role of tourism in creating modern ethnic identity. Each read numerous drafts of this book and guided it through its important early stages. Both David Woodward and David Ward advanced my research through gainful employment and from our interesting discussions. To my teachers outside geography, I want to thank Paul Boyer and Allan Bogue for sharing their vast knowledge of American cultural and economic history, and Jack Kuglemass for introducing me to the methods, questions, and importance of folklore and performance. Jack Holzhueter of the State Historical Society of Wisconsin read an early draft of this book and offered valuable insights into the state's ethnic history. Among my many graduate student colleagues at Wisconsin who helped in innumerable ways, from attending a performance to suggesting a new line of thought, I owe a special thanks to Mike Barrett, Tim Bawden, Christian Brannstrom, Mike Castellon, the late Josh Hane, Tony Harkins, Eric Olmanson, Drew Ross, and Jeff Zimmerman.

More recently, LSU's Department of Geography and Anthropology has provided a nurturing environment for its younger faculty, and I have benefitted enormously from its warm, collegial atmosphere. Bill Davidson, Carville Earle, Jay Edwards, Joyce Jackson, Kent Mathewson, Miles Richardson, and Karen Till (now at the University of Minnesota) made helpful suggestions to oral presentations and drafts of several chapters.

Numerous colleagues outside of Madison and Baton Rouge have contributed significantly to this book with readings, suggestion of sources, and comments on presentations, including Regina Bendix, Christopher Boone, Kathleen Neils Conzen, Michael Conzen, Stephen Frenkel, John Hudson, Gerry Kearns, Anne Kelly Knowles, David Lowenthal, Donald Meinig, Don Mitchell, Timothy Oakes, Rudolph Vecoli, Terry Young, and Wilbur Zelinsky. In many ways, *Heritage on Stage* reflects my earliest interests in cultural geography and here I would like to thank Bob Douglas and Bob Moline at Gustavus Adolphus College and Ted Relph, Aidan McQuillan, Jock Galloway, and Tom McIlllwraith at the University of Toronto for initially triggering ideas of ethnic identity, modernity, and place.

At the University of Wisconsin Press, Mary Braun's unwavering confidence in this project and her skillful guidance of the manuscript through the lengthy publishing process have made writing this book a pleasure. I would also like to thank the manuscript reviewers for their critical and supportive comments.

Finally, I want to express my deepest thanks to my family for all their encouragement that has sustained me during this project. My parents, Douglas and Barbara Hoelscher, provided not only support at every stage along the way, but a wonderful place in Cable, Wisconsin, to write the final draft of this book. I owe a world of debt and inspiration to my wife, Kristen Nilsson, who ever reminds me of the powerful pull of ethnic memory and place.

Heritage on Stage

Prologue

A Place More Swiss than Switzerland

SCHWINGFEST

The September morning sky is overcast and this worries the hotel owner who has spent the past year preparing for today's *Schwingfest*. So many arrangements had to be made, not the least of which is housing the hundreds of foreign tourists and athletes who are straining the village's lodging infrastructure. There are many more visitors than hotel rooms and a "housing committee" is hastily formed to mete out lodgers to village residents who have an extra bed or two. These are not tourists to be sent to nearby Monroe for the night. Most have flown to Wisconsin from Switzerland specifically for the 1993 Schwingfest, or Swiss wrestling tournament, and they want to stay in "America's Little Switzerland."

The Swiss visitors begin arriving at 9:00 AM via yellow school buses that shuttle them from the village's main hotel (which is owned, not coincidentally, by the festival organizer) to the Wilhelm Tell Grounds. Located in a steep-walled valley outside New Glarus, Wisconsin, the festival site is striking. The thick grove that encloses the open meadow is broken only at the outer right boundary where a row of colorful Swiss cantonal flags adds reds, blues, yellows, and whites to the greens of the woods. Ordinarily used once a year for outdoor theatre, the Tell Grounds have been converted into an elaborate grassy wrestling stage, complete with spectator seating arranged four rows deep in a horseshoe configuration. In the middle lie two twelve-foot sawdust rings where the action takes place. At the opening of the horseshoe are the announcer's stand and the official scoring table. The *Schwinger* from Schaffhausen, Winterthur, Hettiswill and eighteen other Swiss communities stand near the announcer's platform and chat about last night's

3

festivities and their American landlords. A young line cook from Zürich who is interning at one of the village's restaurants knows the sport well and keeps track of his favorite Schwinger. Through the crowd he spots Silvio Rüfenacht, the reigning *Schwingerkönig*, or Swiss National Champion, among the competitors: "he's absolutely the best," the cook tells me. "I saw him win the national championships last year. He's very young and they say that he will be around for a long time."

Seventeen even younger looking men stand closer to the scoring table to the left. All but two are local truck drivers, students, and farmers who wrestled for the New Glarus High School team; the remaining pair are from California. The handful of Americans fidget nervously as they await taking on the Schwingerkönig and his Swiss entourage. At the far end of the field, on the edge of the forest, a large brown bull is tethered to a tree. Posing for photographs by the Leica-toting Swiss, later in the day he will be the prize of today's competition.

Two tall Swiss men enter the sawdust ring to the right. Each is wearing a collarless Swiss peasant shirt—one blue, the other red. Around their waists both wear short, heavy, canvas trunks supported by a thick leather belt over their loose pants. After shaking hands and sharing a joke with the referee, each grabs onto his opponent's belt with his right hand. With the left hand, they grab onto the others' rolled-up portion of the trunk on the other side. Then the wrestlers lean forward, bracing each other by interlocking heads and shoulders. Standing with bodies almost at right angles to their outstretched legs, concentration intensifies until, three or four seconds later, the match gets underway. The Schwinger maneuver about the ring in unison. As one hand must be continually on the opponent's belt or canvas trunks, the wrestlers appear to trace laboriously the steps of a complicated dance. Suddenly, one contestant hurls the other into the air and, falling down with him, sends sawdust flying (fig. P.1). It takes only seconds for the hurler to pin the hurlee. He then helps his defeated opponent to his feet and, following tradition, brushes off the coat of sawdust from the other's red shirt.

The skies lighten and so does the hotelier's mood. Dressed in traditional white chef's shirt and neck bandanna tied in the front and topped with a red "Schwingfest" baseball cap, he has been setting up the long banquet table near the end of the valley. As punctual as a Swiss train, the matches stop for the noon lunch break. Contestants and fans queue up in two long lines for bratwurst, chicken, and sirloin steaks. Large amounts of beer and wine are sold at the neighboring stand. Everyone then moves to the adjacent red and white striped circus tent. An electric buzz fills the air along with the heavy cigar and

Figure P.1. "All the best Schwinger are here." Although several local men participate, the Schwingfest features the top national wrestlers from Switzerland. (Photograph by author)

cigarette smoke. The conversations, mostly in *Schwyzerdütsch*, or Swiss German, gradually become louder and more boisterous. A local community singing group finds a small stage at one end of the tent and is joined by a number of the foreign visitors. Many in the crowd add their voices to the well-known patriotic and nostalgic *lieder*. Between songs a retired chemical worker from Basel reflects on the irony before him: "You don't get much of a chance to get this sort of entertainment back home, unless it's for tourists. Here, this place seems more Swiss than Switzerland."

Nearly two hours later everyone returns to the wrestling rings without haste. Before the matches resume, the Schwinger and many of the audience members compete in a side event: *Steinstossen*, or stone tossing. A hefty rock, gleaned from a local river bed, is hurled toward the sawdust rings (fig. P.2). More laughter than awe ensues when many of the older fans, themselves Schwinger of an earlier era, finish their cigarettes, hand their drinks to a friend, and proceed to heave the over-weight shot.

The competition draws to a close. Some of the visitors climb aboard the school buses that will whisk them back to New Glarus for a short

Figure P.2. Steinstossen at Schwingfest. The Wilhelm Tell Grounds are converted to the site of a Swiss style—and Swiss attended—*Schwingfest,* or wrestling tournament. Here, participants heave a local river stone in a competition known as *Steinstossen,* or stone tossing. (Photograph by author)

rest before returning to the Tell Grounds for more food and entertainment later that evening. Some remain in the meadow and swap stories with new friends. The hotelier senses that the day has gone well.

WILHELM TELL

On a hot Sunday afternoon two years later, the festival grounds fill again. Unlike the Schwingfest, today's event makes use of the valley for its intended use: another annual performance (the fifty-sixth, in fact) of Friedrich von Schiller's *Wilhelm Tell.* Audience members stretch out in lawn chairs and blankets placed on the hillside or sit in folding chairs centered in neat rows in front of the area where most of the action takes place. Today's *Wilhelm Tell* performance will be different from those of the past fifty some years as it marks the 150th anniversary of the Swiss-settled village and, as with the Schwingfest, the crowd is comprised mostly of tourists from Switzerland. Many muse over the program, refreshing their memories of the plot line to Schiller's *Wilhelm Tell play;* others read through the four-page cast of characters, recognizing the names they share with people they do not personally know. Though some have seen the famous play in Switzerland, and all have read its patriotic lines in school, for most this is the first time that they will see the national drama; the tourists have come to America to find Switzerland.

Meanwhile the performers, directors, and crew have taken a different path through the woods and reach the other side of the opening. Here, women in cowboy hats and boots tend the dozen horses in the shaded corral. Nearly twenty boys, their Little League baseball season having just come to an end, are being dressed in ruffles and black robes; like butterflies they will soon metamorphose into choirboys. Other children and adults make their way to the large, two-roomed dressing cottage, painted green to blend into the oak woods. There— for a day—an art teacher becomes a thirteenth-century peasant, a dairy farmer is transformed into a noble, a butcher into a priest, a school district superintendent becomes an Austrian villain, a graphic designer becomes an heiress, and a ski shop businessman is recast as a soldier. Costumes vary from the very simple to the ornate and are sized up and fitted by three expert seamstresses. Once in costume, the more than one hundred cast members anxiously pace along the winding, narrow "backstage" path looking for a chat with a neighbor, a peek through the woods to see the growing audience, or just a quiet space to silently recite lines for one last time.

Over the row of black speaker boxes, perfectly accented High German is heard welcoming the audience to this year's production: "Guten Tag und grüezi alle mit einander. Herzliche Willkommen zur Sechsundfünfzigste Aufführung von Schillers *Wilhelm Tell* Spiel. Wir hoffen daß Sie das Spiel geniessen . . . " Soon Rossini's *William Tell Overture* signals that the play is about to begin. The tone softens when three alpenhorn players quietly emerge from the forest and play a pensive, ancient-sounding melody. Their call brings forth the entire cast from the far end of the meadow. A lengthy promenade of goats, sheep, and massive Swiss Brown cattle are driven by men and boys (fig. P.3). The procession then meets up with women and girls waiting near center stage. Friends, families, and neighbors greet each other with a *grüezi* or talk about last night's Milwaukee Brewers game. None of this can be heard by the audience as it witnesses a panorama resembling an accurate transhumance parade. The "mob scene" concludes with yodeling and singing.

Under the hot August sun of this area's hottest summer on record, a retired village lawyer named Ed Willi prepares for his final performance as Baron von Attinghausen. The Attinghausen character has been Willi's for exactly forty-eight years and he knows the part as well as one should after such a lengthy, uninterrupted run (fig. P.4). Willi, a

Figure P.3. Opening scene of *Wilhelm Tell*. Schiller's drama begins every year with a traditional transhumance, complete with Swiss Brown cows from the nearby Voegli farm. (Photograph by author)

Swiss American who moved to New Glarus fresh out of law school, has a reputation for being, as one cast member put it, "a wild man" who enjoys the homemade wine that flows in abundance backstage. This year, however, he is quiet and subdued as he asks two friends to replace the wine in his stage goblet with soda to aid his growing thirst. For the last two weeks Willi has been confined to a hospital room where he received daily blood transfusions for the internal bleeding that was clearly not getting better. There is considerable concern for his health on the part of the *Tell* directors, who called him on the day before the performance to make sure that he knew he didn't have to go on stage. Alternatively, a prompter could stand behind and read his lines for him. His reply to such concerns was typical: "Goddamn it, there is no way in hell that I am not going out there. Just get me there."

Willi does make it to the grounds, leaving the hospital to do so, but looks ashen and sickly during the entire performance. His two scenes are given with considerable emotion. In the opening of the second act, first scene he tells Rudenz that he is dying: "Und so, in enger stets und engerm Kreis, Beweg' ich mich dem engesten und letzten, wo alles Leben still steht, langsam zu. Mein Schatten bin ich nur, bald nur mein Name."[1] At this point the tears that had been gradually welling in the

8

Figure P.4. "Mein Schatten bin ich nur, bald nur mein Name." "I am only a shadow of my former self, and soon only my name will remain." For nearly fifty years Ed Willi played the role of Attinghausen. He is shown here in an undated photograph (circa 1970), on his deathbed as Gertrude (Doris Arn), and Walter Fuerst (Ernest Thierstein) look on. (Courtesy of New Glarus Historical Society)

corner of his eyes streamed down the dying Attinghausen's face. His second scene comes nearly two hours later and is known as the Attinghausen death scene. There is no joking or talking on his part as in years past. Playing the part of a monk, I carry him out into the ninety-seven-degree heat and stand by his bedside while he recites his farewell lines to the audience and fellow *Tell* players for one last time. Willi plays the part well, looking like a shadow of his former self, and, like the character he plays, dies eight days later.

PART I

Place, Memory, and Ethnicity: Theoretical and Historical Perspectives, 1890–1915

1

Invented Ethnic Places

Since its founding 150 years ago, New Glarus, Wisconsin, has come to be known throughout the country, and across the Atlantic, as "America's Little Switzerland." Events like the annual Wilhelm Tell Festival and the recently revived Schwingfest[1] have led the Swiss Consul General Friedrich Vogel to remark that even "in its homeland," the community is "known to maintain Swiss ways and Swiss traditions—it is a reminder that Switzerland is living beyond the Atlantic" (fig. 1.1). One Swiss textbook on the United States notes that the community is "more Swiss than Switzerland," and a regional travel magazine calls New Glarus "a tidy little town that wears its Swiss heritage proudly." As its village board president put it when he welcomed visitors to the fiftieth anniversary performance of *Wilhelm Tell*, "The people of our village are dedicated to the preservation of their Swiss heritage and are proud of their achievement."[2]

New Glarus is hardly exceptional in boasting its ethnic heritage. In 1987 Americans staged over three thousand folk festivals, many of which celebrated ethnic culture.[3] In Wisconsin alone, well over seventy ethnic heritage festivals annually commemorate the state's diverse multicultural origins; every group, from Native Americans to Swedish, Croatian, and African Americans, finds a festive outlet each year to express its unique culture.[4] If we restrict our sights only to Euro-American groups, a map of ethnic heritage festivals across the American heartland reveals an archipelago of cultural displays that corresponds well with the region of heaviest nineteenth-century European immigration (fig. 1.2).[5]

Curiously, these celebrations are occurring one hundred years after Frederick Jackson Turner's famous proclamation that "in the crucible

13

Figure 1.1. Volksfest, 1993. Before a painted mountain scene at a New Glarus *Volksfest*, the Swiss Consul General recalls for his audience that the community "is a reminder that Switzerland is living beyond the Atlantic." (Photograph by author)

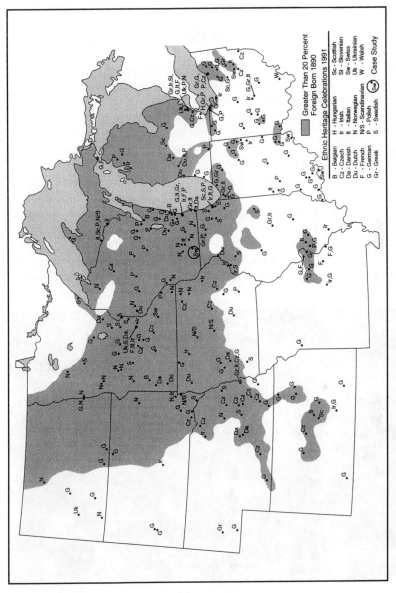

Figure 1.2. Old World Heritage Archipelago. (Source: *Census of the United States, 1890*: Official Calendar of Events for Each Midwestern State, 1991. Map is derived, with modifications, and reprinted with permission, from Hoelscher and Ostergren, *Journal of Cultural Geography*, v.13, 1993).

of the frontier immigrants were Americanized, liberated, and fused into a mixed race, English in neither nationality nor characteristics." Setting the tone for both popular conceptions of, and scholarly research into, immigration and ethnicity for the next seventy years, Turner's thesis contained little room for the ethnic celebrations, museums, monuments, and rituals that have played a role in many individual and community lives since the earliest days of settlement. That role has fluctuated with the changing currents of pluralism and assimilation in American culture, but as the Swiss of New Glarus would surely attest, the connection to an ethnic heritage is far from dead—Ed Willi's near half-century dedication to the *Wilhelm Tell* drama evokes a passion and commitment to "roots," one that is on the rise the world over. At its core, then, this book pursues the paradox that Turner unwittingly invoked by assuming ethnicity's frontier death a century before one small Wisconsin community's recurrent invitation to "discover Switzerland in America" (fig. 1.3).[6]

New ethnic festivals and parades are being organized, heritage plays performed, and landscapes built that reflect the hunger for memory among Americans often considered to be "in the twilight of ethnicity."[7] Across the country, dynamic processes of traditionalization and invention are visibly reconstructing many communities into *recog-*

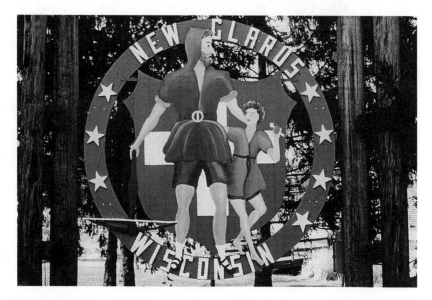

Figure 1.3. "Welcome to America's Little Switzerland." New Glarus, Wisconsin, conspicuously constructs its public memory for visitors and residents alike. (Photograph by author)

16

nizably ethnic places: Leavenworth, Washington and Helen, Georgia are instantly "Bavarianized" by adding half-timbered houses to their Main Streets; descendants of Dutch immigrants in Pella, Iowa enact zoning ordinances restricting new, "non-Dutch" architecture; Solvang, California stands out along the drive from San Francisco and Los Angeles as a quaint "Danish" town; and Louisiana Cajuns stage traditional music festivals with one eye to tourists from New Orleans and Chicago.[8] The striking extent of this conspicuous cultural construction has impressed many observers, leading one to comment that: "Remarkable now is the breadth of ethnic reawakening among countless small communities across rural and small town America as the depredations of decades of dilution and neglect are countered with campaigns to reinvigorate, at least commercially, the ethnic presence on the American landscape."[9] Thus, Michael Conzen notes, "ethnicity has become fashionable" in late twentieth-century America, a point seconded by David Harvey. For Harvey, the revival of vernacular traditions associated with places—of which ethnic heritage is exemplary—is a central component of our postmodern condition.[10]

More than mere fashion, today's exploding interest in ethnic heritage and its conspicuous or self-conscious display represent fundamental shifts in what it means to be ethnic, and thus American. Ethnicity's conspicuous construction joins with its consumption to provide the basis for what Herbert Gans calls the "symbolic identification" of white ethnics. Rejecting the notion that an "ethnic revival" has taken place, Gans postulates that the "roots phenomenon" of the 1970s demonstrates the radically altered content of Euro-American ethnicity. Instead of maintaining their ethnic organizations and group culture, people of third and later generations are more concerned with maintaining a sense of ethnic *identity* and are discovering new ways of expressing that ethnicity in suitable ways that bear little or no social cost: taking part in festivals, eating ethnic foods, and visiting places and museums associated with one's ethnic past.[11]

Mary Waters detects a paradox in this individualistic and voluntary form of ethnicity when she finds that "in spite of all the ways in which it does not matter, people cling tenaciously to their ethnic identities: they value having an ethnicity and make sure their children know 'where they come from.'"[12] The "where" in this case is critical, for I argue that with increasing suburbanization and the destruction of traditional place-based communities, an ever greater demand exists for places conspicuously constructed to impart an ethnic identity—for invented ethnic places. In part a contradictory quest for an antimodernist community and for individuality on the one hand, and in part an easy erasure of the differences between their own situation and that of Americans with non-European identities on the other, today's white ethnics join a global chorus in their celebration of place-based heritage. And the public display of heritage—the willful manipulation of the past for present causes—has become a worldwide industry of staggering dimensions.[13]

This book focuses on a rather under-studied group. Later generation white ethnics, neither marginalized today nor "definitive" Americans through most of their history, nevertheless represent a sizable portion of the American population. After all, 89 percent of white respondents to the 1990 census claimed an ethnic affiliation.[14] Yet, scholarship, with a few notable exceptions, has concentrated almost solely on the immigration phase through the first generation or offers a snapshot of contemporary "unmeltable ethnics."[15] This neglect of historical research is regrettable, for there is ample reason to suggest that the continual invention of tradition among later generation white ethnics is part of a much larger movement and one important consequence of modernity: the assertion of local ethnic identity within a globalized context of shrinking distances. Or, to put it somewhat differently, the advent of Swiss nationals traveling to a "traditional" Schwingfest in rural Wisconsin speaks to the twin forces of modernity: a world torn between globalization and traditionalization, between "McWorld" and Jihad.[16]

The historical reasons for such an upsurge of conspicuous ethnic cultural displays—and for their critical relationship to American culture more generally—are diverse, complex, and serve as this book's principal point of departure. It seeks to strip away the naive and earnest charm of public events like New Glarus' *Wilhelm Tell* play and Schwingfest and unearth their broader significance for the multicultural history of America. This is a history that, as David Hollinger has argued so persuasively, has reached an impasse. Academics have locked theoretical horns over the relevant costs and benefits of cultural diversity and the need for a unified culture.[17] Meanwhile, behind the battle lines of the dramatic culture wars, ordinary people of all socioeconomic classes are quietly performing their heritage, and often at a profit. This book participates in an ongoing dialogue over American identity and ethnic affiliation from a unique perspective: the willful and conscious manipulation of ethnic heritage—in short, its conspicuous construction.

The ever regenerating interest in, and artful public display of, their respective ethnic heritages among Euro-American ethnics raises a host of interesting questions. Where, we might ask, does that interest originate? When a fourth-generation person of Norwegian descent tries to be "Norwegian," how does she derive her notions of what being Norwegian means? How is that "Norwegianness" performed, and precisely by whom? In what ways have the symbols, rituals, and performances of an ethnic identity changed over time? What role does individual agency play in modern ethnogenesis? How is ethnic identity commodified and reworked by the forces of consumer capitalism? What is the relationship between ethnic identity at the level of the community and that of the state and what sort of places and landscapes are created in the wake of this relationship? How do these places and landscapes help shape ethnic identity? Or, to summarize: why and how are ethnic places "invented?"

INVENTING ETHNICITY AND TRADITIONS: SELLING MEMORY AND PLACE

The difficulty in answering these questions stems, in part, from the very ordinary nature of the topic. Heritage displays, though nonordinary events set apart from everyday life, have become so common as to appear "natural" or a given part of modern America. The tradition of celebrating, say, St. Patrick's Day seems as fixed and timeless as George Washington's chiseled granite face on Mount Rushmore. And yet, as Timothy J. Meagher's study of the Irish tradition in Worcester, Massachusetts demonstrates, such festivals are complex affairs that serve diverse political interests and are contingent on both influential individuals and larger societal forces. Neither timeless nor unchanging, ethnic traditions emerge, mutate, vanish, and surface again with ever increasing rapidity.[18]

Equally relevant seems to be an inherent scholarly bias against landscapes and cultural performances tainted with industrial, and especially consumer, capitalism in favor of premodern, "pioneering" societies; or against "fakelore" in favor of folklore.[19] "True" ethnic place would appear to die shortly after immigration, and its successor to be "ersatz" and "inauthentic." Thus, when places like New Glarus are described, they are usually seen as exquisite examples of "hyperreality," as instances—Umberto Eco breezily assures us—where "the American imagination demands the real thing, and, to attain it, must fabricate the absolute fake."[20] An important study on historic preservation codified this view as it proclaimed that such conspicuously constructed places themselves pose serious threats to "authentic" material culture.[21] This criticism of conspicuous architectural (as well as more generally cultural) construction generally relies on a vision of identity

19

as something ancient, unchanging, and inherent in a group's misty past. When change does occur, it would seem to lead down the path toward inevitable assimilation.

This book takes a very different starting point, a position that forces us to reexamine such keywords as *ethnicity, tradition, public memory,* and *selling place.* I begin from the viewpoint that *ethnicity,* for Euro-Americans, is neither a distant memory from the past nor an unchanging and persisting cultural form, but rather a cultural invention. Taking its inspiration from the original work by Werner Sollors, this view sees ethnicity not as primordial or biological—something fixed and given like the color of one's eyes—but as a cultural construction accomplished over historical time that must be continually reinterpreted.[22] In modern settings, ethnic groups persistently create and re-create themselves in response to, first, an ever changing set of social, economic, and political pressures, and second, information about those pressures. This reflexivity of modern life means that the expressive symbols of ethnicity (traditions such as a Schwingfest or *Wilhelm Tell* play) together with the boundaries between groups are endlessly reinterpreted. In other words, the concept of invention historicizes a once static phenomenon and "allows for the appearance, metamorphosis, disappearance, and reappearance of ethnicities."[23] The processes themselves whereby an ethnic group may gain or lose salience are contingent on a host of influences (in part structural and in part due to individual agency) and thus open for investigation.

Michael Fischer prefigures this direction in ethnic research when he notes that "ethnicity is something reinvented and reinterpreted in each generation by each individual and that is often something quite puzzling to the individual, something over which he or she lacks control."[24] That ethnic affiliation is not biological or primordial, but a cultural construct, may be surprising to those who cherish ties of descent. After all, widely held cultural definitions of race and ethnicity take such rigid categories for granted as unchanging and inherited. One is, say, Swiss or Italian, because one's great-grandparents immigrated from Switzerland or Italy. Yet, such taken-for-granted views neglect the flux and choice inherent in modern ethnic affiliation among favored groups. Of course, such choice is not an option for blacks, Hispanics, and many descendants from Southeast Asia, an inconvenient fact that makes the symbolic ethnicity of whites all the more tenuous. Constraint, rather than liberty, persistently defines the experiences of nonwhite, non-European groups, a critical point demonstrated in David Roediger's influential comparison between nineteenth-century Irish immigrants and free blacks.[25] Nevertheless, the tensions and contradictions in celebrating a past so self-consciously created can tell us much about a diverse, complex, and multicultural America.

An equally constructed—and misunderstood—term, *tradition* has been subject to the same kind of deconstruction that makes us reevaluate ethnicity. Despite a lengthy Western history that sees tradition as the stable transmission of inherited cultural traits, recent research is unanimous in regarding tradition as a process that involves continual re-creation. Traditions are "born, nurtured, and grow." Particularly fertile ground for the "invention of tradition" are times of rapid social transformation when old patterns and social formations break down under the uncertainty and chaos of threatening change. With the critical confluence of capitalism and the nation-state, the nineteenth century in particular is seen frequently as a period of mass-producing traditions.[26]

But, as Anthony Giddens notes, all traditions are invented, and the search for "real" traditions versus "invented" traditions follows a circular path.[27] More important (and considerably more interesting) is to consider the instrumental nature of traditions—the use to which they are put and the services to which they are mobilized. Not merely a referent to the past, traditions are an "aspect of contemporary social and cultural organization," Raymond Williams concluded twenty years ago. "It is a version of the past which is intended to connect with and ratify the present."[28] Tradition, rather than simply "the surviving past," is an actively shaping force put to use for a wide array of purposes, primary of which is establishing historical legitimacy for a group, class, or nation-state. Never an arbitrary inheritance, traditions are a selective version of the past which, like all symbols, carry different messages and can be read for multiple meanings.

Even such a "post-traditional" society as ours has need for tradition, for its guardians frequently call upon traditions to generate and regenerate identity and *public memory*. This is no easy task, for traditions (as with ethnic identity) under conditions of modernity offer multivocal readings that are frequently contested.[29] In his *Remarking America*, John Bodnar makes clear that different people will have contrasting views about what constitutes shared tradition and memory. Public memory—the body of beliefs and ideas about the past that help a society or group make sense of its past, present, and future—is focused inevitably on concerns of the present. Those who sustain a public memory often mobilize it for partisan purposes, commercialize it for the sake of tourism, or invoke it as a way to resist change. Tradition, it may be said, is the organizing medium of public or collective memory. As traditions are regularly invented, so will we all have highly selective memories of our shared past. The resulting public memory is a product of that selectivity.[30]

Bodnar's argument on public memory hinges on the divisions between "official culture" and "vernacular culture," a model that this book employs as well.[31] Official culture originates in the concerns of cultural leaders and authorities at all levels of society and promotes an interpretation of the past that reduces the threat of competing interests. Official culture tends toward the dogmatic as it presents an ideal or abstract version of public memory uncluttered by complex or ambiguous terms. Vernacular culture, conversely, "represents an array of specialized interests that are grounded in parts of the whole." First-hand experience in small-scale communities forms the basis for the defenders of this culture which, by its very existence, threatens the sacred and eternal character of official expressions. The success of cultural elites in garnering mass support frequently lies in their ability to appropriate the metaphors, signs, and rituals originally put forth by vernacular culture. Equally important, and frequently underrecognized, is the use of official culture by ordinary people for their own agendas. Ever unstable, this dialectic of public memory is inevitably susceptible to political manipulation and commercialization by both official and vernacular interests.

Indeed, any investigation on the changing role of tradition and memory in American culture must deal directly with the *selling* or *commodification of place*. Commodification—the process by which objects and activities come to be valued primarily in terms of their value in the marketplace and for their ability to signify an image—can take *place* as its starting point. Distinct places and the qualities they imbue, in other words, can be turned into commodities in their own right. Of course, virtually any kind of place can be "commoditized"—most commonly a parcel of land is bought or sold in a market economy—but what is of importance here is the selling or renting of the experience of being in *place*.[32] The most familiar way in which this most modern form of place commodification occurs is with tourist sites and especially those connected with "heritage." As Michael Kammen notes, ever since the 1920s and 1930s, and increasing in the 1960s, the "history of American traditions and 'heritage' has been increasingly intertwined with entrepreneurial opportunities in general and tourism in particular." What is true for traditions at all scales, from the national civic to local community, is equally true for ethnic traditions. Thus, Wilbur Zelinsky is surely correct when he attributes commercialization to the increasing presence of ethnicity on the contemporary American landscape.[33]

Perhaps most central to the relationship of place to commodification is the concept of authenticity. Dean MacCannell sees the tourist as the embodiment of the modern quest for authenticity, the contemporary equivalent of the pilgrim traveling to the spiritual center. As the search for what is real, anchored, and rooted, tourism is the contemporary means by which to overcome the modern condition of weightlessness and rootlessness.[34] Moreover, the problem of authenticity and its relation to power are greatly magnified in ethnic tourism. Not merely a provider of services, the "native" or "ethnic" is an integral component of the tourist spectacle. The ethnic becomes an object of the tourist gaze, an actor whose "quaint" and "different" behavior, dress, and artifacts are themselves significant attractions. But, when the tourist wants to see the authentic, unspoiled Bavarian town, French café, and Nigerian village, she unwittingly alters the very thing that she is there to see. Thus, the authenticity offered to tourists is "staged" and a sense of ethnicity "reconstructed." It should hardly surprise us that neither the ethnic place nor its inhabitants remain unchanged in such a politically charged environment. Or, in the words of David Glassberg, the story of "the invention of a collective sense of place, like the invention of a public history, is part and parcel of the struggle for cultural hegemony."[35]

CULTURAL DISPLAYS AND FESTIVE CULTURE IN PLACE

When tourists visit ethnic places, they expect to see symbols that clearly identify cultural identity. Indeed, a chief characteristic of tourism lies in the tendency of its participants to collect recognizable signs and stereotypes of a place visited.[36] The sorts of symbols most frequently collected include both objects (such as landscapes and museums) and events (such as festivals and performances). Expanding on the work of Regina Bendix to add a landscape element, I will refer to such public presentations of ethnic culture as "cultural displays."[37] A cultural display may be defined as a nonordinary, framed public event or object that requires participation on the part of a substantial group in a community in either its preparation, presentation, or performance. Such public displays are inevitably conspicuous manifestations of a group's festive culture. They are ultimately community affairs that, as Roger Abrahams would have it, "provide *the* occasion whereby a community may call attention to itself and, perhaps more important in our time, its willingness to display itself openly."[38]

Taken together, this constellation of landscapes, museums, festivals, and cultural performances add up to what I am calling the invention of ethnic place. Such a process is an invention in both the sense of the willful manipulation of the image of the place—its conspicuous

23

construction—but also as a way to emphasize the creativity and commitment necessary to develop and sustain cultural displays. It accounts for the growth of easily digested tourist stereotypes and for the resonance such views of ethnic culture may have for individuals. Such stereotypes—though often imposed from outside ethnic communities and at once idealized and romantic—are frequently reclaimed by ethnic groups to create a new niche in a rapidly changing world. Place, as the "geographic expression of the interaction between individual action and abstract historical process," is a unique perspective from which to examine the ongoing tensions, ambivalence, and contradictions inherent in the invention of ethnicity.[39]

Although one may take the tourists' perspective as a starting point, my concern is to visualize cultural displays as closely as possible from their creators' vantage point, while considering consumers of invented ethnic places second. Such cultural displays, whether dramatized in a festival or materialized in a museum, provide important models and mirrors for the invention of ethnicity. Don Handelman writes that "it is in various public occasions that cultural codes—usually diffused, attenuated, and submerged in the mundane order of things—lie closest to the behavioural surface" and are thus most easily read.[40] Moreover, the performative element involved in cultural displays is the crucial medium through which, as Paul Connerton puts it, societies remember. What we envision as our public or collective memory is established through performance in commemorative festivals, by the act of making a landscape, and in the process of reading a travel book. In each case, meaning is achieved through doing—by performance. Characterized by a higher than usual degree of reflexivity, "performances play an essential (and often essentializing) role in the mediation and creation of social communities." Whether those communities are organized around bonds of nationalism, ethnicity, class status, or gender, the study of performance—of *heritage on stage*—forces one to analyze more critically the agents of, and audience for, traditions' invention.[41]

Take one example: in her exploration of the relationship between ethnicity and nineteenth-century invented traditions among German Americans, Kathleen Neils Conzen concludes that cultural displays—the parades, celebrations, and monument and museum building that accompany the "special," set apart occasions of one's life—formed the basis for a distinctive German American ethnic identity. For Germans, Conzen writes, this "festive culture was more than a superficial embellishment of ethnic life. It was the medium within which ethnic identity resided, and was the vehicle with which Germans sought to change

24

the contours of American life."[42] The culturally derived ethnicity that Conzen finds in nineteenth-century Germans was thus a limited one; German Americans became politicized only when their culture became threatened. Yet, for German Americans at least, festive culture's "time out of time" created a real sense of ethnic identity and not simply a set of rhetorical boundaries enclosing a neutral, cultureless interest. Festival organizers and monument builders demanded actual commitment and group organization that Conzen, following Victor Turner, believes fostered a sense of *communitas* for those involved.[43]

But symbols can be read differently by divergent or competing groups. The official and vernacular interests that John Bodnar emphasizes will read the texts of cultural displays in contrasting ways. Thus, we can expect business leaders and parade organizers to commemorate an ethnic past differently than the ordinary people who chose to participate in, act indifferently toward, or protest against festive culture.[44] Contestation and communitas go hand-in-hand in the process of ethnic tradition building.

A few words are in order about the organization of this book. I wish to emphasize that New Glarus's history of ethnic place invention is perhaps unique in its intensity and scale, but surely not in its overall thrust. To make this point and to demonstrate the connections between vernacular and official culture, the level of analysis broadens and contracts by shifting the focus from the level of the community to the state and then back to the community. Significantly, state-level cultural productions beginning in the 1930s and accelerating in the 1940s gave legitimacy to, and borrowed from, the local efforts. Here, the official culture of state-sanctioned research projects, commemorative activities, and popular books becomes enmeshed with the vernacular culture of the local community and, in the process, legitimates those grassroots efforts. The mapping projects of rural sociologists at the University of Wisconsin and the books of popular writer Fred Holmes highlighted the vision of Wisconsin as the "Old World state," a development that places like New Glarus put to good use as ethnic tourism becomes ever more important.

Heritage on Stage combines these customarily disparate geographic scales into one historical narrative. Three periods of ethnic place invention comprise the body of the text, deriving their structure from internal pressures in, and external pressures on, group identity. The first section traces the period from roughly 1890 to the eve of the First World War. Torn by tensions between increasing affluence and a rapidly receding immigrant past, the Swiss community staged elaborate commemorative events designed to minimize this tension. The place, in essence, bound together the increasingly modern community through the performance of festival time and space.

The second section pursues the commemoration of heritage at both the local and state levels during the period between the World Wars. It addresses the way in which ethnic place was transformed from seditious region during the first war into the root and substance of democracy during the second. Two chapters examine this process at the local level, while the third investigates two state-level projects that recognized the contributions of white, European-born ethnics to the American melting pot, and, in turn, made the efforts of local leaders more palatable—and profitable.

The book's third section more directly examines the commodification of that ethnic place heritage. This final period, from 1962 to the mid-1990s, is marked by the full-fledged development of ethnic tourism in the community and a corresponding heightened discourse of authenticity. What passes for "Swiss" in the earlier periods is questioned and new standards are set, all for the sake of consuming an ethnic place identity. I conclude this section, and the book, with a reading of New Glarus's 1995 sesquicentennial commemoration in which the themes that run the course of this study were replayed, performed, and made concrete.

CODA: THE SWISS "COLONY" OF NEW GLARUS, WISCONSIN

Well before there were chalets, schnitzel, and yodeling there were Swiss immigrants trying to eke out a living in southern Wisconsin. One of the central points to any understanding of New Glarus as an invented ethnic place is that its historic roots are as real as the soil the first immigrants purchased in 1845. Thus, unlike Leavenworth, Washington, or Kimberley, British Columbia—theme towns instantly "Germanized" with no connection to an ethnic past—New Glarus owes its earliest years to immigrants whose knowledge of English was only slightly better than their familiarity with American farming practices.

Settled as a transplanted "colony," the community received early recognition as an "unaltered Swiss colony." One columnist for the *Nation* wrote in 1883 of his astonishment that the Swiss of New Glarus "held fast their integrity, in race, language and customs, so largely and so long."[45] And, no less an authority than Frederick Jackson Turner himself held up the village as an example of how "the non-native element [in the U.S.] shows distinct tendencies to dwell in groups." Four years after his presenting his famous essay on the frontier he wrote that "One of the most striking illustrations of this fact is the community of

New Glarus, in Wisconsin, formed by a carefully organized migration from Glarus in Switzerland, aided by the canton itself. For some years this community was a miniature Swiss canton in social organization and customs."[46] Though, by the turn of the century, New Glarus had become, in Turner's view, "increasingly assimilated to the American type, and has left an impress by transforming the county in which it is [found] from a grain-raising to a dairy region," there could be no doubt as to the insularity, homogeneity, and reality of a close-knit and closely-guarded ethnic community. As in most such places, "the Wisconsin Switzerland" was marked by a spirit of "clannishness" and was held together by "the conservative force" of the church. Non-Swiss were seldom welcome during the earliest of years to such an extent that even at the turn of the century, "no Yankee now lives within a ring of six miles."[47]

The migration that created such an insular community was of a special type. Though similar to many transplantations from one overseas community to another, the explicit governmental planned colonization set this—and other Swiss emigrations—rather apart.[48] Chain migration played a role here, too, but the initial emigration took place under the auspices of the *Auswanderungsverein* (emigration society) of Canton Glarus. More than any other European country during the great trans-Atlantic migrations, Switzerland depended on highly organized emigration associations that sponsored the founding of "colonies." Potential immigrants joining such a colony were required to subscribe to statutes and abide by rules of social organization that often included sharing common property and collective decision making on important issues. Vevey and Tell City, Indiana, Bernstadt, Kentucky, and New Glarus, Wisconsin, all owe their beginings to such Auswanderungsvereinen.[49] Formed in 1844 after the failure of many earlier attempts at emigration, the Canton Glarus Auswanderungsverein was created explicitly for the purpose of organizing group emigration to North America.

In the early nineteenth century Canton Glarus ranked as "one of the most industrialized regions of the European continent" and by 1860, more than 220 textile factories lined the Glarnese valley employing nearly six thousand men, women, and children.[50] The hallmark of this early and intense industrialization was twofold: first came a decline in small-scale, family-owned workshops; second, with the increasingly proletarianization of the population, low wages, poor living accommodations and working conditions, long working hours, and child labor became the rule rather than the exception. The number of these new factory jobs fell far below those once available in the former protocapitalist system. After a bitter encounter with the capitalist reorganization of labor in the local textile industry, hundreds of workers found themselves out of work in the early 1840s, triggering the poverty and maladjustment that drove so many from Europe during this century, including those from Canton Glarus. Thus, in the spring of 1845, just under two hundred unemployed textile workers and their families set out for the United States. By August of that year, slightly more than half that original group finally reached the tract of land in southern Wisconsin that had been bought in the name of the Auswanderungsverein.[51]

These early immigrants and the hundreds who followed over the next several decades, therefore, were not expert dairy farmers, much less the cheesemakers who came to typify the Green County, Wisconsin, Swiss. Most did have nominal firsthand acquaintance with small scale cattle raising, but on the whole, the displaced weavers and craftspeople learned their agricultural skills in the New World. It is therefore not surprising that the Swiss, like their Yankee neighbors, took up wheat farming during the early years. The transition from a subsistence, wheat-based economy to dairying came during the 1870s, the "decade of the birth of the cheese industry." That decade witnessed not only the growth of dairy production and the establishment of the foreign-type cheese factories that brought considerable wealth to the community, but also a diversification of the Swiss population, as immigrants from other cantons, most notably Bern, flooded into the New Glarus core.[52] From that core slowly dispersing outward, a steady steam of immigrants created Wisconsin's "miniature Swiss canton," recognized by the nation's leading historian, Frederick Jackson Turner, who was also a firm believer in its ultimate demise.

2

"There Was a Confusion of the Foreign and American"
Public Memory at the Turn of the Twentieth Century

The notion that the small ethnic enclave possessed a past worthy of commemoration occurred to the Swiss colony shortly after establishing its New World base. To a most remarkable degree, that idea persists to this day. Annual commemorations of Swiss Independence, decennial celebrations of the settlement, and yearly reaffirmations to the church have remained significant components of New Glarus's public memory for 150 years. Such reflexive expressions of group connectedness were hardly unique to the southern Wisconsin Swiss. Indeed, scarcely can a community be found in the Midwest—be it settled by Germans, Swedes, Yankees, or a rich combination of groups—that did not set aside time from harvesting corn or managing its shipment to market for a celebration of heritage. New Glarus, in this respect, was typical of many ethnic communities across the Midwest; what was unusual, however, was the intensity and persistence of what one writer in 1905 called the Swiss community's "impulse to celebrate."[1]

By their very nature, such public displays "select out, concentrate, and interrelate themes of existence—lived and imagined—that are more diffused, dissipated and obscured in the everyday."[2] Dominant "themes of existence" at the turn of the century found their way into the public events as a way to grapple with the tensions of making the transition from unemployed immigrant proletariat to wealthy ethnic farmer. At times contradictory and, at other times, complementary, three core themes or symbols emerged that shaped the period's festival discourse: the homeland, the pioneer, and the notion of progress.

29

These crucial "themes of existence" emerged not from a mysterious tradition, but from the social and economic context of the late nineteenth century. As Victor Turner has pointed out, it is crucial to unlock the "field of meaning in which a celebratory object has its potential for arousing thought, emotion, and desire."[3] During this period, two important "fields of meaning" or sets of context aroused a considerable outpouring of commemorative activity. First, tremendous economic forces were transforming the region into the state's most productive agricultural region, giving rise to a most untraditional set of capitalist relations. Second, and in counterdistinction to the first, the lived memory of the pioneering generation was growing ever more dim, leading to concerns over the direction and pace of socioeconomic change. In rhetorical terms, a progressive view of history tangled with a growing antimodernism, the results of which concentrated nostalgic images of a rough hewn and innocent past into a future-oriented, dynamic present.

THE EARLIEST PUBLIC CELEBRATIONS

A Swiss Fourth of July

Less than a decade had passed before the impoverished colonists began commemorating the memory of their departed homeland. J. J. Tschudy, himself an immigrant from Canton Glarus in 1846 and the cantonal government's advisor to the colony, described the event in 1853 as nearly identical to "a day of festival kept in their father-land." Although a number of the oldest (and some of the youngest) settlers had already passed away, "those remaining [were] seen at this festive occasion . . . which served to remind them of their native land." Reaching back into the depths of Swiss—or, more accurately, Glarus—history, the day commemorated "the battle-day which delivered their ancestors from the tyranny of their oppressors on the 9th of April, 1388, when eight hundred men of Glarus defeated several thousand Austrians." That the date of the festive occasion was on July 4 and not April 9 mattered little to the colonists: the "American holiday" proved pliable enough to withstand the colonists' considerable reinterpretation.[4]

It was not only the rhetorical content that the Swiss reworked and appropriated for their own, however. The entire form of the festival departed radically from the customarily staid Anglo-American Fourth of July celebration. Nineteenth-century Germans, in particular, were highly critical of the absence of the festive in American life, the lack of inspiration and richness in public ritual. "The life of the American swings between the market and the church," complained one disgusted German immigrant in 1846, "doing business and praying are the highest moments of the modern republic. Public meetings are his only real festivals." The Fourth of July, in the hands of an Anglo-American elite, rarely measured up to the "idealistic and artistic" stamp of public celebrations to which immigrants were long accustomed in the homeland. Tired of the endless parades and speeches, one German writer complained of the American Fourth in 1856 that "Everything here becomes profane and common. The American people lacks [sic] naturalness, naïveté, humor, and in a country with a free public life, where every activity, effort, every viewpoint can express itself without hindrance, we nevertheless find no real *Volksleben*. The American cannot get enthusiastic about anything; he can't even enjoy himself—for which fact the Fourth of July is a regularly recurring testimony . . . this is a people that doesn't deserve its festival!"[5] This lack of a public culture expressive of the nation's soul, what the German writer described as *Volksleben* and what was so busily being mass-produced across Europe at the time, led to indifference on this side of the Atlantic. Indeed, no sooner had the holiday reached its climactic development in the 1830s than it began receiving all sorts of criticisms about its triteness and the tedium of its long-winded oratory.[6]

By contrast, the newly arrived Swiss celebrated the Fourth with fervor and in a manner to which they were long accustomed. The "naturalness, naïveté, and humor" so noticeably absent among Americans found its place on the Wisconsin frontier. Writing in her memoirs, a local Irish woman describes her experience at the New Glarus celebration. Though the family contemplated "staying home like sensible folk," they decided in the end to travel "the five miles to New Glarus." There they agreed, "we can look at the Swiss, if we can't understand them." And what the family saw on the Fourth amazed them: "New Glarus was celebrating the Fourth of July, but it was a *Swiss celebration*. Gessler was there and William Tell, to shoot the apple from his son's head. There were Swiss wrestlers and Swiss dances in the dining room of the hotel, where a Swiss music box with weights was wound up. In a side room were chairs and a long table, on which stood glasses and pitchers of beer. [There] the dancers repaired to rest and refresh themselves between dances. Round and round the couples would glide while at certain intervals in the music the men would stamp their feet and emit wild whoops."[7] The appearance of mythical figures from Switzerland's romantic past, combined with activities such as Swiss wrestling and dancing, the unusual music, and wild dance set the celebration apart from the sober American Independence Day events that the Irish family chose to avoid.

That the Swiss colonists chose to pattern their festival after a similar recurring event in Canton Glarus would have appeared natural to them in two senses. First, having arrived on the continent less than ten years earlier and living in considerable isolation, they would have been hardly acquainted with American nationalist folklore. It was enough simply to know that the Fourth symbolized a day that, as Tschudy put it, brought forth a "vivid remembrance of a day of festival" in Glarus that is "very similar to the American holiday."[8] Second, and more importantly, the Fourth has always opened up the possibility to express multiple viewpoints. In the nineteenth century, Fourth of July celebrations provided ethnic and working class groups a flexible resource to state their own values, to affirm their ethnic and religious autonomy, and, at times, to voice discontent. As Susan Davis and Roy Rosenzweig have shown in nineteenth-century Philadelphia and Worcester, respectively, the Fourth has long offered nonelites the chance to restate the nation's most important holiday in terms with which they are most comfortable. They also tended to express a vision of a better life—one less regimented and less restrictive—while simultaneously reaffirming their commitment to values of mutuality, community, and collectivity. Community, much to the dismay of the Anglo elite, often took the form of leisure and recreation that threatened the seriousness of the official, patriotic message.[9] In the case of the Swiss, leisure and a commitment to community took place in an arena over which they had control.

Kilbi

The 1853 Fourth of July Celebration bore considerably more resemblance to distinctive Swiss *Kilbi* festivals than to anything remotely "American." Kilbi, a Swiss word for the German *Kirchweihe*, literally means "church hallowing" or "church festival." Annually celebrated in New Glarus since 1850, and formally organized by the Swiss Evangelical and Reformed Church, Kilbi persists to this day. But the form and content of the festival have changed dramatically.

Today the festival is focused entirely on the church and normally consists of a Saturday evening banquet dinner followed by Sunday services. The central event of the Sunday service is a roll call of members who have taken confirmation vows, from the eldest to the youngest, and who recommit themselves to the church. In its more recent renditions, Kilbi has served as a time of homecoming and remembrance for the religious members of the community. Today, Kilbi is largely a religious ceremony, tightly focused on the members of the Swiss Reformed Church rather than on surrounding communities. It is a quiet affair, or, as the local paper put it some years ago, "sedate and abbreviated," with all notions of frivolity and revelry effectively purged.[10] Early in the life of the New Glarner community, however, the secular blended with the religious as Kilbi became the most important annually recurring festive event in the calendar of the entire community, and maintained this role until roughly the late 1920s.[11]

With Kilbi's early history's close church ties, it received official sanction as *the* special church festival in the Swiss Church Constitution of 1859. Noted in the same section as baptisms and weddings, "Consecration of the Church Day, or Kilbi Festival," the constitution read, "shall be on the last Sunday in September, annually."[12] That the Constitution spoke of Kilbi as a "festival" betrays its nineteenth-century function as a spatial and temporal node of celebration that united the sacred and secular. Roger Abrahams notes that this classic distinction (dating back to the sociological work of Emile Durkheim) between the sacred and profane, between "seriousness and high fun," were often juxtaposed and fused. This blend then became a means of attaining community through celebration. But as societies become tied to the marketplace, the two tend to become dissociated. Kilbi is especially interesting, for as we shall see, the sacred and profane coexisted side by side in nineteenth-century Kilbi, just as the market begins to enter the festive space.[13]

Traced back to the Glarus homeland, Kilbi festivals traditionally signified the return of herdsmen and their cattle from summer Alpine pastures. The homecoming provided the opportunity to celebrate the revival and renewal of the church as well as the occasion to revel in customary Swiss activities such as *Schwingen* (Swiss wrestling), music and dancing, and marksmanship. A product themselves of nineteenth-century nation building, each of these three Swiss "cults"—sports, singing, and shooting—were transplanted to the Wisconsin settlement. They became, with special religious services, the central components of nineteenth-century Kilbi festivals.[14]

As in old Glarus, Kilbi in New Glarus became an amalgam of Glarner (and increasingly, Swiss national) festive traditions in one celebration. The religious component took place annually on the last Sunday of September, when the pastor closed the service with the rededication of the church building. The secular element took place immediately afterward, with "a moderate indulging" of target shooting and dancing, but it was on Monday that the festival hit full stride.[15] Combining singing, dancing, marksmanship, and wrestling, the final two days featured leisure and sporting activities (fig. 2.1). As anthropologists have long pointed out, traditional festival typically includes rites of competition—such as sharpshooting and wrestling—where skills required in daily work or military occupations are put on dis-

Figure 2.1. Kilbi Sunday at the Schützen Park. Kilbi at the turn of the century featured Sunday picnics and shooting contests, such as the one pictured here in 1907. (Courtesy of Millard Tschudy Collection)

play.[16] In New Glarus, the didacticism and patriotism of official culture were noticeably absent as days of competition and long nights of dancing, beer consumption, brawling created a period in which the rules of ordinary life were suspended.

The defining event of this "time out of time," and the culmination of Kilbi itself, centered on the Monday afternoon and evening "Kilbi ball." One account offers a nice description of country dance as it was witnessed in 1879: "Strangers come from a distance, and neighbors and friends meet, and renew friendships over loaded tables and foaming glasses. The youth, and, in fact, almost everybody, repair to the village; and music and dance begin about noon and are kept up until morning, at three or more different halls, and all are crowded."[17] From this early portrait, we detect a festival that was perceived as a time for celebration with considerable relaxation of social norms and a unique opportunity for licentious behavior.[18]

The New Glarus Kilbi balls differed from the country dances of their ethnic neighbors in several fundamental respects. First, significant amounts of beer, and to a lesser extent, wine, were consumed, a practice that certainly set the Swiss celebrations apart from the neighboring Norwegians, if not the Germans.[19] One account describes the typical behavior of the men who frequently "retire to refresh themselves with intoxicants, but do not give over dancing, keeping an even keel, all awash though they may be." It would seem, according to this writer, that the Swiss men, "unlike the Yankees and Norse," care little for talking between dances, but would rather "retire to the bar-room and pour things into themselves."[20] Likewise, the music "is played with an exultant elation, almost wildness" noticeably absent from the "stiff, jerky, hoppety motion" of the nearby German dance music. Music was most frequently performed by the local *Männerchor* (men's chorus) or New Glarus' own village band, Stadtmusik, both of which also played at Fourth of July celebrations.[21]

The audience, in turn, responded to the music's "verve, dash, [and] fire . . . with a mighty rataplan of the ample feet of the whole multitude, cheerful, hilarious, exultant, deafening" (fig. 2.2). Indeed, the entire scene, which usually ended at half past five in the morning, struck Wardon Allan Curtis as altogether more boisterous and disorderly than the dances he saw at Norwegian, German, Irish, and Yankee communities: "if you wish to see the old choragic madness as you will see it at no other contemporary occasion, go to a Kilbi ball."[22]

For all the disorder of the festival's performance, its form adhered to a rigid structure that also set it apart from its competitors and was considered "a novelty to Americans."[23] The Kilbi balls were organized and managed for profit. Usually a committee of three "managers" strove to find a suitable dance hall, provide the music, keep whatever order was possible, and collect entrance fees from the men (women, by contrast and as we will see below, did not pay). On most Kilbi Mondays two and sometimes three sets of "managers" competed to create a set of balls with different personalities. Some advertised in a humorous vein, while others attempted to attract festival-goers by depicting lavish, Viennese-type gala balls.[24] All, however, worked within a structure that cut across both ethnic and gender lines.

Though long the province of the tightly sealed community, by 1907 attendance at Kilbi balls no longer remained the exclusive domain of the Swiss. That year, one account estimates that roughly 25 percent of the men were Norwegian, French, and Yankees.[25] Significantly, the women all tended to be Swiss. More to the point, they provided the attraction for the men, who were the admission payers. Wardon Allen Curtis writes that "the managers, in order to draw a crowd, provide transportation for all the girls within a radius of several miles. The management which secures the most and the prettiest girls also secures the greatest number of paid admissions from the men."[26] And where there was mixing among the nationalities, there would also be found—inevitably it would seem—outbursts of disagreement and

Figure 2.2. Turn-of-the-century Kilbi dance. "With verve, dash, and fire," annual Kilbi balls reinforced Swiss cultural identity. (Source: *Century Magazine*, 1907. Drawing by Leon Guipon; courtesy of Memorial Library, University of Wisconsin–Madison)

flares of temper. In this respect, the Kilbi balls were no different than their neighboring non-Swiss dances. "As much occasions of combat as social diversion," Kilbi regularly pitted the Swiss men against their neighbors. Rarely were outbursts more severe than an occasional black eye, but due to "the excitement of the crowd, the music and strong drink," interethnic friction became as common as the dances themselves.[27]

Conflict was made all the more inevitable by the fact that ethnic differences stood out in more ways than language. Here, dress played a central role. The Swiss wore either unique, felt hats with wide, flapping brims or traditional peasant milking caps along with Bernese-style jackets that made the nonwearers (i.e. Yankee and Norwegian men) stand out all the more, thus reinforcing the ethnic boundaries that language and social custom had already erected (fig. 2.3).[28] Werner Enninger writes of the importance of clothing as a marker of identity: "the member of primary groups . . . tend to project their sense of we-ness through collective acts of symbolic identification, such as wearing certain clothing items."[29]

With active participation that cut across all social lines, its licensed relaxation of standard rules through rowdy merrymaking, and its commingling of the sacred and secular, the nineteenth-century Kilbi served to intensify the already tightly-knit community. A powerful example of what Victor Turner calls *communitas*—or the "sense of comradeship and communion," achieved through celebration's loosening of normal, everyday rules—Kilbi festivals regenerated a sense of collective "we-ness."[30] Despite the presence of numerous Norwegians at later Kilbi balls, the event remained largely inwardly focused on the Swiss community. The non-Swiss were attracted, of course, to the universal appeal of intoxicants, rowdiness, and the possibility of courtship—not the "Swissness" of the occasion.

While the early Kilbi festivals created an atmosphere of unreserved merrymaking and were disinclined to memorialize, certain occasions arose in the maturing community that deliberately called attention to significant chapters in the group's collective memory. One of the primary purposes of festivals, Beverly Stoeltje argues, is the "expression of group identity through ancestor worship or memorialization, the performance of highly valued skills and talents, or the articulation of the group's heritage." Such celebrations—and commemorative displays, in particular—are explicitly designed to communicate such messages across and through group boundaries.[31]

These subsequent public displays, then, were distinguished by their commemorative nature. Commemorations of Swiss Indepen-

Figure 2.3. Cheesemaker at a Kilbi ball. Many of the men at Kilbi balls, like the one pictured above, were cheesemakers from Canton Bern. It was customary to wear native costumes at such occasions. (Source: *Century Magazine*, 1907. Drawing by Leon Guipon, courtesy of Memorial Library, University of Wisconsin–Madison)

dence and of the village's founding were distinctly reflexive vehicles for cultural expression. By demarcating celebratory time and space, by programming explicit structures of activity, and by requiring coordinated public participation, the turn-of-the-century commemorative festivals encapsulated an emerging ethnic-American identity. And, where Kilbi generated communitas, commemorative festivals spoke more directly to the fragile tensions inherent in being both Swiss and American.

SIX HUNDRED YEARS OF SWISS INDEPENDENCE: 1891

The first of these grand commemorative celebrations came nearly forty years after that first "Swiss" Fourth of July festival. Like its predecessor of 1853, this festival celebrated Independence Day with all the trappings of Swiss nationalism: folk dance, music, costumes, special food, and competitions, including wrestling (Schwingen) and shooting matches (*Schützenfeste*). Unlike the earlier Independence Day celebration, however, this one took place in early September, not July, and dispensed with any pretense of commemorating American freedom. Rather, Swiss nationalism took center stage as New Glarus celebrated the six hundredth anniversary of the original Swiss Confederation.

That New Glarus would be host to the event was not at all certain. The chair of the event (the *Festpräsident*) came from Monroe, the district's county seat and largest regional center. Nor did either of the two principal orators come from New Glarus; one hailed from Monroe and the other from the neighboring village of Monticello. And, though New Glarus contributed some of the entertainment, Monroe and Monticello each provided more bands and singing clubs. What New Glarus and its immediate hinterland possessed was a virtually homogeneous population of Swiss Americans. As the Festpräsident put it, almost apologetically to the people of Monroe, two factors led the celebration organizers to "choose [New Glarus] for the place of the celebration": it was the place of original colonization; and, relatedly, "because of its being the most exclusively Swiss of all the different townships in which the sturdy and thrifty people had settled in Wisconsin." Monroe, Monticello, and the many other communities in the region might have been home to as many Swiss as New Glarus, but none remained more unabashedly free of competing nationalities than the village on the Little Sugar river.[32]

The village of less than five hundred swelled its ranks more than tenfold on that warm day in September 1891 as between six and seven thousand fellow "Switzers" descended upon New Glarus. Although there was a smattering of "other nationalities . . . the celebration was marked by an immense outpouring of the social and festive Swiss."[33] Kilbi celebrations and annual recurring events such as shooting festivals and homecomings had begun to establish the community as a festival headquarters, but this event secured that status. The impressive scale of the entire event struck those present as something unprecedented and unlikely to recur. Indeed, the immense throng took organizers by surprise as it turned out to be "a much larger crowd than anticipated, and much larger, perhaps, than New Glarus will see again for some time to come."[34]

New transportation technologies rendered old spatial boundaries more porous. With the arrival of the Chicago, Milwaukee, and St. Paul Railroad four years earlier in 1887, the geographic isolation that effectively inhibited contact between the Swiss colony and its wider environment dissipated.[35] A day-long journey from Monroe now became a two-hour train ride. For the price of a one dollar round-trip ticket, most of the celebrants caught the special excursion train in Monroe, or picked it up at intervening points along the way. With no small amount of irony, the commemoration of ethnic insularity was fostered by increasing connections to the outside world.[36]

The "visiting delegation" was met at the train station "by bands of music, the ringing of bells, and the explosion of gun powder" (fig. 2.4). From there, they were escorted by Celebration Marshall R. A. Etter of New Glarus through the village streets. The usually ordinary and somber landscape was transformed into a "gaily decorated scene with Swiss and American flags, banners, garlands, and mottos [with] similarly decorated arches." As the everyday surroundings metamorphosed into a ceremonial landscape, the visitors were impressed with the scene. Indeed, the usually drab streetscapes, neat but unimpressive houses, intractable grid, and smokestacks gave one visitor from Canton Glarus the following year pause to comment that "one may not compare this scene and streets with our lovely Swiss-land."[37] But the 1891 spectacle demanded a landscape reflective of both the event's high fun and weighty seriousness and, with flags, archways, and garlands, the drab and ordinary acquired a distinctive look that differentiated it from the hundreds of similar-looking communities across the heartland (fig. 2.5).

Shortly after the train's arrival, the heretofore "peaceful and orderly crowd" erupted into a noisy confusion when "the body of the immense crowd lost all order and the road was packed with a dense mass of humanity from side to side."[38] Order was quickly restored, however, much to the nodding approval of the program organizers who feared that excessive rowdiness and too much "freely imbibing of the national

Figure 2.4. Festgäste arrive by train, ca. 1900. From the late nineteenth century through the early twentieth, New Glarus became a central gathering place for ethnic brethren. Most arrived by train, as shown here. (Courtesy of Millard Tschudy Collection)

beverage" could possibly cast a shadow over the celebration. One account reported that "beer was there in abundance, as a matter of course . . . and that while those who indulge in the beverage were more than usually exhilarated, there was no intoxication to speak of." The festive occasion had a serious side, and turning it into "a carousal on a big scale" would have undermined its purpose. Fortunately, and to the surprise of many present, the special police who had been hired particularly for that day had an easy time of it and were not forced to make any arrests or break up any fights.[39]

The formal "services" of the commemorative festival (*Gedenkfeier*) took place just outside of the village limits in a natural, outdoor amphitheater and were divided into two components separated by a large group dinner. The first session, considered the "official section" (or, *Offizieller Theil*), consisted of speeches, historical orations, hymns, and patriotic music. The session began with the Grand Chorus performance of the patriotic "Heimat über Alles" ("My Homeland above All"). Bypassing the regional differences that defined identity for so long in Europe (and not least of all, Switzerland) the chorus implored allegiance to the imagined community of the nation state: "Homeland,

Figure 2.5. "Einer für Alle, Alle für Einen." "All for one, one for all"—the drab, everyday landscape was enhanced during festival time. Pictured here is the entrance to New Glarus for its fiftieth anniversary celebration. (Courtesy of New Glarus Historical Society)

homeland above everything/over everything in the world/you alone are our world . . . / unity and justice and freedom for the Swiss fatherland . . . "[40]

Addresses of welcome (*Begrüßungsrede*) from the celebration president in German and from a businessman in English, both from Monroe, immediately followed. Then, after a musical interlude, came sentimental orations from the "oldest federal document" of the Swiss confederation (*Der älteste Bundesbrief, Ewiger Bund der Landleute von Uri, Schwyz und Unterwalden*). At a time before mass communication was widely available, the speaker's stage at such festivals held considerable importance and often constituted the day's highlight. With passion and fervor and to the delight of the surrounding hordes, the pastor from Monticello and a prominent physician from Monroe read excerpts from the "sacred" document that originally signified the union of the three original lands of Switzerland: Uri, Schwyz, and Unterwalden. In the intoxicating romantic nationalism of the nineteenth century, in which such artifacts reflected the "national soul," the text was considered "eternal" (*ewig*). Between orations, the Monroe Männerchor added solemnity with a hymn and then joined the grand chorus to

42

close the session with the sentimental and extremely popular nationalist *lied* "Rufst du Mein Vaterland?" ("Does My Country Call?").[41]

The afternoon session departed from the solemn and quasi-religious tone of the morning with costumes, parades, tableaux, and folk music. Historical oration gave way to historic representation as the "entertainment section" (*Unterhaltender Theil*) introduced an important performative element to the commemoration. A "Grand Parade" led things off with Celebration Marshall R. A. Etter "on horse and in costume," and was followed by two ensigns on foot, bearing Swiss and American flags. Next came four musical bands, and floats carrying Helvetia and Columbia sandwiched between twenty-two "maidens" (*Jungfrauen*) wearing folk costumes of the Swiss cantons and carrying the cantonal flags. A "large number of people" on horseback and foot followed, "dressed in the court, military, ecclesiastical, artisan, and peasant costumes of the Austrians and the Swiss in the time of 1291, including impersonations of the tyrant Gessler, and William Tell and his wife and son." The parade concluded with a herd of fourteen cows, "at the head of them one imported from Switzerland, each cow supplied with a bell of the loudest variety," yodelers, alpenhorn blowers, and—the only representation of local history—a float with a cheese factory in full operation.[42]

The costumed procession moved past the main businesses in town and then returned to the *Festplatz* (festival site) for the highlight of the afternoon's session, the "Grand Tableau." Tableaux, or tableaux vivants, were a popular form of entertainment in the nineteenth century in which costumed figures frozen in place imitated famous scenes from literature, sculpture, history, and painting. Festive occasions across Europe as well as in the United States made frequent use of the pageant form to call attention to national icons as well as to the abstract virtues of a state or nation. Here women—normally excluded from the parade—were allowed the opportunity to participate in the commemoration as silent, allegorical representations of "Columbia," "The Thirteen Original States," and "Justice."[43]

The Swiss of Green County were well aware of the popularity of such entertainment—this came during the entertainment section of the program, after all—and employed the cultural display for their own national celebration. Intended to "represent the struggle for Swiss independence and final victory," the 1891 tableaux made use of well-known scenes and events from Friedrich Schiller's *Wilhelm Tell* to dramatize the struggle. The story of Wilhelm Tell—lodged curiously between history, myth, and legend—stands out as the supreme symbol that all Swiss, regardless of origin, recognize and associate with the nation. This was no less true in the late nineteenth century than it is today. As one recent compendium of essays on the importance of Tell in Swiss life puts it: "no other personality of Swiss history has had such a lasting impact on the Swiss people as Wilhelm Tell. Politicians from the most diverse camps choose him as symbol of their cause and ordinary people venerate him—especially in time of need—as a secular patron of the country."[44]

The tableaux were divided into six sections designed to heighten dramatic tension until the final climax.[45] Following the established form, allegorical tableaux began the program as the daughters of prominent regional merchants took the stage. Helvetia and Columbia, dressed as near twins in long, flowing white robes, led the way, followed by representations of the twenty-two Swiss cantons. Columbia (known also as "Miss Liberty" or, simply, "America") and Helvetia were even closer in spirit than dress. The two allegorical figures, together with Germania, Polonia, and, most notably, Marianne, came to embody nationalist ideals beginning in the late eighteenth century. In each instance, the female figure appeared in classical and conservative guise—as a goddess—representing a stable equilibrium of ethnic and class interests.[46] It is significant that the role of women in the Grand Tableau centered on allegorical virtues, rather than the heroism of the historical characters that were to follow. In doing so—by "playing the part" of the nation-state, and not mythic or lived experience—the women provided a suitable, stable foundation for the scenes that were to follow (fig. 2.6).

That stability flowed over into the next scene where women were joined by men and children, by yodelers and alpenhorn players in alpine costumes (*Alpertrachten*). The homey, domestic display portrayed a "typical" alpine scene (*Alpenscene*) that combined romantic folklore with an idealized version of life in the mountains. Importantly, the tableaux did not attempt to portray the lived experiences of the community, but merely innocent and essentialized folk life.[47]

The next four tableaux took the most famous scenes from Schiller's *Wilhelm Tell* and "brought them to life" (fig. 2.7). Throughout the nineteenth century, local communities in Switzerland invariably included a Tell tableau in their various centennial celebrations, a tradition transplanted across the Atlantic. First came the most famous scene of all: the apple shoot (*Der Apfelschuβ*) at the market square in Altdorf. This was followed by the covenant at the Rütli meadow (*Der Grütlischwur*) in which the compatriots from Uri, Schwyz, and Unterwalden swore allegiance for the first time and produced the sacred document that formed the centerpiece of the morning's orations. Then came Tell's famous monologue in *Die Hohle Gasse*. Played by a Civil War veteran

Figure 2.6. Uncle Sam uniting Switzerland and America. Against a mountainous back-drop and in front of local boys in uniform, Uncle Sam brings together Lady Liberty (*left*) and Helvetia. Women were frequently called on to personify such abstract, na-tionalist virtues. (Courtesy of New Glarus Historical Society)

and Swiss immigrant, Tell "delivered to perfection the monologue of defiance against the Austrian tyrants." The tableaux concluded with the climax of Schiller's drama, the death of Gessler (*Geßlers Tod*). Here, the twenty-two cantonal costume clad young women appeared on the stage and in their midst were "one representing Switzerland as Hel-vetia, and by her side the American Goddess of Liberty." At this junc-ture all others in costume assembled before the grand stage and "three hearty cheers went up from the entire throng for the two republics."[48]

The "entertainment section" concluded with yet another patriotic song performed by the Grand Chorus, "Sempacherlied" (translated as "The Battle of Sempach"). As with the other patriotic songs for the day, the text was printed in the day's official program to encourage participation in the didactic verse from the homeland. The song, soak-ing with the misty pathos of heroism, devotion to the fatherland, and glorification of the ultimate sacrifice, was a fitting way to end such a display of Swiss romantic nationalism. The 1386 battle, long consid-ered one of the greatest victories of the State's "Heroic Age," has con-sistently been given a central place in Swiss history texts. Its principal message, according to one Swiss historian, has been inevitably di-rected toward the "independence of the homeland, and the well-being

Figure 2.7. Wilhelm Tell group. Wilhelm Tell, his son Walter, and the guards who held the Swiss hero captive were all part of turn-of-the-century commemorations, including the 1891 festival. Here, the New Glarus group is part of an undated tableau in nearby Monroe. Note the crossbow and apple with arrow. (Courtesy of Elda Schiesser Collection)

of the family."[49] Of the two virtues, the former dominated the central narrative of the 1891 commemoration.

The detailed homage to the "memory of their mother republic" gave way to the festivity more customarily associated with Kilbi celebrations. Few of the local newspapers even bothered to report the evening's events, although the informal celebration continued into the early hours of the morning: the bands that played the background for the patriotic lieder retuned their instruments for fast hopping waltzes and other dance melodies. After a long day of orations, speeches, and patriotic song, festival-goers were ready to cut loose, as "a few spirits who had imbibed rather freely of the national beverage grew somewhat noisy in the evening."[50]

The 1891 commemorative festival stands out in its attention to the "memory of their mother republic" to the exclusion of local, vernacular memory. Indeed, there was a noticeable absence of rhetoric in both the speeches and tableaux about New Glarus or the emigrants themselves. Rather, the event signified the power of a complex ritual structure developed in Switzerland and of the grasp it still held on the people who left its mountain valleys. The 1891 event in Wisconsin, as in Switzerland, combined a series of unique celebratory forms, including popular adaptations of feudal displays (the alpenhorn and cattle procession), reworking of older forms derived from French revolutionary practice (the female allegories of the nation-state), and recently invented traditions (Wilhelm Tell). But where the six hundredth anniversary in Switzerland marked the final stages of the historical pageant movement, in New Glarus the event signified a beginning.

The work of Rudolf Braun demonstrates that Swiss nationalism developed in tandem with the formation of the modern federal state by modifying and reutilizing such traditional practices as shooting contests, wrestling matches, and folk song. Although each of these practices existed in one form or another for some time, their institutionalization and reformulation altered them radically as they were put to the new use of making the federal state more connected with everyday life. The robust ritual complex that Braun describes—tableaux, the display of flags, parades, oratory, newly created ceremonial landscapes, together with song, shooting, and wrestling—achieved its peak in the late nineteenth century and slowly ossified afterward. Thus, Regina Bendix notes that while the six hundredth anniversary in Switzerland attracted thousands with its elaborate pageantry and a festival play recounting Swiss history in allegories, the event lost much of its intended nationalizing punch.[51]

The reverse occurred across the Atlantic. Not only did the patriotic ritual complex survive the trans-Atlantic crossing, but as it commingled with new American forms, Swiss patriotic displays flourished. They gave rise to a series of ever more dynamic festivals that increasingly spoke to the day's critical "fields of meaning": economic development and the threat such development posed to community and memory.

COMMEMORATING VERNACULAR CULTURE:
MEMORIAL FESTIVALS OF 1895, 1905, AND 1915

The tremendous success of the 1891 observance of the Swiss confederation's birth inspired community leaders to direct their energies toward commemorating an event closer to their own lives and experiences. Ever since 1870, the community observed the arrival of "colonists" to the New World with decennial commemorations, but the events remained small in scale and inwardly focused. In 1885, for instance, former residents from the surrounding region gathered for the event to pay tribute to the men and women who started the families and local community. In the "central place of honor [sat] the venerable remnant of that hearty band of men and women who had broken the way for

47

posterity: a little group of twelve toil-worn, wrinkled, and gray-haired survivors."[52] The pioneers were held in considerable esteem for the significant role they played in maintaining traditional values and for the direct connection to family—posterity—and not prosperity, patriotism, or progress. One by one each founder, with eyes "dim with emotion," stood and "related his or her tale of toil, hardship, and poverty." For the community members and former residents alike, the founders became a symbol of the strength of community, faith in God, and perseverance in the face of adversity. They also became a repository for those "traditional values" that were being challenged increasingly by the many state, regional, and national leaders who, in Madison, Chicago, and New York, respectively, were promoting rapid change and the centralization of political and economic life.[53]

With time and their increasing frequency, the central narrative of the memorial festivals became more complex. A more abstract representation of "the pioneer" or "colonist" began to share the ceremonial space with the actual settlers, although the latter remained important. And, while the Swiss romantic nationalism that dominated the 1891 commemoration never entirely vanished, it now had to compete with the dual narrative of both pioneering tradition and material progress as the central messages put forth on subsequent festive occasions. In 1895, the fiftieth anniversary celebration of New Glarus's founding became the first to express these complex themes.

Progress contra Tradition: The 50th Anniversary in 1895

Following the pattern established for large-scale commemorations in 1891, a dozen overflowing special trains brought festival-goers from the surrounding region. Although Swiss immigrants and their relatives predominated, the village filled with hundreds of people from all the nationalities, bringing the total numbers to between five and six thousand.[54] After assembling at the train depot, the large procession of celebration guests (*Festgäste*) marched behind musical groups, clubs of different stripes (*Vereine*), flag bearers, and a contingent of dignitaries. A troop of young men on horse, each carrying a banner emblazoned with the name of one of the first settlers on one side and Swiss flags on the reverse side, led the promenade to the old cemetery, the site where the colony's leaders, Nicholas Duerst and Fridolin Streiff, first settled (fig. 2.8). Standing next to a "picturesque log hut after the style built here in the early days," County Circuit Judge J. J. Tschudy (himself one of the few remaining pioneers from the 1840s) gave a short speech of gratitude to the colonists. The site, Tschudy reminded the crowd, was important for here was where the "tired pilgrims—with painful feelings of home-

Figure 2.8. Fiftieth-anniversary celebration parade. These young men, momentarily stalled, led the procession to the old cemetery, the site where the colony's leaders, Duerst and Streiff, first settled. (Courtesy of Millard Tschudy Collection)

sickness—received shelter."[55] From there the parade continued to the town hall (*Gemeinderathaus*) where the festival chairman and bank president T. C. Hefty welcomed the guests to the community. Finally, the celebration guests found their way to the familiar wooded grove to the north of town where the main commemoration was to take place.[56]

The memorializing event was then broken into two chief components: the religious portion (*Religiöse Abteilung*) in the morning and the more secular *historische und unterhaltenden Theil* in the afternoon. Presiding over the morning's session was the extremely influential pastor of the Swiss Reformed Church, John Theo. Etter, who had led the conservative congregation for thirty-seven years.[57] His remembrance sermon (*Erinnergungspredigt*)—interspersed with hymns from the various singing clubs—conveyed a central message of thankfulness to God for delivering the pilgrims to the new land. Not content to see the earliest settlers as merely pioneers, but rather as pilgrims (*Pilger*), Etter endowed the colonists with a sacred task: to start life anew. Moreover, "Swiss faith and brotherly harmony" (*Schweizertreue und brüderliche Eintracht*) carried the pilgrims through those most difficult times and were the foundations upon which the honorable colony was founded.

These were terms to which his audience would have been accustomed; what came next, however, was designed to send shock waves through the community. Just as surely as Swiss faith and harmony carried the founders through the most difficult of times, so were they threatened fifty years later. The stern pastor admonished the audience for not living up to the standards of their forebears: "The founders observed better, much better, the Christian principles inconsistent with today's vices of self-centeredness and egoism [*Selbstsucht und Eigenliebe*]." To conclude, Etter entreated the audience to live up to the standards of the founders and obey the creed they brought over to the New World: simplicity, love of frugality, thriftiness, a sense of sufficiency, thankfulness, piety, and above all, warm brotherly love and harmony."[58]

The most welcome communal break for lunch, served at the hotel and in private homes, as well as at the festival grounds, was followed by a second gathering at the train depot and the formation of the historical procession. Here, the arrival of the colonists and the development of industry over the past fifty years were presented in an informal cavalcade. The order of the parade elements—leading with an oxen-drawn covered wagon and ending with a display of modern farm machinery—bespoke the progress of the community. It served, Milwaukee's newspaper noted, as "a forceful reminder of the first days of privation and poverty followed by many years of plenty and prosperity."[59] The cavalcade proceeded from the train depot, past the important businesses of the village, to the festival grounds.

The second half of the commemorative activities resumed with two historical orations, one in German and one in English. The first, delivered by historian and local politician John Luchsinger, focused on the history of the colony and of the many hardships and experiences (*Strapazen und Erlebnißen*) that the colonists were forced to endure during the first years in the new homeland. The second, read by Thomas Luchsinger of neighboring Monroe and "born into the original colony," "exhorted the latter-day colonists to remain in harmony." In both cases the "chief theme of the orators was the poverty, struggles, privations, sufferings, homesickness, hopes and ambitions of the first twenty years of the colony, in marked contrast with the successes, achievements, and prosperity of the present day." The final portion of the commemoration paid homage to the thirty-some surviving original colonists who were formally introduced to the crowd by the master of ceremonies. Afterward, the crowd retired to the numerous taverns and dance halls in town for an evening of Kilbi ball-style dancing and merriment.[60]

The 1895 fiftieth anniversary celebration marked a turning point in the history of the community's festive culture. On the most immediate level, it introduced the large-scale festival form inaugurated four years earlier to the local, vernacular interests of community memory. The celebration highlighted the sufferings of the founding colonists, rather than allegiance to an abstract and distant homeland: the pain of immigration was not easily conjured via allegorical tableaux, as it was an experience of deep personal meaning. A prominent community member had captured this sentiment several years earlier:

[Immigration] is an act of great self-denial, a self-sacrifice, a sort of re-generation, that is forced upon every grown person who undertakes to make a home in the new world. Did you ever notice a foreigner move through the streets of your city or village the first few days after he came across the ocean? Did you ever observe his embarrassment, arising from the consciousness of being different in manners, clothes, appearance, language, in most of all the objects of interest, pleasure or aversions? Did you ever realize that these for-eigners had to conform their whole being, physically and mentally to the new world. They have to abandon their old, form new habits, learn a new language, a new way of living and thinking. This is the gravest of all the sufferings that an emigrant has to endure, the root and foundation of home-sickness, which is the most intense of all sicknesses.[61]

New Glarus's first colonists, Conrad Zimmerman believed, offered this sort of sacrifice, the intensity of which "will never be fully understood by anyone who had not had a similar experience." To further illustrate this "act of re-generation," he recalls the testimony of an older woman in the community who remarked of her suffering: "We often cried until our heads were like *Laegele* and wished were at home again, even if we had nothing else to live on but *Schotte* and *Chrut.*"[62] Importantly, and as all three orators and Pastor Etter made clear, it was precisely this ability to suffer privations together, as a collective, that gave the community its strength.

Progress and distance from these privations, while praised end-lessly, thus carried a double meaning. On one hand, festival organizers began for the first time applauding the progress made by the commu-nity as a specific demonstration to the "many other nationalities pres-ent" that the Swiss should be counted as a model group. On the other hand, this very progress took the colonists even further away from the home so beloved by settlers such as the woman above. And, equally, it pushed the community, however unwittingly, toward "today's vices of self-centeredness and egoism" so abhorred by Pastor Etter.

The Enactment of the Homeland: The Sixtieth Anniversary in 1905

Following the now well-established format of memorial celebrations, the warm August day in 1905 began with greetings and a welcoming of festival guests at the train station to celebrate the sixtieth anniversary of the community's founding. The railroad's twenty-one "packed coaches" carried the largest number of visitors on record, swelling the village's ranks to more than 10,000. Two central events, interspersed between the customary speeches and musical performances, drew the record crowds and showcased what one newspaper called "a confusion of the foreign and the American at New Glarus" and what another, for lack of any other description, merely labeled an "odd celebration": an enactment of a *Glarner Landsgemeinde* and a second series of tableaux that, for the first time, graphically united Swiss and American history with lived memory.[63]

In the Swiss historic (and contemporary) political climate where decentralization takes precedence and statism is viewed as nonfreedom, the Landsgemeinde has become the performative symbol of Swiss freedom. A cantonal assembly through which direct participation of the individual is assured and ongoing self-government is guaranteed, it is the hallmark of the strength of the canton and of individual (though exclusively male until 1971) voter rights. As Benjamin Barber points out, the Landsgemeinde is to Swiss culture what a bill of rights or a declaration of freedoms is to an Anglo-American political tradition: the core signifier of democratic freedom. Unlike a flag or a document upon which a higher court or legislature votes, however, the Landsgemeinde is an inherently participatory and performative symbol.[64] The enactment of such an event would have held special poignancy for the descendants of the earliest colonists. It was in such an open-air meeting sixty years ago in the village of Schwanden, in Canton Glarus, that the emigration plan was first proposed and then acted upon.[65]

The participants of the 1905 New World Landsgemeinde were well aware of its general function in Swiss democracy. One leader from nearby Monticello observed that the day's meeting, the first such event in the community, replicated a "system that represents the principle of self-government more nearly than any other form of government on the globe." Another, in hope of casting the best light on his community, recounted to a reporter from Madison that the Landsgemeinde is "the feature that gives the Swiss people that regard for law and order which is so characteristic of the race."[66]

The New Glarus Landsgemeinde consisted of two components and took place in a "setting as realistic as it is possible to make . . . with scenery that was painted expressly for the purpose, showing a true reproduction of the view presented by the distant mountains and nearby buildings in Glarus, Switzerland." Before the nostalgic, simulated Swiss backdrop, costumed community members re-created the

1845 Schwanden Landsgemeinde by reading the names of the original colonists and the names of the survivors and by "re-enacting incidents of the trip."[67]

The second component used the unique forum to raise initiatives of "matters of local interest." Five measures were proposed and voted upon by the men of the mock Landsgemeinde, including proposals for better rail connections; the construction of an "old settlers monument"; a good roads iniative; lower taxes for farmers; and that limburger cheese, the chief product from the region, be "declared legal tender for the payment of all debts and a medium of exchange throughout the district." This final proposal, coming most likely from a Bernese cheesemaker, added that "refusal to take limburger as legal tender should be followed by [a] sentence that the offender should never be permitted to eat limburger cheese again." Although presented with tongue firmly in cheek, the measure possessed a serious side: it called attention to the widely fluctuating market price for the cheese, a condition that affected nearly every working family in the county.[68]

The second central event of the day, which even created "an atmosphere that was Swissesque," utilized the tableaux format to illustrate scenes of well-known Swiss and American history, as well as the history of the colony. Important members of the community—the bank president, newspaper editor, physician, and leading merchants—donned the costumes that were rented from Milwaukee to silently and motionlessly create the *Lebende Bilder* (living pictures). Wilhelm Tell and his son Walter reappeared, as did the Swiss hero Winkelried (fig. 2.9). They were followed by a scene unique to the Swiss colony: an American historical tableau that featured the landing of Columbus; the rescue of Captain Smith by Pocahontas; Washington leading his forces at Valley Forge; and Lincoln freeing the slaves. Feminine allegory returned as well, with Columbia generously presenting liberty to Cuba. The tableau that stands out, and the one that launched the afternoon program, depicted the lives of the colonists. Here, scenes representing key moments in the migration story stood alongside Wilhelm Tell and George Washington: the cheerless departure from Glarus; the arrival in Baltimore; and, "of special interest," the establishment of *New Glarus* in the Wisconsin wilderness (fig. 2.10).[69]

The mixture of American and ethnic history was a common theme of the historic tableaux of this period. Milwaukee's German-American Day Celebration of 1890, for example, labored to intertwine German and American history by featuring tableaux of German art and history along side William Penn and George Washington.[70] Of central concern here, however, is the increasing influence of the local, or vernacular memory. One need only compare these Lebende Bilder with the tab-

Figure 2.9. Wilhelm Tell in 1905 tableau. Wordless, actionless figures—frozen in time and space—recreate the famous apple shoot scene. Tell is depicted on the far left holding crossbow and the crucial second arrow. Walter stands to his right with an arrow-pierced apple, and the bailiff Gessler is on far right. (Courtesy of Millard Tschudy Collection)

leaux of 1891 that featured Swiss historical scenes to the exclusion of anything local. The same may be said for the Landsgemeinde. Although a nonmaterial relic from the homeland, it was used to reenact *local* history and put forth a concern for *local* affairs.

Taken together, the Landsgemeinde and tableaux provided more than simply entertainment; they afforded festival organizers a valuable educational function. For one, the celebration was specifically designed, as one account put it, "to illustrate how things are done in the canton of Glarus, where there is not legislature and no ruler, the citizens in general hold annual meetings for the purpose of making their own laws, or abolishing or amending them, and it is their power and the part they have in making the laws that develops the disposition to respect the laws after they are enacted." Or, more simply, the enactment of the Landsgemeinde "brought home to the children of the fatherland, a vivid picture of life in the old home country."[71]

The importance of creating such a vivid picture became more acute with the passage of time. Indeed, as the community aged and distanced itself (in time) from the actual settlers' lives, few of the "old gray ones" were still around to attest to their experiences. One reporter summed the situation nicely when he wrote that "the present generation of New Glarus knows but little of the founders of this colony of Swiss people, and the privations that the pioneers were called upon to undergo."[72] Thus did it become necessary to depart from the 1895 festival's form of orations and speeches; more participarty means were required to teach the lessons of the past.

Far from a neutral goal, this educational function carried with it a moral imperative. John Luchsinger, once again, provided a historical background for the Festgäste by recounting the hardships endured in Canton Glarus, the difficult journey, and the arduous few decade in the new world. He reminded his audience that while strides in agricultural production and, especially, the dairy industry have brought increasing wealth, *"it was the poverty of the colonists that held them together thus insuring the success of the settlement."*[73] The eradication of poverty was surely an unmitigated good, but what if it actually diluted that most prized of all virtues—community? The enactment of the Landsgemeinde and the tableaux provided a reminder of the virtues of simplicity and poverty that flew in the face of the progressive narrative that otherwise dominated the day's oratory. Progress fused with an antimodernism—or the desire to reject that progress in favor of an idealized past—to create a charged festive atmosphere.[74]

Figure 2.10. Playing the immigrant, one generation later. This photograph by John Luchsinger depicts participants of the immigration tableau at the 1905 commemoration. This was the first commemorative "living picture" that featured local, or vernacular, memory. (Courtesy of State Historical Society of Wisconsin, neg. WHi [X3] 19002)

A crucial feature of both commemorative ceremonies—and an often neglected aspect of such events in general—was its performative nature. The Landsgemeinde, in particular, involved literally hundreds of men,[75] and they came together in a temporally and spatially bounded space for the explicit purpose of enacting an annual ritual from their collective past. This reflexive cultural expression—standing in a central meeting square and voting on measures of local importance—was one that most would have only known second hand. However distant the voters themselves may have been from the homeland Landsgemeinde, the New Glarus performance provided the means of communicating a vitally important social memory of the ways of life in a country that at one time had been home but was now becoming increasingly distant. It is precisely through the enactment of such commemorative ceremonies that the images and recollected knowledge of the past are conveyed and sustained. No better vehicle of transmitting those images and recollections could have been devised than the Landsgemeinde and historic tableaux.[76]

One final indicator of the commemorative ceremonies' seriousness lies in the heavily patrolled (metaphorical and physical) barriers between the Festgäste and unwelcome outsiders. In particular, community leaders felt that "the program was strong enough to make sure that festival-goers enjoyed themselves." The festival committee, moreover, "did not want the audience to be plagued every five minutes by this or that vendor and wanted to be certain that the visitors could devote full attention to the production without annoyances." They were especially proud that "this element" was not to be found during the entire celebration. The "element" that organizers feared would transgress the steep boundaries of insular ethnic place "were absolutely not tolerated." Indeed, no less than fifteen "gamblers" (or vendors) were discovered in the evening before the event and were promptly shoved on the first morning train out of town. The local newspaper asked, rhetorically, if there was room for such outsiders, and promptly snorted: "No, the celebration was not created to feed this sort of riff-raff!"[77] Such itinerants, whether hocking wares or entertainments, traditionally found a space at festivals. Their banishment from New Glarus reinforced that community's insularity as well as its seriousness of purpose.

The Ossification of the Pioneer: The Seventieth Anniversary in 1915

The only measure advanced by the 1905 "mock" Landsgemeinde to pass from the ethereal to the concrete took the form of a ten-foot-tall granite monument that immediately became central to the community's pioneer iconography. The monument's unveiling, dedicated to the rapidly fading memory of the community's pioneer founders, became the central event of the last prewar commemorative festival. Like previous commemorations, the seventieth anniversary emerged from the complex interaction of progress, the homeland, and the local past; its form, however, differed significantly.

The three-day event nearly did not get off the ground. It was decided only in early July that the celebration would involve more than merely a quiet church service and that local businesses would, for the first time, play a central and visible role. Previously, businessmen participated, but in cooperation with regional farmers, doctors, and cheesemakers as coperformers in musical groups and as co-organizers. Although held at New Glarus, earlier festivals reflected concerns and interests of people from the far reaches of the region. This festival, however, was taken over at the last minute by local business interests resulting in a far more complicated organizational structure than was the case in earlier events. In addition to the executive committee, chaired by the Bank of New Glarus president, other committees included finance, music, stage, decoration, reception, advertising, parade, and amusements. The only board not to be made up of businessmen was the "committee on decoration," comprised of local New Glarus women.[78]

The increased organizational complexity reflected, in turn, a more ambitious many-day affair that tidily segmented the key celebratory components into neat parcels. Considerably removed from the integration of the secular and religious that defined the Kilbi, this celebration cordoned off the oppositions into separate spheres to a greater extent than any other festival. Sunday's celebration, "devoted solely to church purposes," opened the grand festivities with a day of thanks to God. The impressive new church overflowed as Swiss came from far and wide to hear the day-long mix of German language sermons and hymns. At every convenient point, the guest pastors from Sheboygan, Monticello, and Belleville praised the endurance, simplicity, and strength of faith and community possessed by the colony's founders. The church celebration (*Kirchliche Feier*) concluded quietly with evening songs about the sorrows of Swiss homesickness (*Schweizer Heimweh Liedes*).[79]

The following day began with a bang as heavy explosives of dynamite were set off before sunrise. Quick to follow came the "sweet tones of an Alphorn heard all over the village . . . answered by two coronets from the north part [of town] with many Swiss melodies and songs." Visitors began arriving for the first time by automobile, followed by the customary special train from the county seat. Among the more

57

than thirty automobiles came the entourage of Governor Emanuel Philipp. A second-generation Swiss American from Sauk County, the newly elected governor became the first nonlocal person to officially visit and speak at a New Glarus festival. Interspersing Swiss German with English, the conservative politician praised his kinsmen for their diligence and fortitude in making it on their own.[80]

The day, "devoted to memorial purposes," included music, speeches, and a parade in which (again, for the first time) "hardly a business place in the village . . . was not represented." The central event, however, took place early in the day as the crowd proceeded with the formal unveiling of the pioneer monument. Accompanied by the music of two bands and the ringing of church bells, and set against the three flags of the United States, Switzerland, and Canton Glarus, the throng gathered at the monument for commemorative speeches (fig. 2.11). After opening the ceremony with a prayer of thanks, attention focused on S. A. Schindler, the "president of the day," who spoke of the occasion "to meet again with friends and relatives who formerly lived here and to renew old friendships and new acquaintances." The village had begun filling up more than a week earlier with former

Figure 2.11. Seventieth anniversary of the Swiss Colony. The 1915 commemoration featured guest speaker Governor Emanuel Philipp and the dedication of the pioneer monument, shown here behind the Welcome banner. (Courtesy of New Glarus Historical Society)

residents who used the event as an opportunity for homecoming. But Schindler was quick to follow with the observation that "the main purpose of this festival is the dedication of this monument . . . to commemorate the settling of New Glarus by this colony, to express gratitude to these pioneers by their descendants and to turn this monument over to the care and custody of the village and colony of New Glarus." He admonished the community to "hold a protecting hand over it and provide that it will always be kept in a respectable condition and thus be an ornament to the town." Schindler concluded with the hope that "this monument be a new spur to new impulses in the development of this town and enliven anew the spirit of perseverance, of home, of diligence and of faithfulness of their pioneers."[81] Rather than merely looking backward, organizers hoped that the monument would also propel the community forward by suggesting the central messages of the migration story to the festival-goers.

For all the enthusiasm surrounding the monument's dedication, organizers seemed to have had a difficult time drumming up the support necessary for its construction. One speaker recalled the failed first attempt by the village to erect the monument and that, in the end, it was accomplished entirely by private subscriptions at an expense of three thousand dollars, considerably less than the proposed amount ten years earlier. Likewise, Village President T. C. Hefty submitted his halfhearted approval that "the people of New Glarus and surroundings have awakened in a measure at least, to a sense of appreciation and acknowledgment and have . . . responded with their contributions to make possible the erection of this beautiful monument." The bank president concluded by entreating the large audience to allow the granite pioneer's "stern and humble pose be an ever present reminder of its priceless cost, [a] never ending argument for purer heart, for noble deeds . . . and a signifier of our appreciation of the benefits that we now enjoy because of [the colonists'] unselfish sacrifices."[82] Summoning the audience to emulate the virtues of the pioneers, the monument reminded the crowd, in concrete form, the messages implicit in the parade: to recall both the distance and the connection of today's generation to the pioneers (fig. 2.12).

The foreign once again blended with the American as the Monroe Band played a somber "Star Spangled Banner" to accompany the official unveiling of the monument. More well-known songs followed, as the crowd produced rousing versions of "Heimat über Alles" and sentimental *lieder* celebrating New Glarus's founding. Unveiled at the site of the first log cabin and amid the graves of the earliest settlers, the monument—a memorial to the immigration of the Swiss people into

Figure 2.12. "First settlers of the Swiss Colony," 1915. Such floats, a common feature during early commemorations, highlighted both the distance of the present generation from the colonists, as well as their direct connections. (Courtesy of New Glarus Historical Society)

Wisconsin's rugged driftless region—takes the form of an ordinary Swiss male immigrant (fig. 2.13). Looking southward, the direction from which the immigrants came, the man's pose may well be considered "stern and humble" as he bears little resemblance to either the classical sculpture or heroic forms traditionally associated with war monuments.[83] Dressed in everyday work clothes and, significantly, not a folk costume, the granite pioneer stands above the pedestal, on which the names of the twenty-five original colony's male family heads are inscribed.

The very ordinariness of the pioneer monument is indicative of a trend common to commemorations after the Civil War. Vernacular memory, unregulated by governmental and economic elites, encouraged the construction of memorials to ordinary people and ways of life. No longer were such structures dedicated solely to the memory of an individual leader or an abstract virtue; instead, the collective memory of "prototypical Americans" in large and small communities were given concrete form. In the ethnic mosaic of the late-nineteenth- and early-twentieth-century Upper Midwest and Great Plains, the iconography of the pioneer achieved dominance.[84]

The final day of the commemoration severed the link to the past even further. The first of its kind, Tuesday's "Field Day" brought in such popular diversions as baseball games, water fights, races, sports,

Figure 2.13. The pioneer monument as it appears today. A memorial to the immigration of ordinary people, the pioneer monument of 1915 anchored the collective remembering in a tangible and fixed site. (Photograph by author)

and dances. To these events were added attractions such as a merry-go-round, a "moving picture" show, and lunch stands at various places in the village. Perhaps the most "traditional" event of the day was the customary sigh of relief that "despite the fact that all nationalities attended and that drinks flowed in abundance, the festival took place with considerable calm and without incidence." The virulent combination of ethnic mixing and beer consumption, thankfully, failed to produce anything more serious than the usual black eyes.[85]

More than at any previous commemorative festival, the tension between a belief in the virtues of progress, the values of pioneer ancestors, and the memory of a distant homeland came into bold relief in 1915. Numerous elements of the festival's form signified an increasingly modern celebratory structure: the arrival of dignitaries outside the community including the state's governor; the systematic segmentation of the religious from the purely secular events; the intrusion of commercial amusements; and the complicated nature of the monument's fund-raising. And yet, upon further reading, each of these elements could be taken as a means to reinforce nostalgia for the past. For instance, the choice of Governor Philipp as the keynote speaker could be seen not only as an attempt to reach out to state institutions, but also as a way to express pride in being Swiss. Likewise, the inclusion of leisure activities did not come at the expense of either religious ceremonies or somber homages to the pioneers, but rather as additions to the "serious" events.

The three-day festival's content, similarly, reflected this ambiguity. Speeches and songs alternated between admiration for the struggles of the earliest pioneers and awe at the distance today's generation has put between them and the colonists of 1845. This ambivalence took a concrete shape as well. Standing in front of the modern and emphatically non-Swiss looking church, the pioneer monument would seem to be signaling a farewell to the settlement age (fig. 2.13). The large brick church, "an imposing edifice," replaced the old church that had been built by the early settlers. With its plastered walls and "queer belfry" the old church was the only structure in town that bore a distinctive Old World stamp and its demolition took from the community its "only piece of Swiss architecture" (fig. 2.14). The "beautiful, but unique stone church" was built largely by labor donated by church members with the supervision of John Becker, a Swiss immigrant from Canton Glarus. By contrast, the large, Anglo-owned construction firm from nearby Janesville built the new church at a cost of sixteen thousand dollars with no regard for Swiss building traditions. The only notable exception were the stained glass windows, adorned with German text, and donated by the village's Männerchor.[86]

Figure 2.14. The old stone church, ca. 1890. With its plastered walls, "queer belfry," and hilltop location, the old stone church was one of the few structures that bore an Old World stamp. (Courtesy of New Glarus Historical Society)

The timing of the memorial's construction—and, more generally, the event in sum—coincided with the inevitable loss of what the monument was intended to signify. The community was coming to age and, while it was to maintain relative social and cultural isolation for another two decades, the living memory of the immigration was near its end. The evening before the monument's unveiling Henry Trumpy, the oldest of the remaining pioneers, died, giving somber pause to the celebration.[87] David Lowenthal correctly notes that "the memorial act implies termination. We seldom erect monuments to ongoing events or to people still alive."[88] The monument served to anchor collective remembering in a condensed, tangible, and fixed site—properties at odds with the ever changing and intangible nature of public memory. As it turned the pioneer into concrete, the monument ossified the memory of emigration. It was indicative of a general anxiety in the early twentieth century about memory, and in particular of its loss in ethnic place.[89] Trumpy, who had planned to attend the celebration, was one of the last links to the collective past. His passage, and the monument's construction, marked the beginning of ethnic place's conspicuous construction.

FIELDS OF MEANING: THE MATERIAL AND
SOCIAL BASIS FOR FESTIVAL

Importantly, the ambiguities experienced in each festival were being felt at the level of everyday life in the realm of social and economic relations. These tensions, in turn, greatly influenced the rhetorical strategies employed throughout the vast array of commemorative activity. Returning to the themes that opened this chapter, this section more closely examines the "fields of meaning" that furnish the material context for such a complex series of festival displays. Each gave rise to the apparent "confusion of the foreign and American" at the community's commemorations. The rapid economic transformation that turned the New Glarus area into the most prosperous agricultural region of the state provided the first field of meaning for the increasingly elaborate festive rhetoric. Juxtaposed to a modernizing countryside was the village's social insularity, posed to position itself as an antimodern retreat from these changes—the second field of meaning.

These socioeconomic tensions point to a geographical distinction as well, a dissimilarity captured nicely by an unnamed writer during the early years of the century. The anonymous traveler made the crucial observation that although "old-fashioned customs and ideas [generally] linger longer in the country than in the town, curiously, at New Glarus, the contrary is true." The village was the point of memory and tradition for the wider ethnic region:

In the country, on the farms, American ideas and modern methods of doing things are more prevalent than in the village. . . . Strange, indeed, would it seem to one of the original colonists, could he see the comfort and even luxury in which his descendants of the third generation are living. Could he see all the many marks of progress, he would indeed realize that there had been a marvelous change. On the other hand, should he revisit the little village, while he would find many things that are as much modern as he would find anywhere else, he would, nevertheless, find many of the old customs still observed. He would find the old Swiss life still in existence.[90]

The competing influences of a modernizing countryside and the more insular village thus played themselves out on the festival stage, but had a grounding in daily life as well.

More so than virtually any other community in the state at the time, Swiss farmers in the region surrounding New Glarus had ample reason to feel pride in their economic success. Data gleaned form Joseph Schafer's monumental work on agricultural production in turn-of-the-century Wisconsin show the material side to the community's rhetoric of progress. From 1880 and the beginning of the dairying revolution in

Table 2.1 Agricultural Value and Dairy Production in Selected Southern Wisconsin Towns, 1880

Town (County)	Average value per acre, in dollars (rank)	Average value of farm products, in dollars (rank)
Bangor (La Crosse)	18 (14)	959 (3)
Brookfield (Waukesha)	59 (3)	635 (11)
Castle Rock (Grant)	8 (21)	460 (18)
Eagle (Richland)	19 (13)	400 (20)
Empire (Fond du Lac)	50 (4)	983 (2)
Franklin (Milwaukee)	62 (1)	785 (5)
Highland (Iowa)	16 (16)	562 (12)
Lodi (Columbia)	23 (10)	764 (6)
Mount Pleasant (Racine)	60 (2)	740 (8)
Muscoda (Grant)	9 (20)	554 (14)
New Glarus (Green)	*20 (12)*	*680 (9)*
Newton (Manitowoc)	34 (7)	342 (22)
Norway (Racine)	21 (11)	483 (15)
Orion (Richland)	13 (19)	473 (16)
Pleasant Springs (Dane)	33 (8)	1301 (1)
Plymouth (Rock)	36 (5)	751 (7)
Prairie du Chien (Crawford)	8 (21)	334 (23)
Primrose (Dane)	15 (17)	441 (19)
Pulaski (Iowa)	10 (20)	467 (17)
Sevastopol (Door)	15 (17)	380 (21)
Sparta (Monroe)	17 (15)	557 (13)
Sugar Creek (Walworth)	33 (8)	651 (10)
Whitewater (Walworth)	36 (5)	799 (4)

Source: Data calculated from *Wisconsin Domesday Book, Town Studies,* vol. 1 (Madison: State Historical Society of Wisconsin, 1924), 152–53.

Wisconsin, farmers in the township of New Glarus performed ably, but not exceptionally, when compared to their counterparts throughout the southern portion of the state (table 2.1). Their rank of twelfth in average value per acre and ninth in average value of farm products, out of a sample of twenty-three, captures a district reasonably wealthy and productive. However, twenty-five years later, in 1905, the district catapults to first place in comparable figures measuring both wealth and productivity (table 2.2). Indeed, the value of average dairy production per farm in New Glarus of 1905 nearly doubled the average value of total farm products twenty-five years earlier. Making superb use of the new technological innovations and marketing techniques, Wisconsin's Swiss farmers out paced their non-Swiss counterparts in all categories of dairying, the most important agricultural activity in the state. The gap widened even further by 1920 at a rate that the otherwise staid Schafer enthusiastically called "little less than thrilling" (table 2.3).[91]

Table 2.2 Agricultural Value and Dairy Production in Selected Southern Wisconsin Towns, 1905

Town	Average farm income, in dollars (rank)	Value of average livestock production, per farm, in dollars (rank)	Value of average dairy production, per farm, in dollars (rank)	Value of average dairy production, per cow, in dollars (rank)
Bangor	940 (6)	807 (6)	469 (5)	34 (10)
Brookfield	753 (12)	493 (17)	412 (6)	50 (3)
Castle Rock	582 (19)	537 (15)	255 (17)	26 (16)
Eagle	887 (8)	810 (5)	270 (15)	31 (12)
Empire	864 (10)	577 (12)	310 (12)	39 (8)
Franklin	835 (11)	542 (14)	342 (8)	44 (4)
Highland	624 (16)	590 (11)	306 (13)	27 (15)
Lodi	881 (9)	720 (7)	252 (18)	38 (9)
Mount Pleasant	612 (17)	428 (18)	325 (10)	42 (5)
Muscoda	689 (14)	615 (10)	330 (9)	29 (13)
New Glarus	*1642 (1)*	*1586 (1)*	*1075 (1)*	*43 (5)*
Newton	680 (15)	380 (19)	280 (14)	42 (6)
Norway	864 (10)	715 (8)	400 (7)	41 (7)
Orion	705 (13)	638 (9)	330 (9)	31 (12)
Pleasant Springs	1161 (5)	550 (13)	319 (11)	38 (9)
Plymouth	936 (7)	613 (10)	265 (16)	28 (14)
Prairie du Chien	295 (22)	228 (22)	90 (22)	20 (17)
Primrose	1321 (3)	1281 (2)	774 (4)	43 (5)
Pulaski	607 (18)	516 (16)	203 (20)	29 (13)
Sevastopol	517 (20)	258 (21)	175 (21)	33 (11)
Sparta	488 (21)	358 (20)	204 (19)	29 (13)
Sugar Creek	1380 (2)	1153 (4)	798 (2)	57 (1)
Whitewater	1288 (4)	1230 (3)	790 (3)	53 (2)

Source: Data calculated from *Wisconsin Domesday Book, Town Studies*, vol. 1 (Madison: State Historical Society of Wisconsin, 1924), 155.

This tremendous surge in wealth and productivity can be attributed to the coming of the "industrial revolution of dairying" to the region, and to New Glarus township in particular. The organization of dairy associations and boards of trade and Wisconsin's overall improved economic-geographic position in a national and global market (as well as improved transportation facilities at the local level) contributed to the industrialization, or capitalist transformation, of the countryside. The critical element of the transformation, however, was the factory system of making cheese, an innovation that Frederick Merk once called Wisconsin's equivalent of the cotton gin in the South and the reaper in the western wheat states (fig. 2.15).[92]

Although women had been making cheese for local consumption since the earliest days of the settlement, it was only with the introduction of the cheese factory (and the attendant decline of wheat growing) after the Civil War that large-scale dairying fueled the region's eco-

Table 2.3 Agricultural Value and Dairy Production in Selected Southern Wisconsin Towns, 1920

Town	Average farm income, in dollars (rank)	Value of average livestock production, per farm, in dollars (rank)	Value of average dairy production, per farm, in dollars (rank)	Value of average dairy production, per cow, in dollars (rank)
Bangor	2814 (7)	2480 (8)	1418 (8)	85 (15)
Brookfield	2411 (11)	1944 (13)	1318 (11)	133 (4)
Castle Rock	1393 (23)	1376 (20)	751 (22)	65 (21)
Eagle	2526 (10)	2516 (7)	1655 (6)	114 (8)
Empire	3316 (3)	2951 (3)	1843 (3)	118 (6)
Franklin	2398 (12)	2028 (10)	1382 (10)	134 (3)
Highland	2006 (19)	1937 (14)	1044 (16)	74 (19)
Lodi	2393 (13)	1888 (16)	1000 (17)	81 (17)
Mount Pleasant	2156 (15)	1590 (19)	986 (18)	127 (5)
Muscoda	2057 (18)	1859 (17)	1106 (15)	79 (18)
New Glarus	*5338 (1)*	*5329 (1)*	*3901 (1)*	*159 (1)*
Newton	2146 (17)	2022 (11)	1399 (9)	144 (2)
Norway	2158 (14)	1984 (12)	1171 (13)	105 (11)
Orion	2571 (8)	2545 (6)	1650 (7)	106 (10)
Pleasant Springs	2992 (5)	1352 (21)	757 (21)	72 (20)
Plymouth	2566 (9)	1901 (15)	1160 (14)	100 (13)
Prairie du Chien	1394 (22)	1220 (23)	585 (23)	57 (22)
Primrose	3924 (2)	3923 (2)	2677 (2)	118 (6)
Pulaski	2150 (16)	2083 (9)	1219 (12)	84 (16)
Sevastopol	1536 (21)	1290 (22)	836 (20)	104 (12)
Sparta	1827 (20)	1613 (18)	922 (19)	88 (14)
Sugar Creek	3031 (4)	2909 (4)	1842 (4)	117 (7)
Whitewater	2936 (6)	2755 (5)	1777 (5)	107 (9)

Source: Data calculated from *Wisconsin Domesday Book, Town Studies,* vol. 1 (Madison: State Historical Society of Wisconsin, 1924), 155–56.

nomic engine. This significant shift—from household to factory production—reorganized the countryside with profound effects, the foremost being increased and continuous milk production. The dramatic economic transformation that accompanied the ascendancy of factory dairying was only surpassed by its effect on the rural landscape. It required, according to a state agricultural society report in 1883, "only five or six years until cheese making was the main branch of work for the whole farming population" of the New Glarus area. From their modest beginnings with the first factory just south of New Glarus in 1869 until the turn of the century, "cheese factories [came] to dot the landscape." So thick was the distribution of cheese factories in New Glarus that, at its peak in 1905, every crossroad of the township buzzed with the daily deliveries of milk to one of its twenty-two factories (table 2.4).[93]

Yet as quickly as the cheese factory system came to the farmers of

Figure 2.15. Engen Cheese Factory, 1915. Wisconsin's equivalent of the cotton gin, the cheese factory system revolutionized the dairy industry throughout the state. No district was more affected than the Green County Swiss region as it quickly became the state's leader in foreign-type cheese production. The Engen Cheese Factory was typical of many as it employed cheesemakers from Canton Bern. (Courtesy of New Glarus Historical Society)

the immediate New Glarus region, so did it depart; change came rapidly and completely. Green County continued to lead the way in the production of foreign-type cheese for decades to come, but the site of the original cheese factories turned its attention to the even more remunerative enterprise of supplying milk for condenseries (fig. 2.16). The Helvetia Milk Condensing Company chose New Glarus as the site of its new plant in 1910, making it the second to open in the region. The combination of good transportation facilities and the high density of milking cows contributed to the decision to locate in New Glarus. More important, however, was the long-standing connection between the small village and the Swiss community of Highland, Illinois, the location of the Swiss-owned Helvetia Milk. As the New Glarus plant manager once recalled concerning the location of the plant, the decision was made largely due to experiences during festivals: "The confidence of our people was gained through the *Schützenvereine* [shooting clubs] visiting between New Glarus and Highland, resulting in the directors of our company being induced [to] erect a plant [there]." The condensing plant—brought to New Glarus through the ties fostered during festival times—spurred even greater changes in the country-

Table 2.4 Cheese Production and the Number of Cheese Factories in New Glarus Township, Green County, and Wisconsin, 1870–1947

	New Glarus		Green County		Wisconsin	
Year	Factories	Production (1000 lbs)	Factories	Production (1000 lbs)	Factories	Production (1000 lbs)
1870	1	14	2	359	60	13288
1878	12	600	43	2291	370	19535
1885	18	949	173	3204	1000	32478
1895	21	749	200	7796	1330	52481
1905	22	1396	213	10530	1500	109910
1915	3	190	152	13425	2200	234929
1932	1	70	119	12923	2172	313424
1947	0	0	87	18840	1440	535863

Sources: Wisconsin Dairymen's Association, *Reports, 1873–1920;* Wisconsin Department of Agriculture, *Crop Reporting Services Bulletin, 1947;* and Dieter Brunnschweiler, *New Glarus: Gründung, Entwicklung und heutiger Zustand einer Schweizerkolonie im Amerikanischen Mittlewesten* (Zürich: Fluntern, 1954), 66.

Figure 2.16. The Helvetia (Pet) Milk Condensing Co. plant, ca. 1915. The Highland, Illinois, company chose New Glarus as the site of a major factory largely because of its shared Swiss connection. As farmers could receive larger and more regular paychecks from the condensery, cheesemaking in the area effectively came to an end. (Courtesy of New Glarus Historical Society)

side as larger and more evenly timed paychecks, together with increased milk production, accompanied the ultimate demise of the local cheese factory system (fig. 2.17).[94]

While the New Glarus countryside was modernizing and accumulating wealth at stunning speeds, social change in the village advanced at a considerably slower rate, the period's second field of meaning. True, villagers could point to improved sidewalks, a large, modern church building, and an English-language newspaper as evidence of its progress. But on a deeper level, social relations changed more slowly.

Interactions between the Swiss and their neighbors proceeded at a snail's pace, due in no small part to the village's well-deserved reputation for clannishness. From the earliest accounts of the village until its centennial celebration after the Second World War, New Glarus maintained critical barriers to outsiders. Even as late as the 1950s, one area resident recently recalled, salesmen stood little chance of doing any business with New Glarus merchants unless they spoke perfect Schwyzerdütsch, or Swiss German; her father, though from Monroe and himself Swiss American, could not speak the unique dialect and thus was not given access to local markets by community merchants.

Figure 2.17. Henry Hoesley, milk hauler, ca. 1915. Tied to an increasingly modernizing countryside, area farmers and unemployed cheesemakers took up hauling milk to the Helvetia Milk Condensing Co. plant in New Glarus. (Courtesy of New Glarus Historical Society)

Table 2.5 In-Group and Out-Group Marriages in New Glarus, 1851–1950

	Number	Percent
Both partners Swiss or of Swiss descent	920	89.5
One partner of Swiss, one partner of Norwegian birth or descent	80	7.8
One partner of Swiss, one partner of other nationality birth or descent	20	1.9
Both partners of non-Swiss birth or descent	8	.8
Total	1028	100

Source: Derived and calculated from data in Dieter Brunnschweiler, *New Glarus: Gründung, Entwicklung und heutiger Zustand einer Schweizerkolonie im Amerikanischen Mittlewesten* (Zürich: Fluntern, 1954), 94–95; and Swiss Reformed Church records, New Glarus, Wisconsin.

This flour salesman's story is one that was shared by most outsiders seeking to do business in the tightly knit community.[95]

Even more telling are the rates of intermarriage between the Swiss and non-Swiss. Data from the marriage records of the Swiss Reformed (Zwingli) Church reveal that a Milwaukee reporter in 1905 was quite correct in his generalization that "there is little intermarriage of nationalities" in the community, a trend that was to persist for another fifty years. Indeed, nearly 90 percent of all marriages in the community for the hundred year period from 1851 until 1950 took place between partners of Swiss descent (table 2.5).

This remarkably high percentage of in-group marriages for the Swiss of New Glarus must be qualified in two respects. First, these data reflect only marriages at the Swiss Reformed Church (one of two churches by 1950, but by far the most significant). Second (and related), marriages outside the village, which most likely added a slight cosmopolitan tinge to the population, are equally absent from this figure. Nevertheless, given the exceedingly small size of the sole competing church—an Evangelical Associationist (later United Brethren)—and the equally small number of non-Swiss living in the community, this extraordinary figure would seem reasonably indicative of the closed nature of the village. Indeed, due to the "strong prejudice" against the non-Zwingli church, the small congregation was forced to locate some two miles outside the village boundaries. John Luchsinger gives a forthright account of the "other" church congregation, stating that "the Swiss of that day [during and just after the settlement] and . . . of the present day, regarded any one of their nation as degenerate and an apostate who became attached to any other denomination than the

71

Reformed Church; but in spite of opposition, dislike, and even perse-
cution, converts to the new sect were made in small numbers."[96] The
typically increasing rates of endogamy were held at bay considerably
longer among the New Glarus Swiss than other Wisconsin ethnic
groups, thus reinforcing their tight ethnic boundaries.[97]

CONCLUSION

On the eve of America's entry into the Great War, New Glarus had
established itself as the ethnic commemorative headquarters for the
surrounding region. Swiss-Americans, whether born into the community or having arrived via a subsequent migration, flocked to the small,
agricultural village for annual Kilbi festivals and for special commemorative events every ten years. The community welcomed its role as the
"central point of gathering on all holidays and festivals" by providing
commemorative landscapes and by creating increasingly complex festival organizations designed to mediate between diverse interests.[98]
Symbols of the homeland, the pioneer, and progress—at once contradictory and complementary—spoke to the community's fundamental
themes of existence. Most importantly, New Glarus solidified its position as the place of memory and commemoration by providing an intangible, but valuable, asset that was rapidly disappearing in the countryside around them: tradition.

That asset, which one historian termed its "Swissness function"
[*Schweizerische Funktion*],[99] was maintained not by monetary concerns,
but by the internal conflicts of the community itself, by its competing
fields of meaning. The small village, the site of the original settlement,
became a sort of collective hometown for the modernizing farmers in
its hinterland, as well as even less traditional kinsmen who departed
the region altogether. The place, in essence, bound together the increasingly modern and far-flung community through the performance of
festival time and space. In their use of heartfelt historical orations, by
exhibiting artifacts from the town's past in the form of a crude settler's
cottage, in their creation ceremonial landscapes and a historical monument, and by developing parades and performative events such as tableaux and the mock Landsgemeinde, the turn-of-the-century commemorations bore considerable resemblance in form to the civic
celebrations occurring across the country. Their content, likewise, increasingly pointed toward the growing tension between progressivism
and antimodernism found in the more complex historical pageants of
urban areas.[100] The commemorations effectively reminded the more
distant third generation of poverty's virtues while reveling in the diz-
zying thrill of progress.

PART II

Commemorating Heritage by the Community and the State, 1918–1948

3

Ethnicity Discovered and Re-presented

The Localization of Memory Between the Wars

In 1942 the University of Wisconsin, as part of its war effort, released a widely read scientific research report entitled "Wisconsin's Changing Population." Bringing together the expertise of scholars from the fields of political science, economics, rural sociology, anthropology, and geography, the task force sought to find "the solution of problems of immediate, practical concern." As university president C. A. Dykstra noted in his forward, the social scientists hoped to counter the false claims of "totalitarian demagogues" by publishing a scientific account of "our population resources." Prominent among those resources, and distinct from the totalitarian ideal of a racially homogeneous regime, was a state population comprised of many ethnic groups. In a classic expression of cosmopolitanism, the report cited that

Wisconsin today is an outstanding example of that phenomenon which is a peculiar heritage of the United States—a fusion of diverse nationality stocks into one unified commonwealth. Even today, some two or three generations after the arrival of the first immigrants, distinct nationality concentrations can be found. . . . It is in this situation in which the temperaments, aptitudes, prejudices, superiorities—all of which stem out of the values and traditions of culture—are fused into the dynamic reality which *is* Wisconsin. Understood and appreciated, these traits form the bulwarks of democracy."[1]

After displaying more than eighty pages of tables, graphs, and summaries, the bulletin concludes that "foreign-stock" groups have "contributed to the new American culture and have borrowed from it. It is this exchange which makes our present day Wisconsin culture one of the most cherished and powerful assets which the state possesses."[2]

We can begin to appreciate the enormous relief words like this must have brought to readers in hundreds of this report's so-called "cultural islands" when we recall the *100 percent Americanism* of just two decades earlier. With its well known heterogenous ethnic landscape and outspoken pacifist leaders, Wisconsin came under especially harsh fire during the Great War years. One New York journalist summarized the nativist view of many outside the state when he wrote in 1918 that Wisconsin "has not secreted enough digestive juices to dissolve its foreign elements. Scattered around its surface are little communities which retain each its own nationality. Not German alone, but Polish, Bohemian, Dutch, Swedish, Norwegian, Belgian. A traveler among these people finds himself in an alien land. . . . Here and now is Uncle Sam's opportunity to make this nephew state one of the family."[3] Not quite yet one of the family of states and charged with treacherous sedition, Wisconsin soon proved itself with distinction as it forcefully suppressed any identification other than with the nation-state.[4]

The profound transformation of the ethnic between the wars— from pariah to patriot—opened the door for a reevaluation of the role of ancestral heritage in American culture. This review took place, importantly, at two different geographic levels. At the first—or vernacular—level, individuals and local communities slowly emerged from what one historian called their "underground status" to create new spaces from which to express a distinct, local identity. During this period of ambiguity and transition, different groups adapted and reworked dominant society's values for their own, distinct purposes. At the second—or official—level, government officials, scholars, and writers each, along with Caroline Ware, seemed to make the discovery that "immigrants and the children of immigrants are the American people. Their culture *is* American culture, not merely a contributor to it."[5] This official recognition, in turn, helped create the space for ethnic expression at the local level.

The interaction between the local community and the state, or between vernacular and official culture, allowed groups like the New Glarus Swiss to create a new way of memorializing history and inventing place. Ethnic memory was localized between the wars and pressed into the service of larger political and commercial goals. Just as ethnic group values may come to play a strong role in shaping the local "rules of the game"—to embed itself in local political culture— so may the public memory activated by that group come to be seen as representative of a much larger "imagined community."[6] Emerging from severe anti-immigration sentiment, within a short time the once vilified ethnic soon came to personify the state—a most remarkable turn of events. This chapter, and the two that follow, trace this pivotal shift in American identity. Of the next three chapters, the following two examine how representational practices are employed by vernacular culture to legitimize local difference, while the third surveys official cultures' appropriation of those localized representations.

"WHERE DISLOYALTY IN WISCONSIN CHIEFLY CENTRES": MAPPING ANTI-IMMIGRATION HYSTERIA IN THE DAIRYLAND

In many respects, Wisconsin's ethnic experience during the Great War years was not altogether different from the rest of the country. Throughout the nation, states tried to outdo each other in proving their loyalty to the Stars and Stripes. With the rise of nationalism came an unprecedented surge (even for the often anti-immigrant United States) in nativism and an open attack against immigrants of all backgrounds. While previous wars tended to divert nationalism temporarily from nativist channels, World War I, John Higham notes, "called forth the most strenuous nationalism and the most pervasive nativism that the United States had ever known."[7]

Much of the antagonism was directed at German Americans, of course. But to a considerable extent, every group was a potential target, including both "old" and "new" immigrants. A generalized suspicion that the nation could not count on the undivided loyalty of the entire immigrant population gripped many throughout the nation. Condemning loyalties to Europe as "hyphenism," the very notion of group belonging was rebuked in forceful terms. President Woodrow Wilson, foreshadowing today's multicultural debates, helped invigorate this public sentiment by demanding that ethnics shed their dual identities. Addressing an assembly of newly naturalized citizens in 1915, he notified the new Americans that: "You cannot dedicate yourself to America unless you become in every respect and with every purpose of your will thorough Americans. You cannot become thorough Americans if you think of yourselves in groups. America does not consist in groups. A man who thinks of himself as belonging to a particular national group in America has not yet become an American, and the man who goes among you to trade upon your nationality is no worthy son to live under the Stars and Stripes."[8] Across the country, non-English language presses were shut down, community names were changed, foreign language instruction in schools was forbidden and, in the worst cases, public lynching and humiliation were tied to anything not "100 percent American."

"The 58 Percent American State"

Within this national setting, the Wisconsin experience would seem to be especially intense. The potent combination of anti-interventionist leaders such as Robert M. La Follette and Milwaukee socialist leader Victor Berger together with an unusually dense concentration of diverse immigrant groups created a context in which Wisconsin became known nationally as the "Kaiser's state." Senator La Follette, joined by nine of Wisconsin's eleven congressmen, led the congressional vote against Wilson's declaration of war in 1917 and fought a lonely battle for strict neutrality until the war's conclusion. The highly visible politician, joined by his many new supporters among the state's ethnic voters, earned Wisconsin the reputation as the nation's most "seditious state."[9]

Such ill repute churned the stomach acids of the state's prointervention leaders. One Madison newspaper editor explained to his readers that "it is no laughing matter, this reputation for disloyalty." With La Follette in the Senate, socialist Dan Hoan sitting in the mayor's chair in Milwaukee, and 26 percent of the state's vote going to the pro-German socialist leader Victor Berger in the 1916 U.S. Senate election, Wisconsin's bad image surprised none, but flustered the powerful. "Wisconsin everywhere—at home and abroad—is meeting an embarrassing discussion of and comment upon, her loyalty," that same editor grumbled some months later. "Not least of her troubles is the constant defense of her loyalty."[10]

Of equal importance to the "La Follette problem," the publicly expressed ambivalence toward intervention by the state's many ethnic groups deepened antagonisms. Wisconsin's third, and equally derogatory, nickname stemmed from its basic population makeup. Recognized as the nation's "58 percent American state"—a number that reflected the percent of Americans not of foreign birth or parentage—Wisconsin was often seen as an "ethnological laboratory." No other state at the time possessed "such a diversity of racial elements," including so many Germans, Norwegians, Welsh, Cornish, Bohemians, Finns, Belgians, Hollanders, Poles, Hungarians, and Swiss.[11]

Many individuals within these groups—for very good reasons—remained equivocal toward war against countries that had been home only one generation earlier. Foreign-language newspapers of all nationalities habitually railed against intervention as did public speakers across Wisconsin. When American entry into the war finally arrived, most ethnics quietly "supported" the war by buying the obligatory Liberty Bonds and sending their sons and brothers off to battle. Indeed, in response to the hostile suspicion of the nation, Wisconsin surpassed all other states in implementing the war effort once it arrived. It was the first to organize a State Council of Defense and managed to persuade 98 percent of its induction age men to respond to the draft, compared with 92 percent nationally. Wisconsin continually went "over the top" in all war drives, striving to prove its loyalty.[12]

"We Need to Measure the Sentiment in a Region So Overwhelmingly Comprised of the Swiss Race"

Given the weighty pressure to bury ethnic identification, these "patriotic acts" are better seen as performances of self-preservation. From every direction came powerful exhortations to drop dual loyalties, with the consequences for the failure to do so ranging from benign public denunciations to corporal violence. The *Milwaukee Journal* led the way with public denouncements of ethnic loyalties aimed at the spectrum of the state's groups. Throughout the war period, the paper sounded warnings strikingly similar to Wilson's, earning a Pulitzer Prize along the way. In 1917 it declared, "We want no more German-American banks, or Polish-American restaurants, or Italian-American bond companies, or Deutscher Clubs. . . . This is America. America it must be, wholly and unitedly, for all time to come. And, any club, society, company or organization that retards or conflicts with that spirit should change its purpose, close up, or be put out of existence. Appeal to racial or alien ties must be forever banished from American soil."[13] More direct forms of intimidation were recorded frequently across the state: from the "decoration" of a German-American State Bank with yellow paint in Merrill; to the tarring and feathering of a bartender in Ashland; to the suicide of a second-generation Milwaukee baker upon learning of his two sons' induction into the U.S. Army; to the forcing of a man in Pleasant Prairie to kiss the American flag "after he felt the noose of a rope placed around his neck," such incidents became so common as to warrant a mere one-sentence byline in the state's newspapers.[14] During the war, and to the surprise of many, Americans realized that even second and third generations of older groups such as the Germans, Norwegians, and Swiss were still sufficiently "nationality conscious" to oppose unified military action in Europe. Such measures of repression attested to the tensions surrounding dual loyalties as they shattered the illusion of unity among diverse groups.

The depth of ethnic American reluctance to become engaged in the war—one of the critical forces behind such anti-immigrant hysteria—was measured in one of the few official referenda on United States entry into the conflict. The 1917 referendum was held in Monroe, Wisconsin, Green County's seat of government in the heart of "Swiss country." More than 60 percent of the population was Swiss American, as was the referendum's driving force, Green County Judge John M.

Becker. Becker persuaded the city council of the importance of register-
ing the "sentiment toward the war in a region so overwhelmingly comprised of the Swiss race." The referendum question was placed on the
April 3 ballot and read, "Under existing conditions, do you favor a
declaration of war by the Congress?" The citizens of Swiss-dominated
Monroe responded in the most unambiguous of results, as "nays" outnumbered "yeas" ten to one.[15]

While the referendum might have helped to raise the esteem of a
group exercising its political rights, the entry of the United States into
the European conflict three days later surely punctured those short-
lived sentiments. The planned referenda in neighboring Brodhead,
Monticello, and New Glarus—communities that Becker rightly believed would follow Monroe with even more favorable results—were,
of necessity, canceled. Judge Becker himself was indicted and con-
victed on charges of sedition the following year and, though his sentence of three years in prison was later reversed, his career was
ruined.[16]

In response to the Monroe referendum and for its well-known public displays of ethnic culture, the Swiss region in southern Wisconsin
acquired a reputation as one of the state's areas "most infected with
pro-Germanism." Together with the heavily German area extending
from Milwaukee along the shore of Lake Michigan, the Polish core of
Marathon County, and the Slavic lumber workers of Sawyer County, to
name just a few, the Green County Swiss concentration was identified
and mapped as a disloyal center, or a "rotten spotted area."[17] The 1918
"Sedition Map," supposedly based on six months of investigation
work, aimed its sights at dispelling East Coast fears of Wisconsin (fig.
3.1). The map's compiler, the Wisconsin Loyalty Legion, was a private
movement established shortly after the United States entry into the
war with its explicit goal being to "repudiate, in the name of Wisconsin, every disloyal work and deed calculated to misrepresent her and
her people." With its prime objective centering on "eradicating any
and all foreign and non-American racial traits which our emigrants
have brought from [a]broad," much of the Legion's activities were directed against the state's many ethnic Americans.[18] No more effective
way to demonstrate loyalty could be found than mapping the locations
of *dis*loyalty. These "seditious regions," when mapped, could then be
monitored and controlled.[19]

"The Time for Argument Is Past"

The response in places such as New Glarus, predictably, was to push
public displays of ethnic culture underground.[20] Throughout the years

Where Disloyalty in Wisconsin Chiefly Centres

"SEDITION MAP"
PREPARED BY
THE WISCONSIN LOYALTY LEAGUE

The shaded areas show the districts most infected with Pro-Germanism.

For the first time, probably, in the history of the country it has been necessary to map the sedition of a state of the American Union. The Loyalist Legion of Wisconsin, after an investigation lasting about six months, located in certain counties of the State certain centres of pro-Germanism and active or silent disloyalty. It purposes to be only a rough and general indication of the way in which Wisconsin is rotten spotted. There are counties not indicated by the shading where strong sympathy for Germany, and very active backing for La Follette is manifest, but on the whole the counties marked represent the most violent hostility to the President's policy and to the war aims of the nation. These are the counties in which Thompson, La Follette's candidate for the Senatorship, received his heaviest vote and where Victor Berger rolled up a large vote.

Figure 3.1. Sedition map, ca. 1918. The Wisconsin Loyalty League sought to counter the state's un-American reputation by cartographically controlling and containing its "pro-German" element. This map was reproduced in several of the nation's largest newspapers, including the *New York Sun*. Note sedition district #3, the Green County Swiss settlement. (Courtesy of State Historical Society of Wisconsin, neg. WHi [X3] 49718)

before official United States entry into the war, the village's two news-papers, the *New Glarus Post* and the *Deutsch Schweizerische Courier*, maintained official policies of neutrality. In reality, however, both pa-pers, and especially the German-language *Courier*, favored the German side through editorials and reports of English atrocities. The *Courier* regularly carried syndicated columns from other German-language presses that lampooned the Allies and offered a sympathetic portrayal of the Axis powers. Direct criticisms of President Wilson and the esca-lating moves toward involvement were less forthcoming, although con-demnations of "superpatriots" who frequently mistook the Swiss for Germans often reached the front page. In one typical 1917 headline, the *Post* sardonically inquired: "Are Melchoir Schmid and Paul Jackson German Spies?" It seemed that, on the way to a Milwaukee rifle club meeting, the two New Glarners were detained in Milwaukee by detec-tives and accused of "running a secret rifle and ammunition factory in New Glarus." Although the incident involving "our two respected" citizens was brushed aside with irony, a current of frustration and ten-sion ran though this and like stories.[21]

Upon the war's arrival, such mockery ended. Citizens were asked explicitly to put aside their personal feelings about the war and refrain from rocking the boat. Immediately after the U.S. declaration of war, the village's mayor appealed to his Swiss readers, "Whatever our indi-vidual idea or feeling with regards to the war problem, the time for argument is past, and as American citizens we should refrain from disputing the situation either in public or private."[22] Though they bris-tled at the charge of sedition, the New Glarus Swiss, like most "hy-phenated Americans" in Wisconsin at the time, disagreed vigorously with the policy of intervention, a stance that influenced the community deeply for decades.

COMMEMORATIONS IN THE WAKE OF THE GREAT WAR

The events of the war, and the anti-immigrant sentiments that accom-panied it, bore a direct impact on the Swiss American community's festive culture. Public displays of ethnic identity receded into the back-ground for the Swiss of New Glarus, as they did for many "hyphen-ated American" groups in the years immediately following World War I. In some instances, a demonstration of a group's loyalty to the United States became the centerpiece of a commemoration. Organizers of the Minneapolis Norse-American Centennial celebration of 1925, for in-stance, strove to demonstrate their group's loyalty to their adopted country with massive displays of patriotism. In other cases, heretofore frequent public displays of memory faded into the past. Milwaukee's Turners, for example, rewrote their constitution to restrict the festivals and plays that were such a part of their turn-of-the-century life.[23]

In New Glarus, the chief commemoration of this period took a decidedly inwardly focused turn. New Glarners were stung by the bitter memories of the war years and their defensiveness became even more pronounced after Congress passed the Immigration Restriction Act of 1924. As a direct consequence, eighty years after the first immigrants arrived, their memory was commemorated with what the *New Glarus Post* described as "a quiet, homey event." Indeed, the 1925 celebration's departure from previous memorial festivals could not be more pronounced. "No special attractions to draw a crowd" were created. Rather, church services and a small program at the shooting park (*Schützenpark*) north of the village defined the day. At the park, villagers and a handful of friends listened to low key speeches and modest musical entertainment (fig. 3.2). The audience heard the only surviving settler reminisce about the poverty that impelled the emigration, and the day's sole dignitary, former resident and Secretary of State Sol Levitan, praised his quaint neighbors for their assimilability.[24]

Figure 3.2. "It was a quiet, homey affair." In contrast to its predecessors, the eightieth anniversary commemoration in 1925 was a subdued event, marked by a simple day of speeches and songs by the Männerchor. (Courtesy of Elda Schiesser Collection)

The Ambivalence toward "Nonappreciative" Visitors, or the Beginning of Tourism

There had been some contention regarding the scale and meaning of the event. A few influential citizens argued in favor of a larger festival as before, but the majority opted for a smaller-scale celebration. Certainly the shock of noisy anti-immigrant sentiments still rang in the ears of some community leaders. Yet 1925 also marked a moment of "awakening, activity, and acceptance" for a wide array of older stock Euro-Americans across the country. In addition to the immense Norse-American Centennial celebration in Minneapolis, that year also witnessed the creation of a park to commemorate the three hundredth anniversary of the founding of the town of New Sweden, on the Delaware River; the Chicago Historical Society holding exercises in honor of the French explorers Marquette, La Salle, and Tonti; the Norwegian-American Historical Organization making plans to designate October 9 as Leif Eriksson Day, thus preempting Columbus Day by seventy-two hours; German Americans creating a National Historical Society two years later in order to "improve American attitudes towards persons, customs, and things of Germanic origin;" and, at this same time in the American Southwest, an alliance of local Anglos and Latinos transforming Santa Fe's Fiesta into a celebration of "Spanish folk culture."[25]

If John Higham is correct that the strong anti-German nativism of the Great War years—and the vague, if potent, "anti-hyphenism" that grew out of it—was a "spectacular reversal of judgement" toward many groups, only slightly less spectacular was the ensuing reversal. For, almost as quickly as they attained pariah status during the war, so did "Old World people" come to be seen as romantic vestiges from the past. With the politics of identity effectively neutralized by the war, the "contributions" or "gifts" of ethnic Americans to the ever-boiling melting pot became more easy to accept and recognize. Especially after the passage of the 1924 Immigration Restriction Act—legislation that marked the culmination of the nativism that began during the war and brought to an abrupt end a century-long era of mass immigration—a slowly growing reassessment of the immigrant experience was launched.[26]

Some historians, like Philip Gleason, have seen this reevaluation as an indication of the dwindling importance of "the ethnic dimension" in American life.[27] The evidence in Wisconsin points in a slightly different direction, where, if anything, the rhetoric surrounding ethnic life becomes even more frequent and is heard in many different corners of the state. Rather than seeing ethnicity die in 1924, it is perhaps more fruitful to monitor the context of changing rhetoric surrounding group identity. The two faces of the ethnic on either side of the war—as potential traitor, or potential contributor to the American republic—mirror the two sides of a similar concern: how to incorporate diverse people with divided loyalties into a unified state.[28]

A second reason, then, for the smallish scale of the 1925 commemoration stems from a rather different rhetoric. Merely looking for entertaining diversions, a new type of Festgäste threatened the seriousness of memorializing: ". . . in this day of rapid transportation which enables massive crowds to gather at short notice, it is feared that the village would be overrun with visitors during this time, visitors who would not fully appreciate the significance of the anniversary which means so much to the Swiss."[29] The potentially nonappreciative visitors represented a new type of guest to the community—the tourist. Though Festgäste descended upon the village by the thousands in years before, most held some connection to the community, if only a shared ancestry. Previous celebrations had a distinct homecoming quality, but now a new form of visitor was making its way into the community, a visitor whose travel was based not on ties of descent, but on a curiosity toward others.

Perhaps some of the would-be visitors in 1925 had read of the New Glarus Swiss in Wisconsin's first periodical devoted to tourism in the state. Just as the U.S. Congress was passing legislation to restrict immigration, the *Wisconsin Magazine* (the "know your own state" magazine) ran a four-page article entitled "How Our Ancestors Settled in Wisconsin." The article outlined the "story of the conglomerate peoples composing our population." As a point of pride—not shame—the author applauded Wisconsin for its nickname, "the polyglot state, a title it has well merited." Foremost among the half dozen immigration vignettes, and despite their relatively small numbers, "the story of the Swiss is, perhaps, the most fascinating." There followed a detailed account of the planned migration from the snowcapped Alps to the New World, creating a narrative, predictably, of "progress."[30]

If there was resistance toward receiving unwelcome visitors during important times of commemoration, a growing awareness of tourism's possibilities soon became apparent. The same year that the *Wisconsin Magazine* article appeared, the village transformed the Schützenpark into a space to be used by tourists. The site of important communitas-building events, the Schützenpark had long been used by the local Schützenverein for local and regional shooting matches as well as for some of the commemorative activities of early festivals (fig. 3.3). One

Figure 3.3. New Glarus Schützenverein, ca. 1920. Organizations such as the *Schützen-verein* (shooting club) reinforced Swiss collective identity through shared ritual. (Courtesy of New Glarus Historical Society)

year after the last regional Schützenfest, the grounds were turned into a tourist park as a way to persuade some of the new breed of automobile tourists to stop and camp on the edge of town.[31]

In setting up a park removed from the core of the village yet close enough to enable eating at its one or two restaurants, New Glarus was following a path taken by most small towns in the years after World War I. Indeed, just as tens of thousands of tourists took to the road in the 1920s, so were tourist parks or campgrounds organized to control and direct the crowd. By the mid-1920s municipal autocamps became so numerous and commonplace as to be transformed, in the words of one historian, from fad to institution. Wisconsin alone claimed more than three hundred such autocamps in 1923. Every town, it seemed, had a place for tourists to camp. In the case of the Swiss community, no attempt was made to accentuate anything distinctively "ethnic" or "Swissesque," a point reflected in the fact that the park became simply known, by locals and visitors alike, as the "tourist park." Its one "attraction," a replica log cabin, conveyed the hope of catching the eyes of some of those motoring by (fig. 3.4).[32]

The connection between ethnicity and tourism at first seemed to come ambivalently to the burghers of New Glarus. One woman editorialized in a local paper that "it seems odd that a community with so

86

Figure 3.4. Log cabin replica at tourist park. Like many small Midwestern towns, New Glarus opened a tourist park in 1920 to accommodate growing numbers of automobile campers. Here, it is home to the 1925 eightieth anniversary commemoration. (Courtesy of New Glarus Historical Society)

large a Swiss population and known all over the country because of it does not possess a typically Swiss eating place in the entire county." Another writer from Milwaukee expressed dismay at finding only *"American* rural architecture" in the community and at not being able to eat Swiss cheese in the village's newest tavern.[33] Other visitors searched long and hard for quintessential "Swissness" and some even found it, but with effort. One early tourist, hunting for "quaint life" amid the "thoroughly modern" streets, discovered that "the Swiss element still gives the town a richness of color which it must be said the bulk of American towns lack, with their sordid sameness of appearance. Podunk, Iowa, is almost identical with Podunk, Ohio." What impressed this tourist as Swiss was the *Anglo*-designed and -built church that seemed so American in appearance to writers only twenty years earlier. Now possessing an "otherness" that seemed somehow Swiss, the unique scene is described this way: "Turning immediately left again you are on Main Street with a vista that is almost European. The street runs for half a mile through an avenue of trees and fine homes toward a tall red brick clock tower, which had about it a decided 'old country accent'; you might be in Glarus or Schwanden, Switzerland."[34] Without the lived knowledge of landscape in Glarus or Schwanden and by ignoring the commercial blocks on either side of the street, the generic vista could achieve an "old country accent." Local papers were quick to pick this up and, two years later, used the experience to demonstrate to a generation unaccustomed to large-scale commemorations that "when everyone co-operates with those in charge of a big project, things go over big."[35]

"HERE THESE VISITORS BREATHE, IF NOT THE AIR, AT LEAST THE ATMOSPHERE OF THE OLD COUNTRY"

The "big project" centered on the ninetieth anniversary of Swiss settlement in the region. A big project indeed, the 1935 commemoration far surpassed any event before the war in terms of organizational complexity and richness of rhetorical symbolism. Where earlier festivals featured silent and motionless tableaux of historic scenes, the ninetieth anniversary staged a historical pageant, explicitly designed to foster a sense of historical community and continuity through repetition and bodily performance. The historical pageant overshadowed all previous commemorations in transforming history into myth, and the community's past into a shared ritual. And while parades, orations, and special church services accompanied the historical pageant, they only served to reinforce the pageant's central aim of asserting the essential continuity of *local* tradition amid sweeping social and economic change.

Many socioeconomic changes had taken place since the massive prewar commemorations, including the demise of cheese making in the New Glarus vicinity and the advent of factory labor by many area farmers at the Pet Milk plant; increased mobility and a new breed of visitors brought about by the automobile; the cessation of immigration from Switzerland; and the use of English in all school classes. Most disturbing to the self-proclaimed "old guard," however, was the increasing use of English in church services.[36]

Well aware of these changes as well as the threats they posed to community, village professionals feared that insufficient support for a celebration could be summoned. The *New Glarus Post*'s headline, "Shall New Glarus Celebrate its Ninetieth Anniversary?," was met with early indifference. The newly formed Commercial Club—organized and run by the prominent business interests—took the lead in the festival's production. As Larry Danielson found for Swedish Americans in Kansas, such commercial involvement became a feature common among ethnic festivals of this period.[37] For these business interests, curiosity at the organizing meetings, although welcome, was not enough: "idle interest will not put on the celebration, but if everyone in the community will co-operate, maybe the job will not be so difficult." Unity and cooperation were not automatically rendered, but coaxed out of a rather guarded community. "After considerable discussion," a vote was taken and it was decided to hold the celebration over a three-day period that August. The celebration's centerpiece was to feature some sort of ill-defined pageant or Swiss play, which portrayed the life history of the community; but just how that history should be told, by whom, and on which aspects to focus all needed to be decided.[38]

Dr. John Schindler and the Creation of a Historical Pageant

One committee member's devotion to the commemoration was unwa-vering. At the Commercial Club's meeting, the festival's executive com-mittee enlisted Dr. John Schindler to write the historic pageant that was to become the centerpiece of the three-day event. Schindler, the son of the village's first mayor, came from a prominent family that traced its roots back to the original colonization. Although a successful physician by profession, his real love derived from the local history and lore of the region. Schindler saw history in grand, sweeping terms across vast stretches of time and space, ultimately leading to the pres-ent. His progressive view of history was enriched by an admiration for the lives of local residents and the pride that he felt (and believed others must feel as well) for local place.[39]

Fused with that progress came an antimodernism more pro-nounced than in the previous generation's commemorations. Wistfully, Schindler hoped to bring the past back into the lives of the local people and slow down the faceless march of modern, mass society: "Cars whiz by in endless procession, up though the woods, over the hill and are gone. Seldom do the occupants realize that the little road that crosses the main thoroughfare was once the most important road in the state, and that around that road there hangs the romance of the building of Wisconsin."[40] History, John Schindler believed, was filled with romance and dripping with nostalgia. If studied and lived, the use of that history—heritage—could lead to the good life. The past, removed from the forces of modernity, held the key to a better future. No better person could be found to write a similar narrative for the Swiss colonization.

A second factor made the choice of Schindler as author all the more appropriate. Although a community member by birth, he spent a good part of his early life away from New Glarus attending the University of Wisconsin in Madison where he came under the influence of two key members of the nationwide pageantry movement. Thomas H. Dickinson and Ethel Theodora Rockwell were both associated with the University of Wisconsin during the war years (and Schindler's school years) and played considerable roles in the movement's growth. Dickinson, a professor of English and author of *The Case of American Drama* (1915), was a forceful advocate of community historical pageantry as "an early expression of an art impulse springing from the soil." By dramatizing events from the local past and present, the pageantry's naive dramatizations represented an original American form of "social ceremonial." Dickinson advocated the use of folk themes to articulate a political message common to progressive thought: by creating a drama that was "socially constructive," the dramatist should engage in "expanding the purposes of social progress." Socially constructive drama would help a community "discover itself" by "expressing the self-consciousness of the community as a social unit."[41] A student of Dickinson, Ethel Rockwell was equally committed to the art form. While she also theorized about the social benefits of community historical pageantry, the younger Rockwell went on to play a more active role in the actual creation of pageants. In addition to writing historical pageants for Wisconsin's territorial and statehood centennials, in 1936 and 1948, respectively, she also became involved in assisting communities—from Wisconsin to North Carolina—to inaugurate their own historical pageants. She also became involved directly with the New Glarus Swiss community by lending advice for its *Wilhelm Tell* pageant in the late 1930s, the subject of chapter 4.[42]

It is difficult to imagine their paths not crossing, for, even if Schindler did not personally meet either Dickinson or Rockwell, he certainly followed their spirit. The purpose of his pageant, as Schindler put it, would have struck both professionals as wholly on target. The New Glarus pageant, Schindler wrote, "is done chiefly to remind each new generation of the splendid courage, endurance, and co-operation which signalized the rather unique beginnings of our village. It is done to carry the virtues of our past into our present and on into the future."[43] Moreover, Schindler carried the political agenda of the progressive dramatists into his drama. The pageant was not merely to depict a local history, but was intended to "recapture the pioneering spirit of our forefathers by dramatizing the early history of the Colony of New Glarus." With a firm belief in the ability of performance to help the community "discover itself," Schindler hoped that by playing the part of the colonists, present-day New Glarners could learn from them.[44]

Making a Good Impression: Behind the Scenes at a Commemoration

Village business leaders, meanwhile, concerned themselves with the more prosaic business of organizing the festival as a whole. More than a few unexpected difficulties arose. For one, a generational shift in the community had occurred. Fewer residents remained who had actually been born in Switzerland. Knowledge of the homeland and its customs was becoming concentrated among a handful of recent immigrants or second- and third-generation Swiss Americans who, like John Schindler, possessed a historical interest in the colonization. This growing distance presented problems for an accurate folkloric performance, as when, for instance, organizers discovered that nobody possessed a "Swiss Alpine horn" for the event.[45]

Next, participation still required hustling. Newspaper ads ran throughout the spring and summer "asking for further co-operation to enable [the committees] to show just what progress the Swiss did make in ninety years."[46] Finally, there was the problem of language. Although Swiss German remained the language spoken in many homes, both in the village and on the farm, the generation coming to age in the 1930s was the last for whom this was true.[47] The pageant, as it was to encapsulate the traditions and life course of the community, brought the issue of language loss to a culmination, "greatly puzzling" the executive committee. To perform in English would be, in essence, to admit to the ultimate demise of Swiss culture, but to perform in Swiss German ran the risk of alienating the growing number of non-German speaking third- and fourth-generation residents. Though never spelled out explicitly, the executive committee decided to stage the pageant "in the Swiss language mainly," with the assistance of loudspeakers through which key scenes were explained in English.[48]

In the end, festival organizers seemed to have been able to solicit the sort of support they needed to produce the event. Participation came from a broad cross section of the village and township population. The way that participation was utilized in the pageant, however, was not as evenly distributed (table 3.1). The vast majority of the pageant officials and committee members, indeed fully 85 percent, was comprised of the village's business class, with only a handful of farmers and milk factory workers taking part in such organizational work. Their contribution came in the form of performance. Most of the characters in the pageant with speaking roles, as well as those portraying the colonists, either worked on a farm full-time or split that time at the Pet Milk factory. All of the folk dancers came from neighboring farms.

The twofold goal of the pageant—to project a positive impression of the community for outsiders and to reconnect insiders with their pioneer past—appears to have been at least partially accomplished even before the production. Such "restored behavior," Richard Schechner notes, hinges on the rehearsal process. As is frequently the case in drama performance, the lengthy rehearsal period brought the memory of the colonization to life for many of those who participated in the drama.[49] Ninety years had passed, after all, since the arrival of the first colonists, none of whom remained to tell their personal stories of

Table 3.1 1935 Historical Pageant Participants by Occupation

	Number of participants	Village professional[a] (%)	Farmer/milk plant[b] (%)	Unknown (%)
Officials and committee members	20	85.0	15.0	na
Characters with speaking roles	34	32.4	58.8	8.8
Colonists and families[c]	30	6.7	93.3	na
Dancers	20	0	100	na

Sources: *Official Program of the Ninetieth Anniversary Pageant, Presented at New Glarus, Wisconsin* (New Glarus, WI 1935). Occupational data supplied by Millard Tschudy.
[a]Village professions include: furniture dealer, teacher, insurance broker, automobile salesman, pastor, blacksmith, bank clerk, tavern keeper, lumber dealer, factory supervisor, engineer, feed operator, and physician.
[b]During this period, many farmers at one time or another worked at the Pet Milk Condensing Plant, and conversely, many factory workers at one time farmed. For the purposes of this analysis, they were combined into one category.
[c]The data source only lists "household heads," although both men and women, together with their children, performed as colonists. The number 30, therefore, considerably underestimates total participation.

hardship. Moreover, many of the Swiss Americans, though they themselves or their parents might have emigrated, held no familial connection to the earliest colonists, but came from other parts of German-speaking Switzerland or as part of a later migration wave from Canton Glarus. Social memory, it has been argued, is most likely to be found in commemorative ceremonies that are performative. The dozens of Swiss Americans, who night after night finished milking early or raced home from the factory to make it to rehearsal, found this to be the case. "Many people have said," it was reported a week before the scheduled performance, "that they have learned more about the history and development of our community from *hearing the pageant, in practice,* than they have been able to acquire in a lifetime by other means."[50] If Schindler seemed more concerned with "carrying the virtues of our past into our present" for the benefit of community members, the pageant organizers worried considerably more about making sure that link was understood by outsiders.

"This is a Moment for Co-operation, Unity, and Harmony": Messages for an Emigration, and a Depression

A substantial proportion of all visitors were ethnic Swiss. Highland, Illinois, sent a group of 150 Swiss Americans to "spend the week-end

as a friendly gesture from one Swiss village to another Swiss village."
Many visitors were themselves former residents or the descendants of
former New Glarus residents, coming from as far away as California,
Texas, Pennsylvania, and Arkansas. As for many celebrants at the turn
of the century commemorations, for many of these people the event
functioned as a homecoming, a chance to renew ties and celebrate a
shared heritage. "Here, for a few days," the *New York Times* reported of
the festival, "these American Swiss visitors breathe once more, if not
the air, at least the atmosphere of the 'old country.'" And in this way
the festival functioned less as a tourist attraction, and more a place of
memory and a geographic embodiment of tradition.[51]

But other visitors—those with no ties of descent—came with different purposes in mind. Governor Phillip La Follette, for one, seemed
to hardly notice anything remotely "ethnic" about the gathering. For
the state's leading politician, the day presented an opportunity to remind his audience of three thousand that "We must be pioneers today.
The pioneers did not wait for opportunity to come to them—they
blazed new trails that others could follow." Considerations of the Great
Depression were preeminent as the governor repeatedly utilized the
iconography of the village's pioneer monument to drive home his conservative political economic vision. La Follette searched for a usable
past and seemed to find it in the pioneer: "It is up to you and me to
take the experiences, traditions, and principles of the pioneers and
apply them to the problems of today. These [pioneering] principles
have not failed—they are as true today as when they were formulated,
but we have not always remained true to the principles." The governor
then pleaded for state and federal governments to "stop paying people
to be idle." Every American should have "an opportunity to earn his
own living by useful work."[52]

La Follette's was an age-old tactic; namely, to draw selectively from
the perceived virtues of the past to fight today's problems. But, coming
from the governing elite, with no ties to the community or their local
heritage, his speech rang hollow. The individuality of the work ethic
was emphasized at the expense of any group cohesion, a message that
flew in the face of the festival organizers. For La Follette, pioneer iconography did not sit alongside an ethnic vision, but came at its
expense.

Fred Ott, the festival's chairman, found it considerably easier—and
much more important—to reconcile the pioneer and the ethnic. Recall-
ing questions about the group's loyalty less than twenty years earlier,
the local furniture dealer reminded the audience of the festival's real
purpose. New Glarus celebrated to "keep alive the Swiss spirit.
Though we are loyal citizens of the United States, we are still Swiss

and believe that we have a heritage that is worth preserving." Ott entreated the crowd abandon from environmentalist notions of identity, and to recognize that a people's spirit rested not on geography, but, instead, on folklore: "We have no mountains or deep valleys which may re-echo our songs this afternoon. It is, however, not the landscape which makes a country, but the people, who live there. . . . Our lives must reflect the real Swiss spirit—undaunted courage, unswerving determination, the spirit of independence, love of liberty and highest loyalty to God and *our countrymen*. If we can re-awaken or strengthen this spirit in you this afternoon, we will feel that our efforts have not been in vain."[53] In the end, Ott's values differed little from those of La Follette, except in one crucial respect. Whereas the governor's "experiences, traditions, and principles" were found in a generic pioneer, for Ott that same spirit was found in an ethnic, and more specifically, a *Swiss*, pioneer. Pioneering could be—and for the purposes of the ninetieth anniversary commemoration, had to be—communal.

The central event of the festival, of course, was John Schindler's historical community pageant. Depicting various chapters in the community's collective memory, the pageant portrayed the organized transplantation from the Swiss canton of Glarus through to the early years of the twentieth century in Green County, Wisconsin.[54] The cast included the early colonists (i.e., pioneer men) and their families, dancers and yodelers, Civil War veterans, idealized female representations of Switzerland, the United States, and Uncle Sam (fig. 3.5). As much as possible, descendants of the early settlers were recruited to play the role of their ancestors. A virtual throwback to the popular historical pageants of the progressive era, the New Glarus celebration of *community* and *tradition* offered ethnic folklore as the region's "authentic" culture. Local community cohesion, it was hoped, could be strengthened by accentuating locality's unique, ethnic identity and its place within a wider historical narrative (fig. 3.6).[55]

True to the "formula" of similar pageants across the country, the New Glarus spectacle seemed to echo one pageant coordinator's vision of an event in which "the place is the hero and the development of the community is the plot."[56] Indeed, the audience was supposed to feel awe at the heroism of the place creators and, in turn, take their example into their own lives. Reworking the rhetoric of romantic nationalism for vernacular purposes, the author summarized the twin themes of place and its development, saying, "The growth of the colony has been steady until the present time, and it is a far different village that would greet our founders today. But the advantage we have over the founders in a material way, would be more than outweighed by the advantage which they possessed in things of the *soul*."[57] At this time of economic

Figure 3.5. Helvetia at the 1935 historical pageant. Miss Olive Becker won the honor of representing Switzerland as "Helvetia." (Courtesy of New Glarus Historical Society)

Figure 3.6. Characters in historical pageant. The historical pageant of 1935 reenacted the immigration story of ninety years earlier. With a cast comprised of both immigrants and third-generation ethnics, the pageant reinforced the theme that the Swiss were the real winners of the West. (Courtesy of New Glarus Historical Society)

crisis and national hardship, a hero came in the form of a local place where ordinary men and women eschewed material extravagance to live simpler, better lives.

The story line progresses upward, retelling a fable of solidarity and unanimity overcoming individualism, cowardice, and opportunism. When dissension threatens the emigration at an early stage and some argue for turning back, one mock colonist reminds the group not to forget that, "We are Swiss whom hardship can not stop. This is a moment for co-operation, for unity, and harmony."[58] Sounding more than a little like the festival organizers ninety years later, this passionate plea for unity recurred throughout the narrative. Upon reaching the New World, the pageant's mock colonists heard the same message repeated: "Stay together, Swiss with Swiss, and wherever you go, you will live and do as they do in Switzerland. You will keep alive the things which are Swiss."[59]

The pageant proved to be an unqualified success. The three performances attracted nearly 6,500 people and was marked, according to the local newspaper, by "a spirit of co-operation." Though organized largely by the Commercial Club, the Monroe newspaper noted that "probably the most noticeable thing about the celebration was the absence of the 'Commercial spirit' customarily found at a community celebration." The re-presentation of community, so often held in contrast to commercialism, seemed to hold more importance than making money.[60] Its middle-class organizers, almost all of whom were prominent members of the village's business community, breathed a sigh of relief in the event's "absence of 'rowdyism.'" Curiously, the scheduled carnival curiously failed to show up. Rowdyism and elements of the carnivalesque, so much a part of festival, would have posed a potential threat to the propriety and seriousness of the event and might have mocked the earnestness of the complex messages being sent forth.[61]

And just what were those messages? First, that ethnic heritage sat squarely in the middle of American cultural life. The pageant depicts the Swiss immigrants as the real winners of the west, as civilized openers of the frontier to white settlement. The only non-Swiss in the play—a quintessential "American" who doesn't speak Swiss—is lampooned for his crude manners and ungrammatical speech. Dressed "roughly, with no shirt, but only ragged underwear," the American pioneer chews, spits and talks in "drawling English." He is lazy as well as racist as he puts his nose to the ground, takes a couple of big whiffs and declares, "By darn, seems to me I smells Injuns." By contrast, and with no small amount of ethnocentricity, the Swiss pioneers are industrious, thrifty, and, despite their poverty, well-mannered and better dressed. Their relations with Native Americans are those of amicability and deference, attesting to a frontierism seemingly better than that of the classic American version.[62]

This leads to the pageant's second theme: the values and traditions of the Swiss can be of considerable use in dealing with the age's great problems, both economic and political. Indeed, beyond their work ethic, the alleged *Swiss* pioneer values of patriotism and the inherent love of democracy could be used to highlight the moral chasm between freedom and oppression. Arguing for the "Buckskin Shirt" of the pioneer to combat the "Dictatorship of Italian Blackshirts," the "Fascism of German Brownshirts," and the "Communism of Russian Redshirts," the pageant's author positioned Swiss pioneers as prototypical democrats. The celebration of ethnic pioneers enabled the community to place itself within the continuity of American history and yet simultaneously to express pride in being different and not completely welded into the larger, homogenizing national culture.

This strategy was both adaptive and defiant, for the final—and really most important—message of the performance is a veneration of local attachments to place and vernacular culture. It argued that the real heroes of the story are the ordinary people being born, living difficult lives, and encountering and overcoming obstacles. This was accomplished through hard work, of course, but even more importantly through "co-operation, unity, and harmony." The construction of peaceful historic community spoke to both heartfelt pride of local residents as well as ethnic leaders' very real concerns over the decline of an integrated, harmonious ethnic community.

What rowdyism and the carnival threatened to do was to break the seamless web of community and cooperation of this cultural presentation. The desire to present an illusion of consensus prevented the pageant's author from including other, more complicating factors in the village's history, including well-known social and geographic schisms brought over and transplanted from the earliest immigration; discrimination against non-Zwingli denomination churches; an "epidemic of cross burnings" in nearby Monroe and well-attended Ku Klux Klan meetings just south of town; and the increasing industrialization of the countryside coupled with the attendant weakening of the traditional small-scale and decentralized Swiss cheese factories.[63] To present such an unflattering history would puncture the pageant's affirmation of community and social cohesion and imperil the performance's celebration of local attachments to place and the ever receding ethnic past.

PRESERVING A RELIC COMMUNITY IN PLACE: THE SWISS HISTORICAL VILLAGE

As plans were being made for the 1935 commemoration, several community members vocalized the need for a permanent site of memory dedicated to the early colonists. The ninetieth anniversary celebration articulated the current necessity of remembering the communal values of the early colonists and of demonstrating the continuity of descent as tangible connections were fading from everyday life. If anything, the event inadvertently demonstrated the distance between the immigrant and third generations. It was thus felt that a museum created in the spirit of the Swiss pioneers and exhibiting Swiss folk life in "living style" might make the connection between past and present more palpable.

"There Is No Money for Buying the Swiss-Men-History": Making Ethnic History Local

The first tangible steps toward creating a museum dedicated to the collective memory of the area's pioneers came not from New Glarus, but from neighboring Monroe. In the fall of 1937, two years after the production of his pageant, John Schindler was joined by a contingent of history buffs in organizing a county historical society, and to "preserve the history and records of the county as well as those relics and antiques which are still left from the pioneer settlers." One hundred miles to the south, in Rock Island, Illinois, a recently formed historical society of Swedish Americans received an appeal for similar action. Only four months earlier, Marcus Lee Hansen told members of the Augustana Historical Society to grasp the historical moment and begin the preservation of their group's history. Immigrant historical societies such as these could waste no time in collecting the documents and recording the important experiences of the many millions of Europeans who came to the United States in the nineteenth century. The time to rescue memory from oblivion had to begin immediately, while the third generation was awakening to the legacy of its grandparents. Or, to paraphrase Pierre Nora, memory's rescue assumed critical importance because there was so little of it left.[64]

Unlike the Swedes of Rock Island, the Swiss of Green County did not have a professional historian to guide and cheer on their organization. This is one reason why, despite the best intentions on the part of its early organizers, the Monroe-based organization foundered. To be sure, well-intended ideas bounced around the room, some of them quite ambitious, including the conservation of official records and the compilation of township and family histories; the marking of historical trails; the preservation of the growing number of abandoned cheese factories; and the collection of the locally found relics. Foremost was the creation of a historical village in Monroe dedicated to the "predominant groups" that settled the county, the "Scandinavians, Irish, Swiss, and New Englanders." Each nationality, it was hoped, would develop and maintain its own separate unit in a shared space designed to demonstrate harmony between groups.[65]

More significantly, New Glarus successfully siphoned interest to their own local history, a move that brought about the demise of the county group. During the first three years of the Green County Historical Society, membership among New Glarus residents grew at considerably faster rates than in Monroe or other communities (table 3.2). Indeed, by 1939, New Glarus had enrolled almost as many members as Monroe, a city many times it size. This surge in membership was interrupted in 1940, when records indicate that no new members from the original Swiss colony joined. This interruption was not due to a general waning of interest, for Monroe residents continued to join at roughly similar levels to years past, but from a diversion of interest to the local community. In essence, New Glarus used the county historical society to launch its own historical preservation effort.[66]

The emerging focus on *localized* ethnic history in New Glarus is demonstrated further by the estrangement of the community from the

Table 3.2 Green County Historical Society Membership by Community, 1937–1940

	Monroe		New Glarus		Other communities[a]	
	Members	% increase from previous year	Members	% increase from previous year	Members	% increase from previous year
1937	16		7		0	
1938	32	100	18	157	5	500
1939	57	78	54	200	8	60
1940	130	128	54	0	8	0

Source: Green County Historical Society Treasurer's Reports, 1937–1940, New Glarus Historical Society.
[a]Includes Monticello, Brodhead, and Albany

larger national ethnic organization that purported to represent a larger, descent-based collective. The Swiss American Historical Society, organized in Chicago during a 1927 Swiss singing festival, came on the heels of innumerable such organizations.[67] Like many ethnic organizations of this period, its founders determined to shed the best light possible on its diverse members, to make their "contributions to American life" known and respected. The urban, middle-class organizers felt that, in the general public's eyes, their group had melded with that of the Germans, and a separate identity had to be forged. The businessmen leaders, as their constitution made clear, were annoyed and concerned "by the fact that every outstanding person of Swiss origin was claimed by some other nation" and wanted to make certain "that Switzerland and no other country should get credit for the accomplishments of her former sons and daughters in the United States."[68]

The solution to this dilemma was the publication, five years later, of a book with the self-explanatory title *Prominent Americans of Swiss Origin*. In wickedly dull detail, dry compilations of seventy-two Swiss American theologians, soldiers, statesmen, pioneers, physicians, industrialists, and scientists were presented in no apparent order.[69] Although the book was intended to raise the consciousness of the Swiss contribution to American life, the result was a stockpile of hundreds of unsold copies. The "uniformly meager results" occurred because nowhere in the tome could be found the lives of ordinary people with whom readers could personally connect. The Chicago-based society, of course, saw things differently. As a way to explain the poor sales, one board member complained: "we all recognize the fact that our compatriots in this country are not very literally [*sic*] minded and that is perhaps chiefly the reason that they do not buy our book in greater

numbers."[70] Articulating the official culture of the national organization, one member condemned the simplemindedness of ordinary
Swiss Americans: "I do not want to say . . . that they have no interest
in literature. But I guess the reading of newspapers is all they do in
reading and so about history even about our honored late Swiss there
is 'nothing doing.' So it seems to me that there is no money for buying
the Swiss-Men-history. From earlier experience I have learned to let
them [i.e., the uncultured Swiss] alone. It is to [sic] bad that it is so,
but it cannot be changed."[71] With such a patronizing tone and lack of
interest in their "compatriots in this country," it is little wonder that
the early years of the umbrella organization attracted such meager
attention from places like New Glarus. The official culture represented
by the Chicago-based organization contained little of interest for the
locally-based vernacular culture, making the "Swiss-Men-history" easy
to ignore.

"We Want the Community Back of Us": Esther Stauffacher and the Beginnings of a Historic Village

Within a year of the county level organization's beginning, a handful
of the village's most influential citizens met to form a local chapter in
New Glarus, with complete autonomy from the county and national
organizations. Its defined purpose joined antiquarian, preservation interests of the county level organization with explicit pedagogical goals.
The New Glarus Historical Society's constitution called upon its members to "collect and preserve objects of historical value and to *cultivate
an interest* and understanding of such articles as well as the *traditions
and heritage of our background.*" Merely obtaining relics was not enough;
arousing interest in Swiss heritage among village residents became the
organizers' driving aim.[72] It soon became apparent that the organization's means of preserving and transmitting heritage would differ considerably from the Green County group. Leaders quickly jettisoned
the idea of a multiethnic museum and, in its place, called for an ethnically pure "old Swiss Village."[73]

Membership in that new group cut across all economic sectors of
the society. Farmers, bank workers, lumber yard owners, and teachers
made up the core leadership. These people were a far cry from the
Chicago businessmen and Madison university professors who comprised the Swiss American Historical Society. They were the sort of
community members that John Bodnar, in his work on ethnic commemoration, calls "ordinary people." Their concerns were rooted in
the vernacular: more interested in honoring pioneer ancestors than abstract American founding fathers or equally distant "Swiss-men," the

local historical society privileged the personal and vernacular dimension of memory over the official one.[74]

Nevertheless, it cannot be said that they spoke uniformly for the entire community or that as "ordinary people" they accomplished their objectives without needing to convince other "ordinary people" of the task at hand. As a subgroup with interests not always shared throughout the community, their achievements were brought about only through continuous, and at times contentious, effort. Their goal, after all, centered on *cultivating* an interest in their heritage. Cultivation implies creation, or at the very least, preparing the conditions for the transmittal of heritage that is on the verge of being lost or forgotten.

Significantly, a large proportion (more than half) of the key players were women, a feature common among early "patrons of tradition." What Karal Ann Marling points out for colonial revivals is equally true for tradition-building work more generally, namely that "women were the primary custodians of the American heritage in its tangible [or material] manifestations."[75] And of those women, one in particular stands out. Esther Streiff Stauffacher, a third-generation Swiss American farm woman, dedicated the final years of her short life to creating a tangible, material reminder of the region's ethnic heritage. Though described as a "typical New Glarus farm wife," her position of community leadership and her relentless push to awaken slumbering filiopietistic sentiments were anything but "ordinary." A "common woman" leading an "uncommon life," Esther Stauffacher, more than anyone else, is responsible for the creation of the state's first ethnic museum (fig. 3.7).[76]

Stauffacher possessed strong powers of persuasion and, equally important, the legitimacy to back up her earnest words. The larger community's initial interest in creating an "old Swiss Village" seemed lukewarm at best, making these qualities all the more important. The first problems focused on soliciting workers to get the project off the ground. Stauffacher hoped that "work . . . will be done by the entire community as in the early days when each settler felled trees for the community house." But when volunteer labor from the community failed to materialize, outside intervention became necessary. Labor for the "community" house's construction came in the form of the National Youth Administration (NYA)—a national public works project—and despite the pious rhetoric of "community spirit."[77]

The next consideration centered on the location of the proposed "old Swiss Village." Original plans called for open-air museum's construction at the recently purchased village park, but opposition came fast and direct. Esther Stauffacher, who had already lined up the building materials and was ready to have the masonry work begin, voiced

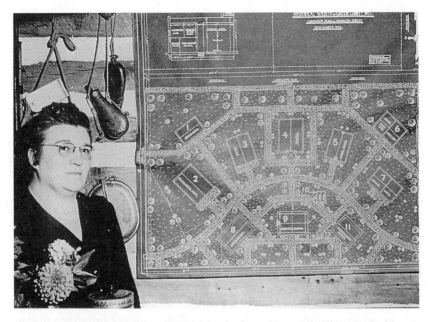

Figure 3.7. Esther Stauffacher with plans for the Swiss Historical Village. Backed by the legitimacy of being a respected community leader, Stauffacher directed the construction of a permanent monument to her Swiss pioneer ancestors. (Courtesy of New Glarus Historical Society)

her exasperation: "Suddenly 'like a blot out of the clear sky' shall we say—propaganda was started, such as 'Why spoil our park?' or 'No room for something like that in our park.' Soon we decided to our regret [that] we perhaps had better look for another place to build. We felt we would be letting down our forefathers if we dropped the idea and would allow ourselves to weaken and give up SO EASILY—because of the opinion of a few who probably haven't given the idea any thought or consideration whatsoever."[78] These were strong words in a society that acknowledged conformity as the supreme virtue and permitted disagreement only behind closed doors and unknowing backs. And in the end it worked, for one month later the heirs of a large lot on the edge of town donated the property to the Historical Society in return for the payment of delinquent taxes. The estate—whose transfer was arranged by John Schindler—seemed ideal; its large size and open space would allow gradual development, one unit at a time. Moreover, its location at the far edge of the village made it seem removed enough so as not to annoy its detractors.[79]

Once materials, labor, and property had been acquired, "the local historical project," at this time officially called the Swiss Historical Village, was "now ready for action." For her part, Esther Stauffacher seemed less interested in relics or attracting tourists than in vigilantly neutralizing memory's demise: "I believe our forefathers could feel proud if they could know that we are honoring them by reviewing and studying and preserving the things which went to make up our heritage. We must not forget all the hardships these pioneers suffered." The importance of establishing a connection to the past through a visibly recognized, and materially grounded, image was reaching a critical stage. Living memory of the hardships of the Swiss pioneers was fading rapidly, a fact that gave Stauffacher considerable cause for concern: "The sad part of it all is that our unusual history is fast becoming lost to the younger generation and unless the community supports our project 'up on the hill' for a museum and Swiss village there is no doubt that in time most of our history and traditions will be forgotten."[80]

Forgetfulness was not the only obstacle Stauffacher and her fellow guardians of tradition faced; indifference was worse. The Historical Society president pleaded that if only "we all put our shoulders to the wheel, boost instead of always try to discourage those working on the ideas, we will . . . have something which will surely be a fine monument to our forefathers. We want the community back of us."[81] Part of the problem, as Stauffacher saw it, was rooted in "idle gossip" that complained about the "outside help" from the NYA instead of the use of the "local boys." Her response to the Main Street-type prattle was biting and unforgiving: "If some of those folks would work as hard for the project as they work against it, we could have the whole village up in a short time. We have asked time and again for suggestions, [b]ecause we know that it is our, yours, and my project. But it sort of seems like some folks glory in seeing others make mistakes so they can say 'I told you so.'" Sounding more than a little like Carol Kennicott in *Main Street*, Sinclair Lewis's novel of small-town life, she pleaded for "loyal support," rhetorically asking, "Why is it always such a difficult task to arouse interest in some of these finer things for our Community? . . . It [always takes] some foresighted folks to put them through, regardless of the few who always want to put the damper on such things. I do hope that all of you will show just a bit more interest."[82]

"Tradition and Heritage Are the Qualities of Life," or, Cashing In on the Past?

In the end, enough support came to complete the first "community house" by 1942. Approximately six hundred people—including the president of the State Historical Society of Wisconsin and the director of the State Historical Museum, in Madison—attended the dedication of the first building of the Swiss village that September. Held on Kilbi Sunday, the event turned into a full-blown "Historical Folk Fest" complete with folk music, costumes, and dance "typical of the various nationalities who pioneered Green County." Songs by the New Glarus Yodel Club complemented other performances, including early American dances; Irish fiddling and jigging; Norwegian folk tunes; and Edvard Grieg violin solos.[83]

The multiethnic nature of the cultural display reflected the continuing involvement of the county historical society that, essentially, sponsored the Sunday afternoon event. It also reflected a confusion as to precisely whom the Swiss Historical Village was to represent. Some newspaper articles continued to refer to the *open air museum* as "for all groups." One Monroe article explicitly stated that, because the original blueprint plans show that there is to be a Norwegian dwelling, "the village is not to be typically Swiss, but rather for all nationalities." Even in New Glarus, some believed that the "first building . . . is to be a community house for all nationalities." Many more in the Swiss-centric town, however, referred to the community house as the "Swiss house." More telling, the planned Norwegian house was the only structure *not* built from the original plans. The area's pioneers, everyone connected intimately with the project presumed, were *Swiss* pioneers and that a museum should be built in their memory went unquestioned.[84]

In the end, New Glarus's Swiss Historical Village had its roots in the fervent efforts of a handful of committed enthusiasts (fig. 3.8). Neither simply "social elites" nor just "ordinary people," the prime movers sprang from the community to form a heritage-oriented subset. They were cultural leaders, but also representatives of vernacular, local interests. Well aware of the critical role of human agency, Esther Stauffacher scoffed at the suggestion that "New Glarus" built the Swiss Historical Village: "I was amused at the 'New Glarus undertook' [notion]." She knew well, as did other tradition guardians, that communities or cultures undertake nothing of the sort; rather, certain individuals within the community make and sustain traditions, whether material or strictly performative.[85]

Within the subgroup, competing interests emerged that—while never directly threatening the museum's existence—stood at odds and in relative tension. For the first time, the prospects of "commercializ[ing] on our traditions" became a distinct possibility. To be sure, merchants profited from the thousands of turn of the century Festgäste who poured into the village during periods of high celebration. As mentioned, it was hoped that the "tourist park" might catch a few auto-crazed campers just beginning to hit the newly paved state high-

Figure 3.8. Community House shortly after its completion. Located on the edge of town, the Swiss Historical Village had its beginning with the Community House. Pictured, *second from left,* is the replica pioneer house's designer, Jacob Rieder, and, *far right,* is Esther Stauffacher. (Courtesy of New Glarus Historical Society)

ways. But only in the early 1940s did it become necessary to give "all commercial aspects every consideration" in the construction of the museum; and, only at this time did it become feasible to appraise the geohistorical position of the community as a potential "nucleus for capitalizing upon our natural [i.e., ethnic] assets." Tourism, as business leaders commented, was becoming a "billion dollar National business" and they hoped that adding a museum might help the town lure a handful of those dollars.[86]

Over the next several years, business-oriented members of the historical society believed that "the excellence of our heritage and tradition," would divert some travelers to New Glarus and Green County. The Norwegian American Museum in Decorah, Iowa, was held as a model of the sort of museum that "proved [to be a] valuable asset." And when national attention came in the form of publicity in *National Geographic* and *Time* magazines, local attention to the museum came more readily. As one businessman put it, "all this points, again, to the unusual interest others take in our village, and how our responsibility lies to supply that interest with adequate hospitality"—and with things to see. Businessmen who straddled the boundaries between business and heritage were the first to make the connection to tourism, a connection that, in less than two decades, was to completely envelop the community.[87]

Other leaders expressed hesitation over cashing in too quickly and blindly on heritage. "Commercialize on our aspects of tradition—our heritage—yes, as it is necessary to do so," Esther Stauffacher wrote on the eve of the museum's opening. But, she also warned that with such commercial interests, "to become too casual or effusive is to become destructive." In her view, "the motivation impulse to such establishment had deeper inspiration, much further reaching aspects than commercial valuation alone. Tradition and heritage are *the* qualities of life" (fig. 3.9). The museum, she felt, should maintain a simplicity that speaks for itself, for to create a place of memory too flashy and overly dedicated to the passersby would "assume a quality of polite tyranny." It threatened to become, in Mike Wallace's words, "Mickey Mouse history." Such tension between tourism and the presentation of the past, or heritage, is endemic to the enterprise.[88] Heritage, for people like Stauffacher, is sacred, and peddling it affronts decency. Placing the grave markers of their ancestors directly in the Swiss Historical Village

Figure 3.9. The Männerchor at the Swiss Historical Village, ca. 1952. Funded by the Wilhelm Tell Guild, the Historical Village's second building—the log church—has been home to annual services of the Swiss Reformed Church. The Männerchor customarily leads in hymns, reflecting the museum's sincerity and quasi-religious temper. (Courtesy of New Glarus Historical Society)

Figure 3.10. Relocating the past at the Swiss Historical Village. By placing the grave markers of their ancestors in the Historical Village, local guardians of heritage demonstrated the high seriousness and deeply felt connection to the past engendered in the site of memory. (Photo by author)

demonstrated the high seriousness and deeply felt connection to the past engendered in the site of memory (fig. 3.10). For others, heritage offers the material for an exceptionally profitable industry.[89] This friction between cultural prostitution and cultural preservation cannot be avoided, though in the early 1940s it remained largely hidden from view, emerging thirty years later only with tourism's full development.

CONCLUSION

Out of the complex messages of both the 1935 folk production and the Swiss Historical Village museum emerges a view of ethnic culture in the late 1930s that sought to refashion a more inclusive definition of what it meant to be American. For groups like the Swiss, ethnic identity was neither obliterated through an abstract Americanization, nor passed down as an unbroken tradition, but rather discovered and represented through cultural performances and sites of memory. The celebrations of both ethnic and American ideals made use of classic Americanist rhetoric in an attempt at cultural legitimation within a civic culture that had previously remained exclusive. By proposing their compatibility with American goals and values at the same time as proclaiming their uniqueness, the New Glarus guardians of tradition deployed localized culture to crack a seemingly all-powerful American cultural hegemony. Indeed, it would appear that such memory work actively shaped a discourse of ethnicity between the wars that became decreasingly antagonistic and gradually more open to ethnic and cultural difference.[90]

In both cases, the New Glarus Swiss created a notion of culture that was as place specific as it was ethnic specific. In performance as well as material construction, they planted their values, aspirations, and myths deeply into local culture and, in doing so, helped reshape that culture in their own image. That localization of ethnic memory was accomplished only through considerable effort and with the leadership of key individuals, pointing to the critical importance of human agency in tradition's creation and sustenance. By overcoming indifference, by selecting an "appropriate" view of history, and by forging a common ground between heritage and a nascent tourism industry, the Swiss immigration story became *the* place history. Over the next few years that story would broaden and deepen through the invention of a tradition that, even more radically, transformed the community and, eventually, the state. With the successful establishment of the *Wilhelm Tell* tradition, memory's loyalists found an even more powerful outlet from which to express and mold their ethnic identity.

4

"Replanting the Swiss Scene"

Wilhelm Tell Returns to America's Little Switzerland

Three years after the ninetieth anniversary pageant of 1935, the *New Glarus Post* printed a lengthy open letter intended to remind villagers of yet another upcoming festival. Although ostensibly addressed to a specific part-time community member, the anonymous author's real target was the entire village. "Dear Sir," the writer began,

Switzerland's hero of song and story will relive again and re-enact the exploits for which the world hold[s] him famed in the open-air production of Shillers' "William Tell." The production, to be presented in the original German, is patterned after the Interlaken version, presented annually in the Swiss canton once native to Tell. It is particularly fitting that "William Tell" be *re-introduced* to residents of the site of the original Swiss colony in America. Its presentation provides a happy medium for telling the Swiss of the accomplishments of their great land and at the same time offers an outlet for the . . . talent of the residents of the New Glarus community. It is a fine thing that you are doing in *replanting* the Swiss scene in settings appropriate to it. [The play] provides a bit of the old world in the new.[1]

Unwittingly, the anonymous letter writer illustrated the points he or she had tried to assert. The misspelling of the playwright's name seemed bad enough, but to mistake the Interlaken region as part of the "Swiss canton once native to Tell" betrayed a glaring unfamiliarity with the epic myth, a myth that goes to the heart of Swiss national identity. Apparently, the third generation Swiss Americans really *did*

110

need an accessible medium through which to learn about their "great land." The letter also intended to arouse enthusiasm for an ambitious—if potentially disastrous—*community* project; it was not written to thank the community interloper who was viewed by many with suspicion. The project *could be* a means of personal, group, and artistic expression, *if* the letter's readers would commit to it.[2]

In the end, many did commit to the project with a passion that nobody could have predicted. If longevity is any measure of that commitment, the sixty years' ongoing production of Schiller's *Wilhelm Tell* alone has come to demonstrate a near inexplicable devotion to the performance. It is hardly an exaggeration to suggest that the annual staging of *Wilhelm Tell* replanted the Swiss *scene*—in all the meanings of that word, from situation and ambience to vista and drama—in the small Wisconsin community, changing it forever. That same conspicuously constructed ambiance has become increasingly important for both community members and for those many thousands of tourists who now visit every year. Where earlier larger-scale public displays of ethnic culture recurred intermittently, organizers of *Tell* hoped, from the very beginning, to "*make* this one of the traditions of New Glarus."[3]

"Replanting," "re-introducing," "making": all these words, coming from the pens of those involved with the first 1938 production, point toward the creativity and inventiveness behind the New Glarus *Tell* tradition. To speak of *Tell* as an "invented tradition," however, is not to say that the cultural display is considered phony or untraditional by the people involved behind the scenes and in front of the stage. Likewise, I do not mean to suggest, by the high degree of self-consciousness implied in the term "invention," that the tradition is somehow less pure or "authentic" than its nineteenth-century tableaux counterparts, and devoid of meaning.[4] On the contrary, I want to argue just the opposite. The annual performances of *Wilhelm Tell* have played a fundamental role in defining what it means to be ethnic American in this most conspicuously ethnic place.

This chapter, then, examines the interconnectedness of ethnic identity and tradition creation at a critical time and place. The genre established in the wake of this association—a folksy cultural display of Germany's high culture icon—allowed a bridge into the nation, while still permitting ethnic distinctiveness. By enacting Schiller's *Wilhelm Tell* in German during the second war this century saw with that country, the Swiss community simultaneously demonstrated both New Glarus's "different heritage" as well as its compatibility with American ideology.

111

WILHELM TELL'S CENTRAL ROLE IN SWISS
NATIONAL CULTURE

The legend of Wilhelm Tell in Switzerland has been long considered a national founder's story. In a country that prides itself as the world's oldest democracy and is renowned for its isolationist nationalism, such a story assumes no small importance. So powerful and all-encompassing is the myth of Tell that it exemplies the major symbolic properties of the United States Constitution, George Washington, the American frontier, Abraham Lincoln, and Davy Crockett. Tell is vested with establishing the modern confederation of Switzerland. His quiet leadership and master shot, coming on the heals of the famous oath on the Rütli meadow of 1291, supposedly inspired the confederates of Ur-Schweiz to overthrow their Austrian oppressors. Also, the mountain hero exemplifies the virtues of the state itself: self-reliance, honesty, hard work, and bravery. Regina Bendix summarizes the importance of the Tell myth for modern Switzerland in this way: "If there is a single symbol that all Swiss, no matter what language they speak, will recognize and associate with the nation, it is William Tell." Situated precariously between history, myth, and legend, Tell has become a sort of "secular patron saint" of the country.[5]

Given the overwhelming significance of Wilhelm Tell for Swiss national identity, the lengthy debates and investigations surrounding his existence should come as no surprise. Yet even the most casual observer is struck by the seeming endless number of studies into the historicity of Tell. As early as 1907, one archivist was able to compile a 189-page bibliography on the Tell legend; today another scholar tries to find his way through the "labyrinth of 200 years of Tell research."[6] While thirteenth- and fourteenth-century documents indicate that Swiss peasants did swear allegiance to each other in the face of foreign oppression, the name of Wilhelm Tell does not appear in any documents for nearly two hundred more years. Most scholarly accounts now argue that oral tradition either turned an individual who had existed into a legend or created a character who personified key elements out of the nation's past. So successfully did the elements of liberty, heroism, and politics merge that Tell and his association with the establishment of Swiss confederacy became the story of Switzerland's modern, democratic birth.[7]

The historical reluctance of many Swiss to accept Wilhelm Tell as more legend than history demonstrates the power of the myth. Responses to critical reappraisals have been largely negative and denigrated as unpatriotic. Max Frisch's satirical retelling of the legend from the perspective of the ailing bailiff Gessler, for instance, received decidedly mixed reviews. More recently, an exhibition reevaluating the seven hundredth anniversary of the Swiss confederation at the historical museum in Lausanne questioned the foundation story of 1291. Even worse, it "relegated Tell to the realm of folklore." The exhibit's creator, Werner Meyer, created a furor as he debunked Tell: "They [the Tell legend's early authors] invented all the great events and then put the invented figure of Tell into this landscape." For such heretical (but accurate) views, Professor Meyer received a notebook full of angry letters—including a death threat—as well as abusive calls. Many Swiss, it would appear, are not quite ready for such a critical historiography.[8]

Friedrich Schiller and Wilhelm Tell

The most famous rendition of the Wilhelm Tell story came not from the pen of a Swiss author, of course, but in the writing of the great German dramatist, Friedrich von Schiller. For centuries before Schiller, versions of the Tell legend found their way onto outdoor stages in the Protestant and Catholic regions alike. As early as 1512 a rendering known as *The Old Urner Tell Play* featured Tell as the central character who united the three cantons and who, when forced by the sadistic bailiff Gessler, shot an apple off the top of his son's head. This early peasant or folk theater persisted in remote areas and was strengthened by growing sentiments of nationalism in the late eighteenth century. The widespread formation of shooting contests, a practice that generated memories of Tell and his achievements in marksmanship, contributed to such early peasant theater. Thus, when Schiller began his last completed play in 1803, he was covering well-known Swiss terrain.[9]

That he never actually saw the physical terrain is the first item that attracts most critics' attention. This is most likely because Schiller's drama—with its radiant patches of "local color" in the form of Swiss expressions, its intricate regional detail and picturesque Alpine settings—gives the sense of geographical intimacy. The aging dramatist determined to familiarize himself with the landscape and people of a country he never visited, a point frequently made in his correspondence. Goethe, his friend and inspirer for the drama, thus described Schiller's method: "He began by pasting as many maps of Switzerland as he could find on the walls of this study. Then he read descriptions of travels in Switzerland until he knew every nook and cranny of the scenery of the Swiss uprising. He read the history of Switzerland; and after he brought together all the material, he sat down to work, and quite literally did not get up from his place until he was done with *Tell*."[10] In his many letters from this period Schiller repeatedly stressed the importance of local color in the play and his conviction that the historical Tell legend "should grow organically out of the sublime landscape setting of the Swiss Alps."[11]

Schiller's foray into the mythic historical geography of Switzerland provided him with a powerful setting to address very contemporary concerns. In the wake of the Napoleonic invasion of Switzerland (the first foreign army to do so in the modern era) and the accompanying seizure of Bern and the all-important mountain passes, the drama offered a national past much preferred to a dreadful present. The play brushed aside a dismal recent history of disharmony and weakening national unity, as Schiller's *Tell* provided Swiss leaders with a usable past. By setting his drama in the thirteenth century, Schiller's glorious past successfully accomplished the two key political tasks of the time—revolution and national integration.[12]

Conservative loyalists in the German states, not surprisingly, frequently censored Schiller's drama after its first successful performances in Weimar in 1804. German audiences, on the other hand, adored the play and flocked to performances in numbers that made it by far the most popular central European stage show of the late nineteenth century. For the middle classes, *Tell* was the lively, swashbuckling drama of personal freedom by their most beloved playwright. In Switzerland itself audiences saw the play in a different light, as *Tell* resonated with emerging sentiments of romantic nationalism and *Heimatbewuβtsein* (feeling or awareness for the homeland), a fervor ardently encouraged by federal Swiss leaders. As Eric Hobsbawm puts it, Schiller's *Tell* became one of Europe's mass-produced traditions and played a critical role in Swiss nation building. Significantly, a growing number of these performances began occurring outdoors as a way to equate national identity with the "purity" and "simplicity" of nature. Such is the direct origin (combined with an early eye toward tourism) of the annual Swiss Interlaken outdoor production of *Tell*, begun in 1910, and from which a Long Island bank clerk drew his model for a New World folk production.[13]

INVENTING A TRADITION

Enter Edwin Barlow, Stage Left

Everyone connected with the play in its early years credits Edwin Barlow for the initial idea and the ability to motivate community members to stage the outdoor production (fig. 4.1). As one early participant put it recently, Barlow's "interest in his Swiss heritage and the stage [s]parked the vision for the Wilhelm Tell play in New Glarus."[14] That vision was the product of Barlow's unique background, one that straddled both sides of the Atlantic and enabled him to become an ethnic

Figure 4.1. Edwin Barlow as a young man. Related to New Glarus by birth, but with years of experience in Switzerland, Edwin Barlow perfectly fit the role of ethnic cultural broker. A substantial inheritance enabled Barlow to pursue his dream of world travel, exotic collections, and eventually bringing the *Wilhelm Tell* drama to America's Little Switzerland. (Courtesy of New Glarus Historical Society)

cultural broker par excellence. Such cultural brokers most often appear when a group is undergoing change in its status and its relationship to a "host society." In the wake of the pressures during the war years to Americanize, and their more recent reassertion of local ethnic identity during the 1935 anniversary pageant, the Swiss community appeared to be searching for such a go-between leader who could effectively bridge both worlds. Although some recent immigrants trickled into the Swiss-dominated region, none possessed the economic ability or intellectual leadership to move effectively between the local, the national, and international scales of memory. Louis Wirth grasped the unique qualities demanded to fill this position during his studies in Chicago at roughly this time, noting that such brokers influence their fellow ethnics by "recovering, disseminating, and inspiring pride in the group's history and civilization and pleading its case before world public opinion."[15] Nobody seemed more poised to argue for the Swiss—to the world, and themselves—than Edwin Barlow.

Described aptly as a community "insider/outsider," Edwin Barlow had relatives in the village, but spent very little time there.[16] Born in Milwaukee to the unusual and unwelcome combination (at least in 1885) of a Swiss mother and a Yankee father, the orphaned Barlow spent most of his childhood on a Green County farm where he was raised by his aunt and uncle. He moved away from Wisconsin upon reaching adulthood and became a clerk at the Long Island Savings Bank. After serving briefly in World War I, he returned to New York where, his obituaries claim, "he turned to the theatrical field and followed that profession for several years as director and producer." The extent of his involvement in professional theater is questionable, however, and most likely a product of a wishful imagination.[17]

More significantly, the bank clerk spent this time cultivating a friendship that would change his, and New Glarus's, life history. In 1924, at the age of thirty-nine, he was "adopted" by a fifty-two-year-old "New York social registerite" who provided Barlow with the lifestyle he had been craving. Clara Bosworth Mather was presumably "a distant relative on the Barlow side of the family" who took an unspecified interest in the bachelor and enabled him to begin extensive European travels, including a nine-year residence in Lausanne, Switzerland. Freed from the cares of work at the Long Island Savings Bank, and with the unprecedented opportunity to soak up Swiss culture, Barlow traveled extensively. It was now that he began amassing a wide array of artifacts that would eventually form the collection that filled his future New Glarus residence. It was also during this period that Barlow most likely saw the Interlaken version of *Tell* that had such a profound effect on him.[18]

These foreign travel experiences, so utterly in contrast with his common, rural upbringing gave Barlow a demeanor that set him apart from his former neighbors in New Glarus. Flaunting his "worldliness," the self-proclaimed "man-of-the-world" was viewed with suspicion by many in the small Midwestern town who saw his theatrical presentation of self as "weird" and "out of the ordinary." (fig. 4.2). Some called him *"Papagei"* (parrot), or "parrot nose," for the facial feature he considered "haughty and sensitive." Whatever his eccentricities, Barlow also possessed an "unflagging, unflappable energy." With that energy and the assurance that naturally accompanies a thick wallet came a confidence and unswerving insistence on doing what at first blush might seem impractical.[19]

Precisely when Barlow first concocted the idea to stage an outdoor production of the Schiller drama is difficult to determine, but it is clear that he followed the events of the Wisconsin Swiss community throughout the 1930s from his Long Island home. His occasional visits to Green County were brief sojourns, but in the wake of the successful ninetieth anniversary celebration, the cultural broker found his inspiration. The historical pageant, he wrote, "has given me courage to lift my voice . . . for the benefit of unborn generations, and the glory of the land of your origin." Why not, then, direct that "wonderful spirit" to create a living, performative monument to ethnic memory, a "Folkfest" to be performed year after year in the tradition of the finest outdoor dramas of Europe?[20]

Over the next three years, Barlow spent increasingly more time in New Glarus, supervising the construction of a chalet structure—the village's first, and soon his second home—and reacquainting himself with the community that he had left at such a young age. He also began planting the idea of a Wilhelm Tell play in the minds of a handful of key village leaders who might find value in such a production. The first person he approached was Esther Stauffacher, the woman so instrumental in constructing the as-yet-unrealized Swiss Historical Village.[21] That he should turn to Stauffacher demonstrates the interloper's skill at reading the community and knowing how to negotiate its local political apparatus. Barlow himself would have gotten nowhere on his own, but with the support and legitimation of one so well connected and respected as Stauffacher, perhaps the "crazy idea" stood a chance. Indeed, his outsider status was crucial to his vision. With no vested economic or political interests in the daily life of the village, and appealing to his authority as a knowledgeable "world traveler," the outsider could take risks and incite an enthusiasm that no local resident, no matter how well connected, ever could.[22]

Figure 4.2. Edwin Barlow in theatrical drag. The community insider/outsider brought his flair for the dramatic to the Wisconsin community. He is photographed here in an unknown, undated role in a production of *The Two Orphans*. Although he was never a professional, his love of the theater contributed to the establishment of *Wilhelm Tell*. (Courtesy of New Glarus Historical Society)

"Wilhelm Tell Had a Painful Beginning"

Little direct action was taken on the "Folkfest" idea for another year. Behind the scenes, however, and with the support of people like Esther Stauffacher, Barlow sweet-talked, cajoled, and coaxed important members of the community into staging the drama that summer. The initial resistance to an idea that seemed so "out of the sky" and disconnected with New Glarus's own history appeared, at first, insurmountable. Although a silent Wilhelm Tell had appeared in an earlier generation's *tableaux vivants,* nobody had ever considered staging the entire three-and-a-half-hour drama before.[23]

In a revealing interview a few years later, Barlow himself recalled that at first few people seemed interested in his "Folkfest idea." The indifference of the community that he had lifted onto such a high pedestal pained the nostalgic Barlow. "*Wilhelm Tell,*" he relayed, "had a painful beginning" that nearly caused him to retreat to the safe confines of Long Island. Indeed, this was precisely the problem. The possibility of retreat in the face of failure was Barlow's option, but not one that existed for the deeply rooted Swiss of the region. This simple fact was pointed out to the "world traveler" by the one person with whom he had a direct, familial connection. When the "preparations were in such confusion and the people in so great an uproar," his aunt "begged him to give up the notion of a community project." Barlow recalls that he nearly did when she scolded him to "remember, you can go back to New York after this thing is over, but I'll have to live here."[24]

The threat posed by *Wilhelm Tell* was a danger faced by every community when creating a public display that purports to speak for the identity and underlying reality of that community. The culture created by such displays is always a *public* culture that eventually leaves the carefully contrived grasp of its creators. Consequently, such public culture can expose, accentuate, and tear open internal divisions within a community that had been carefully concealed beneath the surface. Social ties within small towns—more often cobbled together through personality than social class—are particularly fragile and contingent. These ties must be nurtured and protected from the divisiveness that can pose a danger to the very survival of the community as a coherent entity. New Glarus, for all its communal rhetoric, was not immune from the ever-present cleavages that run through such places. Due to the sheer scale of the undertaking—requiring the committed efforts of hundreds of people—and because the story represented *the* central founding myth of the homeland, *Wilhelm Tell* assumed a weighty seriousness. If staged successfully, it could reinforce community; if it failed, community could be undermined.[25]

Barlow did manage to garner enough support for this "Folkfest" during the spring of 1938 to consider staging the drama late that summer. His growing network consisted of a representative cross-section of the community, including businessmen, lumber and feed dealers, farmers, clergy, and teachers. Creating a broad base of support was a crucial maneuver that was to legitimize a fundamental aspect of the production: *Tell*, to be successful, had to be a community production which, in turn, became one of its chief selling points. By involving such diverse strands of people, the play could appear as if it sprung naturally and organically from the soil. It could give the illusion of consensus.

Creating a Festival Structure

A group of the most committed among Barlow's followers met at the Swiss church's community house that July to formalize plans for two performances later that summer. Called to order by a lumber dealer named Gilbert Ott, the thirty-three people sought to formally "discuss ways and means of presenting a 'William Tell' pageant." Although this is the first recorded gathering of Tell enthusiasts, the meeting's minutes indicate that other meetings, discussions, and rehearsals had been taking place for some time (fig. 4.3). With the date of the first production little more than a month away, those committed felt that it was important to organize in an official manner.[26]

Methods of financing headed the agenda, and five suggestions were offered, including the selling of advanced tickets and the securing of monetary backing "in case the project does not pay for itself." Economic solvency, if not profit, was crucial for the festival's success. A financial disaster would have spelled *Tell*'s doom well before any considerations of its artistic merit.[27] Next, in good Swiss fashion, executive and finance committees were formed, as well as a temporary umbrella organization. With Barlow in charge of casting, the other committee members included those already tapped for major roles in the production. And as the performance date neared, further committees were created, including those for seating, ushers, parking, costumes, property, tickets, first aid, and concessions.[28]

These backstage functions were left to the increasing number of community participants as Barlow focused his attention on the casting and directing of rehearsals. Many of the already committed cast members assisted him by recruiting family members for lesser roles, who, in turn, contacted other family and neighbors. In this way a tightly-knit web of familial and community relations—*not* formal auditions in which people would be judged and evaluated in public—worked to

Figure 4.3. Edwin Barlow directing *Tell*. With a style and manner that set him apart from his adopted community, Edwin Barlow introduced *Wilhelm Tell* to the Swiss village. Here he is shown in Bavarian costume at an early rehearsal of *Tell*. (Courtesy of New Glarus Historical Society)

fill out the large cast. In the end, seventy individuals were assigned named roles in the first production, with up to forty more assisting as peasant women, men, and children in the appropriately named "mob scenes." If we then add the additional thirty or so who were involved solely in the backstage functions, roughly 140 people, or nearly 10 percent of the combined village and township population, participated in the 1938 performance.[29]

The logistics behind the scenes seemed endless. First, there was the matter of finding a suitable location for the festival. The Interlaken *Tell* served as the model, so it was decided immediately to stage the play outdoors.[30] The unglaciated region surrounding the community offered a number of long, wooded valleys that might serve as suitable backdrops for an outdoor performance. As in Interlaken, an outdoor production could reinforce the ruralness of the setting and add to its "folksiness." This natural simplicity quickly became a significant point of distinction, organizers relayed to national pageant coordinator Ethel Rockwell: "The only scenery we use is that what nature so correctly provided," *Tell* loyalists boasted. Of course, the farm valley eventually chosen was a working farm, demanding hours of labor to transform a cow pasture and corn field into a "setting of rustic simplicity."[31]

Next, concerns of varying complexity required attention for the production's success. Obtaining horses—fourteen in all—for the apple shoot scene was a priority, as was renting costumes, folding chairs, and a sound system. Swiss Brown cows had to be somehow brought in from the Voegli farm, two miles away, as did goats and sheep for the opening transhumance scene. Finally there were the rehearsals themselves. Although cast members had been learning their lines to varying degrees for the better part of the summer, practices intensified for the final ten days before the September 4 opening. Barlow's insistence that the extensive lines flow together harmoniously kept the performers there from afternoon until late into the night, a demand especially onerous for farmers who had to finish milking early in order to make it to practice on time.[32]

"Stick to Traditions After We Once Start Them"

To nearly everyone's amazement, the two 1938 performances of *Wilhelm Tell* proved an unqualified success. It was not so much the surprising fact that fifteen hundred people attended the first performance; nor was the two hundred dollar profit from the occasion cause for the greatest celebration, though that certainly helped everyone bask in the afterglow of festival. More than this, what most impressed organizers was the simplest fact of all: that enough of the community had rallied around the cause to make it happen. The *Post*'s columnist expressed her view of the play in this way: "One of the finest things about this whole production, is the co-operation shown by the people of the community. Everyone in this community undertaking found something to do and did it willingly and cheerfully." Of course, there were many more who chose not to participate than those who became involved, a convenient point of omission. And there were those who wished "not to be involved in any manner." Enough individuals did support the project, however, to legitimize its claim to be "of the community."[33]

The play, even at this early stage in its career, seemed to offer as many interpretations as there were seats for the growing audiences. One Swiss American from Illinois saw the play in the light of American patriotism: "What a wonderful demonstration this drama is of the principles and simple prerequisites of democratic rule!" A Madison, Wisconsin, visitor, on the other hand, was most touched by the connection of the performance to her own Swiss heritage and identity. Moved by the performance's "Swiss spirit" (*Schweizergeist*), Ida Mauer felt that the play demonstrated that *true* community hangs together because "we all stem from the same old homeland." A Chicago resident found the antimodern rusticity most important, noting that she "liked the sincerity and simplicity of the production" (fig. 4.4).[34]

For New Glarners, *Tell* also provided multiple readings and experiences. Some endured the heavy commitments of the weekend and the weeks leading up to the festival as community and family obligations, and, as quickly as possible, withdrew their support. Others, similarly, used the occasion as an opportunity to leave New Glarus for the weekend and the crowds that began invading their town. A significant number, however, used the opportunity to rediscover their ethnic roots, their "Swissness," through a ritualized performance. Many of these third generation Swiss Americans began to question seriously their own private assumptions regarding assimilation and Americanization. Katherine Theiler, for one, was struck when "one boy asked whether he could be in the *Wilhelm Tell* play, and was told by his father, 'No, you can't be in it because you can't talk German.' 'Well, I could learn, couldn't I?' was his reply." Following many ethnic leaders in condemning the loss of language, she then editorialized that "maybe some of us haven't been giving our children the opportunity to learn to speak Swiss German when he would have found them willing to learn."[35]

In the exciting aftermath of the successful performance, leaders gathered weekly to discuss the festival's future. From the very beginning, Barlow and the handful of people closest to him aspired to make an annual ritual of the performance. "It is hoped that this presentation can be given yearly and be enlarged every year," it was noted one

Figure 4.4. Reawakened public memory. Though the audience for *Tell* interpreted the performance in multiple and often competing ways, for many, as with "Muetti" (or Mother) Ingold, it served to reinvigorate public memory of her native Switzerland. She is shown, *on the right*, with a friend in folk costume and with Canton Bern's chief symbol, en route to a performance of *Wilhelm Tell*. (Courtesy of New Glarus Historical Society)

month before the first performance, "until it has grown to mean a great deal to this village of New Glarus with its Swiss traditions and to Swiss people near and far."[36] This conspicuous construction of tradition was seconded in a letter to Ethel Rockwell in 1939: Henry Schmid believed that everyone wanted to "keep it [the play] here at New Glarus and build up a tradition or shrine (if it may be called such) where the Swiss people and others can return every year."[37] A permanent organization with the notably antimodern name "Wilhelm Tell Guild" was entrusted to perpetuate the tradition and scores of other suggestions that soon followed. The study of *Wilhelm Tell* in the area's schools and the cast's obligatory attendance of church services the Sunday of the festival were carried out for several years. Other suggestions, such as the one in which "everyone in town especially those in business places should be dressed in Swiss Costume the days the play is presented," were summarily ignored. Most important, the originator of the festival offered the following words of advice: "Stick to traditions after we once start them, do not get professionals in, work these things out by *ourselves*, keep the play in *our* own community."[38]

Barlow, the theatrical interloper from Long Island by way of Lausanne, not only successfully made himself part of the community (the following year he was even described, for the first time, as "a grandson of Fridolin Streiff, one of the village founders"),[39] but also astutely foretold the key to maintaining the tradition he had just invented. If it was to survive as a tradition, Barlow noted at a Guild meeting, *Tell* had to function as the collective responsibility of the community.[40] This was crucial for two distinct but related reasons. First, the festival would lose its legitimacy *within* the community if it were "professionalized." The only way to obtain involvement was to position it as a community affair that held the promise of individual and group renewal. Outsiders might make it a "better" performance, but they would be on their own, without the support of the farmers, milk haulers, and feed dealers who made the play work. Second, and no less important, bringing in professionals would tarnish the view of the festival as indigenous or homegrown. It would lose its "folksiness" and appearance of rooted authenticity and thereby lose its legitimacy *outside* the community. These seeming unrelated tensions were, in fact, inextricably tied together, and achieved a resolution with such skill as to appear "natural" or preordained.

MAINTAINING A TRADITION

"Above All, Boost the Play, Don't Knock It": Legitimation within the Community

Overcoming the ever-present threat of indifference and inertia posed the first serious challenge in sustaining the newly created tradition. The first indication that such a danger might threaten the integrity of the production came shortly after the first performance. As one of the ways to bring "the younger generation" into the play, high school seniors were encouraged to study Schiller's *Tell* and take part in a forensics contest in which passages from the play would be recited from memory. Although eight students took part, with the five dollar prize eventually going to the son of one of the two Wilhelm Tells, the Guild was disappointed with the results. It seemed that "very little interest was shown by the public."[41]

As a way to combat such an attitude, appeals were made in seeming endless repetition for local involvement and community support. Herbert Kubly recalls that Barlow incessantly "ran around town, drumming up support. It was really hard to say 'no' to him. His enthusiasm was infectious and, although I think that turned some people off, lots of folks got wrapped up in it all." Similarly, the *Post* ran editorials calling for support that appeared with the regularity of the changing seasons, entreating skeptical readers to "above all, boost the play, don't knock it."[42] The editor was tired of all the "criticism . . . that there has not been co-operation in putting on these performances. Of course, there has been co-operation or everything would not go as smoothly as it did." As evidence, she pointed to the several dozen ways that people cooperate, "not only on the part of the actors, but also of those behind the scenes."[43] Though such a defensive statement betrayed the lingering suspicion of the festival, it also contained a great deal of truth. Barlow and his fellow organizers recognized the legitimizing importance of bringing as many people as possible into the *Tell* tradition and to making sure that it spoke to their sense of memory and identity at the personal, family, and group scales.

Thus, the first half dozen years of *Tell* saw an increase in the production's complexity and organizational scale, drawing most heavily on the community's large and unpaid skilled work force. Women took the lead in creating the costumes that became so integral to the *Tell* ambiance. It was clear that the generic rented costumes from Milwaukee, in addition to being expensive, possessed little of the sense of "authenticity" that Barlow believed would bring his pageant distinction. Bridging the gulf between folk art and high culture, Barlow solicited the creation of a complete series of cast costumes based on notions of "authentic Swiss design" and designed from color plates at the New York Public Library (fig. 4.5). Next, the Biblioteca Helvetica (Swiss Library), also in New York, furnished renderings of the Swiss cantonal *Trachten,* or folk costumes. The "usherettes" made their debut in these Trachten in 1940, adding not only "accuracy and authenticity" to the play, but also important participation opportunities for nearly two dozen young women (fig. 4.6).[44]

Women, again, played the key role in the next addition to the *Tell* tradition. In 1940, the same year that women became involved more directly on stage as "usherettes," Barlow solicited a professional dancer with the Bern Opera to teach the community "traditional" Swiss folk dancing. Roesli Witschi, who was in New York directing dances at the Swiss Pavilion at the World's Fair that year, found that the waltzes, polkas, and schottisches known in New Glarus were "only remnants of the fancier and more exact steps the old, old Swiss did centuries ago." Such commonplace dances would not do for such a high culture tradition as Schiller's *Tell.* The dances of the "old, old Swiss"—never known in Wisconsin—had to be "revived." Working with the 21 "usherettes" and younger girls of the "wedding dance scene," the Swiss dancer introduced the traditional dance steps that are still used in precisely the same manner today, sixty years later.[45]

Finally, the Guild worked strenuously to bring in volunteers for the

Figure 4.5. Costume design for the character of Stauffacher. After the success of the first *Tell* performance, Barlow contracted with an art student to design costumes for the immense cast. Her designs, as in Stauffacher's costume here, were derived from patterns found at the New York Public Library. (Courtesy of New Glarus Historical Society)

127

Figure 4.6. Wilhelm Tell usherettes, ca. 1970. Making their debut in 1940, the usherettes brought not only "accuracy and authenticity" to the play, but also a significant opportunity for women to participate in the male-dominated *Wilhelm Tell*. (Courtesy of New Glarus Historical Society)

scenes that involved hefty numbers of peasants, goat and cow herders, and soldiers. All told, more than 250 people participated in the 1940 production and fully 290 were "front stage" during the following year, amounting to a 100 percent increase in participation in *Tell* during its first three years. Indeed, if all the people involved backstage were added, fully one third of the village and township population worked toward staging *Tell*.[46]

The Swiss Are More American than Americans: Legitimation outside the Community

The first four years of Tell witnessed, in addition to a doubling of participation numbers, a threefold increase in attendance, reaching more than five thousand by 1941.[47] These first performances all took place in the original German language, a point of pride for organizers, but one that required some justification. The first year's program explained, "The play would lose much of its charm were it to be given in the English language." Cast members of the German-language version today reiterate this theme as many see the English translation as "clunky" and "boring." By contrast, the German script, as one contemporary *Tell* player asserts in a way that is typical, "has a poetic ring that comes from the fact that it is great literature. Listen to the rhythm, to the way that the words flow together. This is the work of a German language master and cannot be translated into English."[48]

Despite this feeling, in 1941 one of the two German performances was dropped and, in its place, an English-language translation was added. The timing calls into question the reason for the change. On one hand, and from a very practical perspective, the Wilhelm Tell Guild clearly saw the possibilities of attracting larger audiences to the performance if they expanded it beyond German speakers. But equally important, heightening tensions between the United States and Nazi Germany recalled the anxieties of just twenty years earlier when all public displays of ethnic culture were suppressed. The Swiss Americans found an ingenious medium to maintain their German play and, simultaneously, fit into a growing American anti-Germanism.

The first way to accomplish this rhetorical sleight of hand was to offer the play in English as well as German. One week before the 1941 performance, the local publicists noted that "in previous years the Swiss of New Glarus have been content to please themselves with their 'Wilhelm Tell,' presenting it in the original tongue. But this year, believing the old play has a new message for all, they will give one performance in English."[49] The new message held that *Tell* could speak to an American, and not merely a Swiss, patriotism. By adding an English version to the *Tell* weekend, organizers felt confident that they could reach a wider audience and thereby position the festival as more fundamentally "American."

Second, the Guild launched a blitzkrieg of programs, brochures, and press releases designed to liken the play with American revolutionary history. Press photos were released that made the connection between the thirteenth-century Austrian bailiff, Gessler, and his twentieth-century counterpart explicit and unmistakable (fig. 4.7). The text to the 1942 brochure offers only a hint of the broad efforts to legitimize *Tell* as an *American* drama. Likening their four-year-old festival to a sacred ritual, the Guild conveyed a heavy seriousness of purpose:

Again America is fighting to preserve a free world and democratic ideals— the same kind of freedom and democracy which the Swiss have preserved in their tiny nation for six centuries. The parallels between Switzerland and the United States are many in their desire for freedom. Both nations fought against heavy odds to win their liberty in the first place. Both have had to fight again and again since then to preserve that liberty. So it is now the two nations feel closer bound than ever, and that feeling is nowhere stronger than in the village of New Glarus. . . . For it is in New Glarus that every year, like some *holy*

Figure 4.7. A Swiss farmer playing an Austrian tyrant. The parallels between the *Wilhelm Tell* story and the war against an aggressive Nazi Germany were heightened by Paul Grossenbacher's evocative portrayal of the Austrian tyrant Gessler. Grossenbacher, a Swiss immigrant farmer, said that he "played the role just like Hitler." (Courtesy of New Glarus Historical Society)

reaffirmation of the democratic spirit, the townspeople and near-by countrymen join in a great and colorful outdoor drama, Schiller's 'WILHELM TELL,' which tells the story of the Swiss freedom fight.[50]

So far had the *Tell* players dissociated themselves with anything German that the Sunday performance was given in the "Swiss" language. Similarly, a medallion commissioned by Barlow in the early 1940s demonstrates this link between Switzerland (and, by implication, New Glarus) and the United States (fig. 4.8). Separated by the flags of each country, the "founding fathers" of both are depicted at opposite ends of the medallion next to the dates of their respective "declarations of independence," 1291 and 1776. This sort of nationalist rhetoric had been with the drama since its inauguration in 1938, but had previously played, by comparison, a diminutive role.

The press picked up this viewpoint with remarkable uniformity, demonstrating the ideological chasm between 1918 and 1942. The *Wisconsin State Journal* called the Swiss heritage "a liberty-loving heritage," the *Capital Times* spoke of the "Swiss fight for freedom," and the *Chicago Tribune* noted that the English-language version was necessary

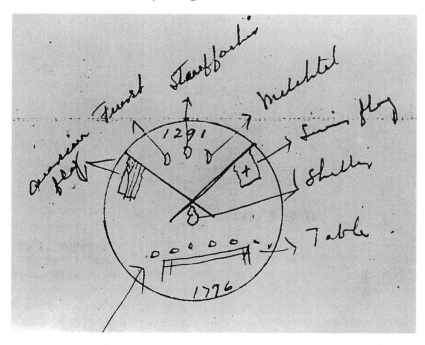

Figure 4.8. Proposed medallion for Wilhelm Tell Guild, ca. 1940. The Tell medallion, though never completed, was to unite Swiss and American history by highlighting the shared revolutionary history. Note the central place for *Tell* playwright Schiller. (Courtesy of the Elda Schiesser Collection)

this year because the "noble celebration of a nation's struggle to free itself from tyranny carries such poignant and ominous overtones as to make its presentation more valuable than ever." Finally, the effort to "Swissify" and Americanize Schiller's play received a dramatic reaffirmation by the Nazis themselves. By banning *Wilhelm Tell* in Nazi Germany, Hitler and others afforded the play's New World version a status of "national cultural freedom ritual." *Tell*, its organizers' claimed, was a more American story than the Anglo founding of New England.[51]

Tell Speaks for the Group, and the State

By enacting *Wilhelm Tell*, the community placed their particular group's heritage into the marrow of American national ideology. The articles of incorporation for the production's official organization made this point clear; the chief reason for the festival was "to stimulate a greater interest in Democracy ... in the State of Wisconsin and anywhere in the world."[52] Indeed, the connections between Schiller's drama and American Revolutionary history and its salience for a world on the verge of another world war could not be more explicit. *Tell's* 1941 program, for example, declared: "To this day, Switzerland is a free country, one of the few in Europe, and it is this message of liberty and freedom from oppression that the people of New Glarus are trying to show by their 'Wilhelm Tell play.'"[53] Thus conceived, Schiller, first in German and then in both English and German, could be used to demonstrate the tolerance and power of democratic pluralism in the face of Nazism's ruthless repression of ethnic minorities and, in turn, position Swiss culture as *the* archetypical American culture.

And yet, this is telling only half the story. The second reason for the festival also can be found in the articles of incorporation: "to perpetuate the memory, character and traditions of the Swiss people for the welfare and common good of the residents of the community of New Glarus."[54] The annual ritual of performing *Wilhelm Tell* reinvented for many in the community their own Swiss heritage. Reporting that the children "as well as many old folks in the cast, learned more about Wilhelm Tell and Swiss history the past few weeks than in the rest of their lives," the local paper questioned if the community had not too hastily discarded local traditions twenty years earlier. Newly created traditions, it recognized, could provide a means for community cohesion and stimulate public memory.[55]

Looking back across the span of sixty years, Millard Tschudy, who was involved in the earliest *Tell* productions believes that *"Tell* was a means, an almost sacred web, for people to find their identity. It gave the early performers a sense of cohesion that was missing from their lives and gave people a chance to mingle with folks that they might not otherwise talk with. *Tell* really defined what it meant to be Swiss." Lila Kubly Dibble—who played the part of Bertha von Bruneck for that first performance—recalls the role of the play for people like her and of Barlow's key contribution: "We were just starting to get excited about being Swiss and that we were just a bit unique. Mr. Barlow was a leader in this area, so we followed." Clayton Streiff seconds this perspective. Streiff, who had performed in some capacity in every performance since 1938, recollects that "we really didn't know what it was to be Swiss until *Tell.* It taught us Swiss culture, that's for sure."[56]

This high degree of reflexivity or self-consciousness became a crucial element of the cultural pluralism fashioned between the wars in communities like New Glarus. As with the historic pageant only a half dozen years before, this conspicuously constructed ethnic identification became a rallying point around which cooperation and unity, however manufactured, could gather. The *Wilhelm Tell* play became an ideal genre with which to build a connection to the state—indeed, to position the group as the small-scale archetype of the state—while simultaneously permitting ethnic heritage and difference to thrive.[57]

Critical to the genre's success was its unique blend of ethnic folklore and high culture. Parochialism, Werner Sollors suggests, is usually associated with the ethnic writer while national or international fame with high culture. Thus is the poet Carl Sandburg, the son of Swedish immigrants, excluded from ethnic culture because of his very popularity. So with Schiller, an international writer of first order and pillar of high culture.[58] It is in the simultaneous *Americanization* and *ethnicization* of Schiller—the reworking and adapting of his best known drama, which appeared on nineteenth-century New York, Berlin, and Zürich stages—and pressing it into the service of a local public memory, that the festival achieved its greatest rhetorical power. As *Wilhelm Tell* became both American and ethnic, so did the cultural performance create a space for ethnic memory and the official culture of the state to coexist.

UTILIZING A TRADITION

No matter how high the lofty-minded rhetoric of renewal and regeneration flew during these early years, just beneath hovered a second reality. Although most people involved in the original *Tell* do not recall it as a tourist attraction, early records indicate that organizers focused a large portion of their considerable energy on bringing visitors to the community. Two weeks before the 1938 performance the *New Glarus Post* editorialized, with a fair degree of wishfulness, that if things were done properly, "New Glarus could be made to be a center for tourists interested in what we have to show them."[59] That this was to happen eventually is due to the twin factors of the community's own successful efforts at publicity and the fact that the community possessed a quality, a sense of place, that tourists wanted. That quality—a sense of an authentic retreat from the modern world—sat in uneasy tension with the all-too-modern efforts to publicize it.

"*Tell* Has Become a Veritable Trade Mark of This Swiss-Settled Community": Tradition-Powered Publicity

That the *Tell* tradition was created with one eye toward tourists should come as no surprise. Its Swiss model, after all, had itself been invented in large measure to facilitate tourism in Interlaken roughly three decades earlier. Quick and fast assumptions as to the festival's meaning stemming from this simple fact can be misleading, however. As Regina Bendix has demonstrated persuasively, the Interlaken *Tell* in Switzerland, despite its touristic origins, also imbued its participants with a sense of group cohesion in the face of a tourist invasion that itself helped inspire. Invented traditions, even those primarily directed at tourists, can have an ameliorating effect on its conspirators.[60] This introductory caveat is important because the New Glarus *Tell* players saw no incompatibility in their aim to renew Swiss and community culture with the possibility to make the village "a center for tourists." As it was put at the time, "some commercialism must be acknowledged to make possible the expense entailed in keeping the play authentic in detail."[61] It was a fine line they walked, however, and a misstep in either direction could spell the tradition's doom. Too open a display of commercialism would undermine the "authenticity" it had to purchase. Likewise, without publicity the Guild could hardly expect to attract the crowds needed to keep the drama moving forward.

Henry Schmid, a third generation Swiss American businessman, took the lead in the festival's publicity efforts (fig. 4.9). Though involved "front stage" in several character roles, Schmid's most important and longstanding contribution came backstage as secretary and treasurer of the Wilhelm Tell Guild from 1939 to 1969. Schmid, more than most, saw the potential of luring large numbers of tourists to the community and went about attracting those visitors with vigor and resolve. While the tradition's inventor might have warned about hiring professionals for the performance, Schmid felt no compunction in working with promotional specialists behind the scenes. Three years after the first performance, the Guild treasurer hired a professional publicist for the crucial 1941 season, a move that helped insure the "Folkfest's" success. That he had never been to New Glarus—or knew next to nothing about it or the Swiss culture of the village—mattered little to the Illinois-based advertising agent. In their efforts to "sell the play," Schmid created the informational material and his publicist took responsibility circulating it. Over the course of the summer, they distributed more than twelve hundred informational leaflets throughout the region and an equal number of press releases to regional and national newspapers[62] (table 4.1).

So successful was Schmid's "selling" of the play, that, by 1945, *Wilhelm Tell* and the replanted Swiss scene had taken on a national scope. As we have seen, organizers of *Tell* hoped that their production—and the conspicuously ethnic place that stood behind it—would attract attention from a wider than regional audience. Eight years after the first performance, this had been largely achieved. Thus, by the time of the community's one hundredth anniversary celebration, in 1946 (it was delayed one year due to the war), the *Capital Times'* chief arts and culture correspondent felt certain that "since its inception [*Wilhelm Tell*]

Figure 4.9. The "front-stage" side of promotion. Henry Schmid, a third-generation Swiss American businessman, took charge of "selling the play." Though he is shown here in a publicity photo as the character of Fronvogt, his main contribution was behind the scenes as he created a sophisticated advertising system for the anti-modern presentation of ethnic place. (Courtesy of New Glarus Historical Society)

Table 4.1 Advertising for *Wilhelm Tell,* Distribution of Informational Leaflets, 1941

	Number	Percent
Milwaukee Road railroad	500	39.3
Chicago Motor Club	200	15.7
American Automobile Association of various Wisconsin communities[a]	200	15.7
American Automobile Association of Madison	100	7.9
Wisconsin Conservation Department, Recreation Publicity	100	7.9
American Automobile Association of Lake Geneva	100	7.9
Milwaukee Travel Bureau	50	3.9
Various other locations[b]	22	1.7
Total	1272	100

Source: Lawrence N. Eldred to Henry M. Schmid, Elgin, Illinois, August 1, August 11, August 18, 1941, correspondence folders; Wilhelm Tell Archives, New Glarus, in possession of Elda Schiesser.
[a]Includes Kenosha, La Crosse, Green Bay, Wausau, Eau Claire, Oshkosh, Antigo, Three Lakes.
[b]Includes independent travel bureaus in Illinois and Colorado, and the U.S. Department of the Interior.

has become a veritable trade mark of this Swiss-settled community and has helped spread the name and fame of New Glarus nationwide" (fig. 4.10).[63] That trademark carried with it a self-confidence that bordered on chauvinism as the Swiss made the transition from a defensiveness about their own uniqueness to a position that argued for their centrality in American history. Two episodes in the effort to reach a national audience, one ultimately futile and the other successful, demonstrate the group's increasingly secure position engendered by the tourist publicity surrounding *Tell.*

In the winter of 1945, a handful of *Wilhelm Tell* enthusiasts represented New Glarus at a countywide meeting, the purpose of which focused on the creation of a special postage stamp commemorating the one hundredth anniversary of the region's first Swiss settlement. Such a stamp, to be used nationwide at the conclusion of the war, would benefit the community in two contrasting ways. First, it "would do much to spread the fame of Green County" and second, the stamp would "pay fitting tribute to the 108 bold-hearted and liberty-loving Swiss colonists." From here, opinions diverged about why such a small group should be so honored. In Monroe, the county seat, where a greater ethnic heterogeneity was becoming increasingly important, ad-

Figure 4.10. *Tell* becomes a trademark of the community. Less than ten years after the first production of *Wilhelm Tell,* the invented tradition has come to symbolize the core values of the community. In the village's centennial commemoration of 1946, *Tell* takes center stage. (Courtesy of New Glarus Historical Society)

vocates stressed the cheese industry and the representativeness of the Swiss for all immigrant groups. The appeal for such a commemoration, the Monroe paper argued, "cannot be construed solely as a narrow gesture honoring a small racial group here," but rather was to stand for all the "hopes and industry and dreams of European emigrants who could not find fulfillment of those dreams and ideals in 'the old country.'"[64] Such a cosmopolitan view gained little ground in the more self-consciously Swiss New Glarus. The commemoration, leaders there argued, should highlight the Swiss story alone and boost that one group's important "contributions to the cultural and economic advancement of our country." Reflecting this position, the final design touched on the important cheese industry, but as a stepping stone to imagined Swiss culture of folk costumes, alpenhorns, and glaciated mountains (fig. 4.11).

The effort received endorsements from such patrons of official culture as Governor Walter S. Goodland, *Old World Wisconsin* author Fred Holmes, the *Capital Times* editor William Evjue, the president of the University of Wisconsin, Edwin Fred, and U.S. Senator Alexander Wiley. By far, the most important patron was the state's other U.S. senator,

Figure 4.11. Legitimating ethnic place through the postal service. The 1945 commemorative stamp, though never issued, highlighted the community's increasing self-consciousness in the wake of *Wilhelm Tell*. (Courtesy of New Glarus Historical Society)

Robert M. La Follette, Jr., who personally shepherded the request to the U.S. Postmaster General.[65] Despite this high-level espousal on behalf of the Swiss and the ground swell of local support in the form of petitions, the request was denied, due in large measure to high costs at the end of the war. The increasingly confident Swiss cried "foul" and "discrimination" at the snubbing. One annoyed resident, Emma Becker, wrote that she "did not believe your department appreciated the importance of this community and its Swiss population," while Edwin Barlow fired off an angry letter to Senator La Follette lamenting that he was "not the fighter your father was." With ample sarcasm, he charged: "of course, we are not being discriminated against, but if you have the courage, please come and *tell the Swiss that*."[66] Despite the attempt's failure, it shows the growing self-confidence and exceptionally high degree of reflexivity in the wake of nearly a decade of successful *Tell* performances. It also demonstrates the distance in social standing that white ethnic groups like the Swiss enjoyed in contrast to less fortunate "racially" defined groups.

The second, and in this case successful, attempt to publicize vernacular heritage to a national audience came one year after the commem-

138

orative stamp disappointment. Since 1939, the Guild tried unsuc-
cessfully to attract the attention of *Life* magazine,[67] but in 1947 *Time*
magazine and, more importantly, *National Geographic,* spent several
weeks "deep in the heart of Swissconsin." *National Geographic* focused
its celebrated cameras on dairy farms, a cheese factory, and on many
of the principal people involved in the performance.[68] The resulting
story and photographs could hardly have been more flattering, and
are seen to be directly responsible for the record crowds that attended
Tell the following year. Photographs alternate between the modern and
the rustic, but with a heavy accent on the latter. In them, New Glarners
are depicted as up-to-date small-town inhabitants who rely on modern
farming techniques. Whether pictured in front of the pioneer monument, in the tavern, or on the farm, they look and dress like other
Midwesterners. Indeed, the landscape itself is pure Americana, as the
village's "main street . . . is not much different from any other Midwestern town of a thousand population." But quickly the costumes come
out of the closets and "Swiss background and tradition crop out
aplenty for anyone willing to explore its quiet streets." Photographs of
the "warbling Switzer" (fig. 4.12), the Wilhelm Tell family, and usherettes add to the allure and invitation for a visit. The article can be read
as a lengthy promotional vehicle for the community and its "crowning
achievement in the preservation of Swiss tradition . . . the William Tell
Pageant." Sandwiched between articles on "Sailing the Inside Passage"
and life in Korea, authentic, traditional, and untainted folkloric culture, it would seem, can be found here in the least likely of places,
Sinclair Lewis's *Main Street.*[69]

"REVERSING THE TREND TOWARD MODERNIZATION, STANDARDIZATION, AND COMMERCIALIZATION": GIVING TOURISTS WHAT THEY WANT

The antimodernism for which thousands of *National Geographic* readers
were searching could not be purchased and, accordingly, all publicity
efforts had to be carefully hidden from view in order to not taint this
fragile vision of authenticity. Word of the festival, it seemed, came not
from the determined efforts of Schmid and an Illinois press agent, but
from the sincerity of the local culture and the authentic connection of
the play to local history and geography. One newspaper editor, envious
of *Tell's* success after three years, noted that "anything tinged with
commercialism could not command the attention the *Tell* play does."
Another saw the festival as a "community hobby" whose prime virtue—during the summer that the Guild began its heavy publicity campaign and on the heels of attendance figures reaching past the five
thousand mark—was the fact that it was "unspoiled by heavy tourist

Figure 4.12. *National Geographic* discovers American ethnic place. Tavernkeeper Ernest Thierstein, depicted here as the "Warbling Switzer," appeared in *National Geographic* after the flood of publicity engendered by *Tell*. For more on Thierstein, see chapter 6. (Photo by Joseph Baylor Roberts, *National Geographic Magazine*, June 1947. Reprinted with permission.)

traffic or commercialism." Still another lauded the fact that, unlike city dramas, *Tell* inherited its charm "not [from] professional actors, but villagers who portray the characters in the familiar story."[70]

Moreover, this near obsession with the absence of obvious tourist life was seconded by many of the middle-class playgoers themselves. One second-generation Swiss American wrote to Gilbert Ott in 1941 that "the most impressive fact [of the *Tell* play] to me was your utter lack of attempted commercialism. This bespeaks of your sincerity and that of your fellow townsmen in your efforts to present something really worthwhile to the public and yet not capitalize on it." Yet with unintended irony, the admiring fan from Illinois lamented that "unfortunately, you do not advertise this venture sufficiently" and suggested paying more attention to publicity. Another tourist, a Japanese student living in Madison, was moved, similarly, by the "frontier spirit" and "purity" of the play, while yet another felt that the "traditions and folksiness" of *Tell* deserved the fifteen-hundred-mile trip that he and his wife made to see the festival.[71]

At the same time that they embarked upon a heavy and sophisticated advertising campaign, *Tell* organizers were well aware that a good deal of the pageant's tourist appeal lie in its antimodernism. That very first production called attention to this fact. "The world seems to go faster and faster," the *New Glarus Post* editor explained. "Every year cars are made to go more miles in more hours. Every year, the streamlined trains go faster and faster. . . . We seldom stop to think about the changes that have taken place in less than a hundred years." Other writers agreed, and applauded the Guild's efforts to "revive" its sacred traditions. In this way—by "recapturing the old Swiss spirit" that made it distinctive—the community would be seen as "reversing the trend toward modernization, standardization, and commercialism."[72]

This reversal of modernism seemed to come from three related factors. The first stemmed from the very ordinariness of the cast, a point repeated endlessly in early newspaper accounts. The *Chicago Sun*, for instance, found the play's power to be derived from the "farmers and merchants, housewives and children who learn the lines, grow the 'old country' beards and embroider the costumes."[73] Second, the outdoor setting in the rural, Wisconsin countryside offered a geographical basis to the antimodernist sentiment. In Schiller's drama, the human relationship with nature lies at the root of Swiss resistance to Austrian occupation and, as such, a landscape assumes a spiritual presence (fig. 4.13). Indeed, one may argue that landscape is not merely present to increase theatrical effect, but has a pivotal role, linked to all aspects of the drama's plot, themes, and characterization. Especially in the Rütli scene, where confederates from Uri, Schwyz, and Unterwalden swear

Figure 4.13. The apple shoot scene in *Wilhelm Tell*, ca. 1938. Like its Interlaken, Switzerland, model, the staging of *Wilhelm Tell* took advantage of a natural, outdoor setting. The driftless area valley heightened the sense of naturalness that is a central leitmotiv in Schiller's drama. (Courtesy of New Glarus Historical Society)

allegiance, the link between Swiss peoples' constant contact with the natural world and their love of freedom and democracy reaches its apex. Swiss folk performers of *Wilhelm Tell* have on some basic level understood this central fact and, correspondingly, have frequently staged the play outdoors. In New Glarus, as in Switzerland, the outdoor setting heightened the play's sense of "nature" and its "folk aspects," blurring the boundary between both.[74]

Third, and most important, the tradition was seen to reestablish indigenous heritage to a "real" place. The festival gained resonance in the view of many tourists because of its seemingly authentic connection to the community's own past. As one Wisconsin paper put it, "Spectacular as the pageant is, with its colorful processions, dramatic speeches and tense situations, the reason for its great success lies in that it is such an interpretation of the New Glarus people's own lives: the rich heritage of freedom given them by their ancestors, the preservation of their ideals, the solidarity they feel in their transplanted homes, the perpetuation of their industries."[75] There is considerable evidence that many of the visitors expected to find this vernacular connection for their own lives. Ethnic memory rather than the play's didactic patriotism dominated, as few letters collected by the Guild during these early years mention the patriotism taken for granted by newspaper editors and publicists.[76]

The audience, however, was far from a uniform, homogeneous mass and received the "messages" of the play in diverse ways. Some found it boring and didactic, while for others *Tell* was "sublime." Though the highbrow interpretation of *Tell* outnumbered the competing lowbrow reading, enough of each found their way to the festival to give it a sense of instability.[77] One visitor, George Willet, felt tremendously moved by the pageant's sincerity, but was rattled by the "inferior minds" who "failed to understand the pathos and tragedy" of the production, saying, "I would have enjoyed it more if it had not been for a 'wise-cracker' right behind me, who evidently thought he was at a county fair, and wanted the entertainment louder, and funnier. It is to the credit of the New Glarus people that they put on the drama without comic relief, as near to the classic Schiller text as possible. But our country is full of people who are not happy at a play unless it is a slapstick comedy."[78] A high seriousness marked the play from its inception, a seriousness that contained little space for comedy or irony. One of the Guild's perennial suggestions—reiterated the year before Willet complained of his lowbrow fellow tourists—was to minimize the "restlessness of those in the cast which draws attention away for the scene of action." In the end, it was this sincerity of purpose that brought together tourists searching for an antimodernism or their ethnic roots with the performers who sought to reinvigorate their own slumbering ethnic identity.[79] What was unknown to tourists—that the drama was invented by a wealthy outsider from Long Island—was unimportant to the increasingly symbolic and ever more reflexive ethnicity of the New Glarus Swiss.

CONCLUSION

By 1950, annual productions of *Wilhelm Tell* had become a firmly established tradition. Its instantaneous creation only twelve years earlier had all but vanished behind a screen of antiquity. So vast was the distance between tradition and its origin that the Monroe paper saw *Tell* as a "spectacle carried over from the Old World."[80] The spectacle was, from the very beginning, what may be described as an instance of "folklorism," that is to say, the utilization of folk culture (here, the Tell legend) for a secondary, manipulative framework (in this case, performances for audiences of noncommunity members). Such consciously created and framed folk culture has been customarily disparaged as "commodified," "fakelore," or "culture by the pound." The notion of the invented tradition—a phenomenon that *Wilhelm Tell* represents par excellence—implies a self-fashioning and representation that is often deceptive and exploitative. Indeed, it is not difficult to find examples of duplicity and exploitation on the part of elite or official culture.[81]

Yet, the success of *Wilhelm Tell* in replanting the Swiss scene hinged not on performers duping an unsuspecting and gullible audience, but on the performance's resonance with individuals and groups on both sides of the stage. Heritage, David Lowenthal argues, is *always* a fabrication, inevitably both creative art and act of faith. To view the use of the past as either correct or incorrect misses the dynamic and instrumental nature of public memory. In other words, the "mistaken dichotomies" implied in the authentic folklore/disingenuous fakelore model, fails to account for the *Tell* career of someone like Clayton Streiff, who never missed a performance in English or German during the play's fifty-eight-year history (fig. 4.14).[82] Taking on different roles as he aged—including Wilhelm Tell—Streiff found, as many have, that the annual festival assumed something of a rite-of-passage quality. Planning, rehearsing, and performing *Wilhelm Tell,* an invented tradition of recent origin, eventually did what older traditions accomplished so successfully: it helped fulfill a shared need for celebration and communitas. Public dramas like *Wilhelm Tell,* Barbara Myerhoff suggests are "definitional ceremonies, performances of identity." As cultural mirrors, they offer a critical medium for presenting public or collective memory. But, "like all mirrors, these reflections are not always accurate. They may also alter images, sometimes distorting, sometimes disguising various features." Despite (or, perhaps, because of) their inherent duplicity, public dramas such as *Tell* are the means people employ to "see" themselves.[83]

Equally important, the performance set apart the Swiss from mainstream American culture as it simultaneously created an integral space within that wider national culture. After seeing the first 1938 production, J. P. von Grueningen felt that the village and its connection to Swiss culture "must have something sacred about it for the inhabitants of the town and for all knowing visitors who go there. To many it may be a matter not of mere show going but something of a pilgrimage." While not every visitor was so moved as Grueningen, enough were to give *Tell* the feel, or *Stimmung,* of a secular pilgrimage.[84] Fabricated from the cloth of Switzerland's best-known myth and cross-stitched as high culture and folklore, as both American and ethnic, *Wilhelm Tell* provided an ideal medium through which Swiss Americans asserted their centrality and their "difference" within a previously restricted civic culture.

As a powerful, multifaceted symbol, *Tell* articulated the patriotic exigencies of the day as well as the reassertion of an identity that was

Figure 4.14. Clayton Streiff with alpenhorn. Clayton Streiff had never missed a performance of *Wilhelm Tell*, in either the German or English productions, since the drama began in 1938. Here he is pictured in 1994 with the alpenhorn he purchased at the local drug store and which he played weekly for tourists. (Photograph by author)

gaining currency across the country. One longtime New Glarus observer recognized the importance of this larger context when he noted that staging the drama was possible because "doing that sort of thing was in the air."[85] Wisconsin at this time provided an especially receptive milieu for such a cultural display. The next chapter explores that context.[86]

5

Provincial Cosmopolitanism
Mapping and Traveling Through Old World Wisconsin

In the summer of 1945 Giradela Wackler felt homesick for her native Wisconsin and longed for a return visit. More than twenty years had passed since her family moved to Oakland, California, and it seemed that the memory of her home state was becoming cloudy. This granddaughter of Swedish immigrants remembered Wisconsin as American as apple pie and a far cry from the richly textured way of life she recently encountered while touring Sweden. Now—and to her great surprise—she discovered that a trip home might register the same feelings of Swedishness that she experienced in "the old country." Writing to the author of a book that she had just read and which led to this discovery, Wackler noted that "we have traveled abroad quite extensively but never realized there was so much of 'abroad' right in our native Wisconsin" (fig. 5.1). It hardly seemed possible that the state she had left so long ago possessed so many rich Old World customs and traditions. Not just the Swedes, but the Poles, Germans, Finns, Swiss, Belgians, and so many others apparently maintained customs that one found in California only with considerable effort, "if at all." The third generation Swedish American concluded that she would waste no time before visiting "the many interesting spots with their foreign backgrounds you describe [about] old world Wisconsin."[1]

Giradela Wackler was not alone in her surprise discovery of the state's conspicuous ethnic places. Communities themselves were busily discovering and re-presenting their ethnic identity as a form of cultural legitimation within a previously hostile environment, as a way for individuals to establish connections with a rapidly fading collective memory, and as a means toward community renewal. But these local cultural displays—a Norwegian museum, a Swedish Midsommer

147

Figure 5.1. Traveling abroad through *Old World Wisconsin*, ca. 1944. Many readers of Fred Holmes's *Old World Wisconsin* were surprised to learn of the state's many persisting ethnic places and traditions, such as this Swedish kaffeeklatsch, documented, photographed, and romanticized in his 1944 book. (Courtesy of the State Historical Society of Wisconsin, Fred Holmes Collection, neg. WHi [X3] 51274)

celebration, a Swiss performance of *Wilhelm Tell*—did not take place in a vacuum or without critical support from institutions, writers, researchers, and politicians at the state level. Rather, as groups like the New Glarus Swiss or Wackler's Swedish neighbors were becoming more assertive in demonstrating and performing their heritage, so were powerful cultural leaders at the state university, historical society, local newspapers, and regional publishers also realizing the extent and persistence of ethnic cultural displays. The eminent Wisconsin historian Milo M. Quaife, upon reading the same book that amazed Giradela Wackler—Fred Holmes's *Old World Wisconsin*—summarized this growing awareness well: "most surprising is the astonishing number of European nationalities and races who have established colonies in Wisconsin, in which the cultural atmosphere of the homeland is preserved more or less fully."[2]

This chapter widens the lens of analysis by shifting the focus from the level of the community to that of the state. It seeks to account for the growing presence of ethnic places in the late 1930s and early 1940s that so surprised Giradela Wackler and Milo Quaife. At exactly the

148

same time that the New Glarus Swiss replanted their heritage through festive occasions and museum work, influential leaders at the state level legitimated, borrowed from, and modified those efforts by their own recasting of ethnicity. The role of state level educational, governmental, and scientific bureaucracies in constructing cultural meanings is central to the way in which public memory is shaped, altered, and utilized. Numerous scholars in Europe and the United States have demonstrated how elite supporters of the nation and state frequently impose master narratives on "vernacular culture" as a way to consolidate control over a citizenry: George Mosse's work on the representation of the past in Nazi Germany, Michel Foucault's connection of systems of knowledge and power in France, Ian McKay's quest for the construction of innocence and the folk in Nova Scotia, and John Bodnar's reading of twentieth-century American commemorations all point toward a hegemonic "official culture" with a common interest in promoting social unity.[3] Like the culture of the vernacular, official culture finds its public expressions in a variety of media or forums. But where local ethnic culture, as in the case of New Glarus, may commemorate heritage through unpaid performance, leaders working at the level of official culture generally possess greater resources. They have access to government grants, advances from publishers, and monies raised through large private donations. The resulting historical imagery tends toward consensus and homogeneity.[4]

Despite these differences, and frequently overlooked in studies which focus exclusively on the governing elite, considerable interaction often takes place between these different levels of cultural expression. With interaction usually comes confusion, for as official culture often draws on local expressions, so does vernacular culture make use of official culture for models and parodies.[5] This chapter deepens the understanding of this dialectical relationship as it examines, first, the role of scientists, then popular writers, in shaping popular conceptions of ethnic place. The image that emerges from these two disparate media points toward a vision of acceptable difference, of a cosmopolitanism sufficiently provincial to satisfy the state and to provide the imaginative material for a local, and thoroughly tourist-oriented, ethnicity.

MAPPING ETHNIC CULTURE: GEORGE HILL AND THE WISCONSIN NATIONALITY PROJECT

Wisconsin Rural Sociology and the Study of Locality Groups

In many ways what Chicago is to urban sociology, Wisconsin is to its rural counterpart: both places were among the first used as scientific "laboratories" for examining the lingering effects of ethnic background on a heterogeneous population. The Chicago school of sociology—led by William I. Thomas, Robert E. Park, and Ernest W. Burgess—formulated the most influential scholarly theories of immigrant adaptation to life in American cities. Their fundamental contribution, as Stow Persons argues, lies in several concepts bearing on immigrant adjustment: migration as a process of disorganization and reorganization; the "race relations cycle"; symbiosis and ecological succession. Each of these concepts relied on the accurate and precise mapping of ethnic group distribution in the Chicago laboratory as well as more schematic mapping, most notably in Burgess' famous, and endlessly reproduced, zonal model. Add to this mapping the less well known, but equally important, Hull-House maps of Jane Addams's research staff, and Chicago emerges as the first testing ground for urban ethnic studies.[6]

Just to the north, in Wisconsin, researchers at the state's university were attempting to make a science out of the study of the adjustment to rural life and, equally important, to apply that science to the benefit of state residents. As with their urban counterparts in Chicago, ethnic studies comprised an important part of the large corpus produced by the University of Wisconsin's Department of Rural Sociology. From the arrival of Charles Josiah Galpin in 1911, the formal establishment of the department and the appointment of John H. Kolb as chair ten years later, and the appearance of such key scholars as Arthur F. Wileden, Ellis L. Kirkpatrick, and George W. Hill over the next decade and a half, Wisconsin gained a reputation as the major center for the study of rural life in the United States. In many ways the rise of rural sociology as an academic discipline in the 1920s through the early 1940s mirrors the development of the Wisconsin department.[7]

During these early years in Wisconsin, research concentrated on identifying and characterizing the important rural types of group arrangements. These included the family, voluntary associations, and special interest organizations. The most important group, however, was distinctly spatial in arrangement and became known as the "locality group"—the community and the neighborhood. With his first locality group study of Walworth County, Wisconsin (1915), Charles Galpin presented a landmark analysis of the rural community as comprised of both town and country people. Equally significant, he created a method for locating and mapping the locality group's region, a technique with which it now "became possible to view the areas of rural association as discrete [and mappable] units." Robert Park, for one, immediately recognized the utility of identifying locality groups and credited the Walworth County research as the stimulus for his first studies of ecological areas in the Chicago region.[8]

The second fundamental work on locality groups in rural society came from John Kolb in his study of rural neighborhoods. The presence of country neighborhoods had been long recognized, but it remained for Kolb to draw on Michigan sociologist Charles H. Cooley's general primary group concept to derive a method of formally identifying active neighborhoods and establishing their boundaries. Kolb defined a rural primary group as "that first grouping beyond the family which has significance and which is conscious of some local unity." This fundamental point—namely, that rural communities center around discrete groups in which propinquity plays a decisive role— spawned a host of studies that concentrated on institutions such as the high school, libraries and hospitals, and churches.[9] From here, it was a very short distance to the recognition of a unique type of locality group—the nationality, or, as it was beginning to be called, the ethnic group. This research had to wait until the arrival, in the mid-1930s, of a young assistant professor, George W. Hill.

"Yes, the Cultural Background of Our Various Nationalities Does Make a Difference"

George Hill brought a new dimension to the study of locality groups. The son of Finnish immigrants on Minnesota's Iron Range, Hill was among a growing cadre of scholars across the nation for whom identification with an ethnic background was a component of both personal life and scholarly research. Like historians Theodore C. Blegen, Carl F. Wittke, and Marcus Lee Hansen, literary scholars Alfred Kazin and Lionel Trilling, and fellow sociologist Louis Wirth, George Hill's own ethnic heritage became intertwined with his scholarship.[10] Although Hill never considered his work to be Finnish American, his personal experiences and the contacts that he cultivated among Wisconsin's Finnish community opened him to the possibility of the lingering effects of nationality background. He spoke and wrote Finnish fluently, maintained an active interest in Finland's political situation (particularly during the crisis precipitated by the Soviet invasion) and, with John Kolehmainen, coauthored the definitive history of Finns in Wisconsin.[11]

But George Hill's interest in the state's nationality groups extended well beyond his own Finnish roots. As he began working within the emerging tradition of Wisconsin rural sociology, Hill believed that he stumbled across a key link that Galpin, Kolb, and others had hinted at, but neglected to approach directly. "Does the *cultural background* of a rural people," Hill asked, "have any influence upon the prevailing type of farming, ratio of farm tenancy to farm ownership, value

151

of farm land and buildings, tax delinquency, relief acceptance, and other related sociological phenomena?" For Hill, "cultural background" was synonymous with ethnic or ancestral heritage. The answer came from the collective voice of Wisconsin's county agricultural agents with such clarity as to make the original question seem rhetorical: "Yes, the cultural background of our various nationalities does make a difference."[12]

Shortly after his arrival in Madison, the young assistant professor turned his attention to this "nationality question" where it became the center of an extremely large, government-funded research project that was to have three distinct dimensions: (1) a retabulation of the 1905 Wisconsin State Census; (2) a map showing the distribution of ethnic groups in the state; and (3) a larger, crowning monograph that would make use of the data gathered in the previous two components. The state, it was believed, offered an ideal testing ground or "laboratory of human relations" for the central issues of the day.[13] The first two dimensions of the Wisconsin Nationality Project were completed; the monograph, however, was jettisoned.[14] As the project was expected to offer important findings into problems of federal relief and, later, combating fascism, so, following Foucault, its archaeology suggests insights into how an official culture created a sense of acceptable ethnic difference during a period of crisis and turmoil.[15]

Fields and Faces in Wisconsin: Two Ways to Highlight Diversity

The first phase of the Wisconsin Nationality Project owed its beginnings to the New Deal's Works Progress Administration (WPA). Between 1937 and 1942 three project grants brought together more than fifty full- and part-time "professional, educational, and clerical persons to conduct a study of Wisconsin nationality groups."[16] That initial study aimed to retabulate and summarize population data from the Wisconsin State Censuses. As project director Hill and his colleagues immediately realized, the customary county-level data proved too coarse a unit of analysis. Even the briefest survey of settlement patterns in the state revealed that "there is more variation within counties than between counties which necessitates information on a minor civil division [i.e. township, or "town"] basis."[17] Although the censuses of 1885 and 1905 were initially chosen, it soon became apparent that retabulating even one data set would stretch the project's limits. Thus, Hill dropped the 1885 date and concentrated on 1905, a point in time when Wisconsin's foreign population was "just one generation old [and] the most accurate source of information for our present nationality groups."[18]

Three graduate students were enlisted to serve as project supervisors, each of whom applied the project's data to their own dissertations or research papers,[19] and a WPA research coordinator was hired full time. Under them, a staff of seven certified WPA clerks oversaw the immense task of reviewing the manuscript census rolls at the state WPA office, retabulating 400,000 of those schedules by townships, coding that retabulation onto Hollerith cards, feeding the cards through electric counting sorters, making the final computations with four merchant calculators, and finally, typing the end calculations into tables. In the end, eleven bound volumes containing more than four thousand tables were created.[20] If the creation of population tables was the Wisconsin Nationality Project's final product, it could have been admired for the employment that it provided for dozens of university researchers and clerical staff as well as for its early use of historical census data. But Hill's concern with the distinctly geographical question of location called for accurate maps of the distribution of the state's many persisting nationality groups.

The purpose of the mapping, the Project's second component, was not merely illustrative. It represented a vital step in uncovering the relationship between cultural background and variations in social, economic, and political ways of life. Hill knew from experience what he now wished to demonstrate cartographically, that "different nationalities . . . develop certain social values and attitudes particularly their own [that] tend to crystallize into social heritages and, as such, condition the behavior of nationality groups in their new cultural *settings*. Hence, to the soils map and to the type-of-farming map, we are going to add one more—a nationality map of the people."[21] The statistical tables might have shown that sociological and economic behavior differed from locality to locality in direct proportion to the change in "dominating nationalities," but alone remained unwieldy and unconvincing to the nonexpert. A map locating these culture areas, on the other hand, would make the connection between nationality and "ways of life" unmistakably clear.[22]

Two sets of maps produced at the level of the county furnished the raw material for the nationality map. The first set of compilations used the 1905 retabulations and mapped the "dominant nationality" by township, defined as "one in which more than 40 percent of the family heads are of one nationality by birth or parentage." A second set of compilations was then created, using data from fieldwork undertaken in 1938 (fig. 5.2). This time, township boundaries were ignored and an effort was made to "place exact limits around any distinct nationality group."[23] Here, Hill contacted county agricultural agents, postmasters, doctors, judges, and school superintendents across the state. In some cases, standardized questionnaires were mailed, in others interviews

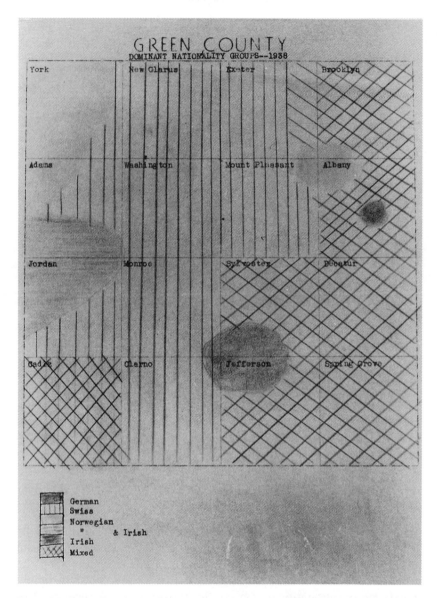

Figure 5.2. Green County dominant nationality groups, 1938. This rough map is derived from interview and ethnographic data collected by George W. Hill and provided the basis for his statewide studies. Each county was thus mapped and compared to similar maps using 1905 retabulated census data. (Courtesy of University of Wisconsin Archives)

154

were conducted, many by Hill himself. In each case, the informants—
all of whom had intimate knowledge of the county in which they were
located—were asked to draw the boundaries of the area's dominant
groups.[24]

Next, composite maps were drawn from the 1905 and 1938 data
(fig. 5.3). Hill believed that a comparison of the patterns would reveal
the persistence or disappearance of the nationality groups and thereby
imply their dominance over local culture.[25] Although they employed
radically different sets of data (historical census data on the one hand
and qualitative interview data on the other), Hill combined both into
one final overlay map. Hill released this map as a foldout supplement
to a scientific research report that brought together experts from across
the University. "Wisconsin's Changing Population" (1942) soon became
the standard reference on the state's demographic composition and a
key text in the shifting attitude toward cultural diversity. In total,
11,000 copies of the map were printed by Rand McNally in Chicago
and distributed as the bulletin's centerpiece (fig. 5.4).[26]

Entitled "The People of Wisconsin According to Ethnic Stocks," the
full-color map graphically demonstrated the thesis that Hill and his
colleagues had been developing for the past half dozen years: that
Wisconsin was, at its most fundamental level, a rich tapestry of persisting ethnic groups. In a lecture on the state's ethnic composition
entitled "Fields and Faces in Wisconsin," Hill noted that this work
tried, "in a common sense way to state the cultural resources of Wisconsin as something to which *every group* has contributed."[27] Scanning
the 14½-by-16-inch map from the lower right to the upper left, pockets
of Germans in the eastern regions give way to a complex mixture of
Norwegians, Poles, Bohemians, and so on. In some areas concentrations take up entire counties; in others, a fine scale of multiethnic complexity works its way down below the township level. No one group,
and especially not the Yankees, dominates. The map *seems to prove* that
an intricate mixture of cultures defined the Wisconsin scene.[28]

"Wisconsin Is Living Proof That Democracy Can and Does Work"

George Hill's study of Wisconsin's nationalities was conducted not out
of mere interest, but for its "implications." As both a dispassionate
science and a tool for addressing contemporary social concerns, Hill's
rural sociology shared a utilitarian approach favored by progressive
social scientists across the country. He clearly recognized the importance of this functional vision as the Nationality Project was intended
to "show whether the present socio-economic patterns [in Wisconsin]
can be influenced through a better understanding of the cultural fac-

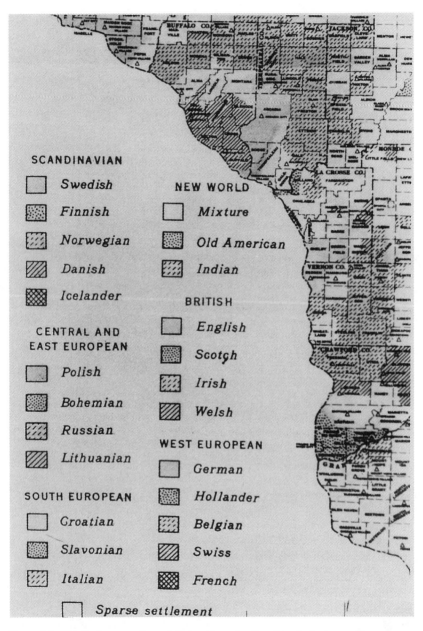

SCANDINAVIAN

	Swedish
	Finnish
	Norwegian
	Danish
	Icelander

CENTRAL AND EAST EUROPEAN

	Polish
	Bohemian
	Russian
	Lithuanian

SOUTH EUROPEAN

	Croatian
	Slavonian
	Italian

NEW WORLD

	Mixture
	Old American
	Indian

BRITISH

	English
	Scotch
	Irish
	Welsh

WEST EUROPEAN

	German
	Hollander
	Belgian
	Swiss
	French

| | Sparse settlement |

Figure 5.3. Detail of composite map 2: location of nationality stocks in rural farm population, 1939. This close-up of western Wisconsin demonstrated the fine detail of Hill's ethnographic research. (Courtesy of University of Wisconsin Archives)

156

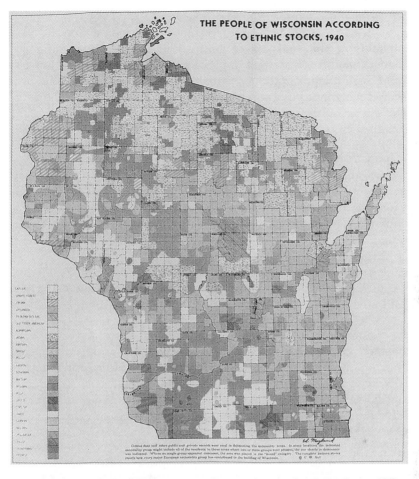

Figure 5.4. "The people of Wisconsin according to ethnic stocks, 1940." George Hill's 1942 map shows the state to be comprised of a mosaic of persisting ethnic groups. The full color 14½-by-16-inch map made use of the 1905 state census and field work to document the location of twenty-five groups at extremely fine levels. (Courtesy of the University of Wisconsin Archives)

tors."[29] The socio-economic patterns that needed influencing shifted somewhat during the course of the project's five year life, though the shift was never a complete one. The move was roughly from agricultural economics to cultural politics as ancestry became the causal factor for every variable from farm tenancy and soil erosion to the glue that binds together the state's democratic institutions.

Begun during the Great Depression, the Nationality Project's primary object was to uncover the factors behind diverging responses to the national economic crisis. Those factors, it was becoming increasingly clear, could not be attributed solely to location within the state, suitability of soil, and size of family. Rather, some other influence, some *cultural* factor, seemed to play a more significant role. How else could one explain variations of wealth, productivity, and condition of farms belonging to two sets of people in the same county? The variation among farmers in Price County, for instance, was astounding. Although they settled the land in roughly parallel fashion, in virtually every measurable index, Czechs outproduced their Finnish neighbors. In light of this and similar data, Hill posed the question this way: "With approximately the same type of natural resources, similar working conditions, and about the same accessibility to markets, [why] do two-thirds of the people succeed in making an independent living while the remaining third fail? Clearly, factors other than the purely economic and geographic must account for the difference."[30]

Through many conversations with county agents and his own direct experience, Hill believed he had found an answer: "The varying rate of tenancy in the state home ownership, tax delinquency, farm values, crop culture, local government, social organization and other phenomena are all strongly dependent upon nationality patterns."[31] This positive correlation between nationality patterns and such socioeconomic characteristics led directly to the question of public assistance, the heart of the depression-era research. After conducting an extensive survey with county agents across the state, Hill found, perhaps not to his surprise, that "the need for relief" varied greatly among different groups in the state. "Old stock Americans," or Yankees, fared least well with only 15 percent being "very successful" in staying off relief, a figure considerably worse than Germans, Czechs, Norwegians, and Poles. Equally important and in contrast to the Yankee, these "newer Americans" were "inclined to view the land as a precious acquisition, to be cherished as a home and to be handed down to the children unencumbered."[32] Ethnic stocks, Hill concluded in a way reminiscent of New Glarus's increasing self-promotion, enriched the state as they proved to be *better* agriculturalists, fertilizing Wisconsin's economic landscape rather than depleting it.

As the United States prepared itself for a different type of emergency in the early 1940s, so did Hill's Nationality Project respond to the new political situation. Though he never strayed too far from questions of rural public assistance and material variations among groups, the rural sociologist became more interested in the political implications of the state's heterogeneous population mixture. In particular, the cultural diversity that Hill and his associates had been documenting for the past four years was brought into service as a way to refute Nazi claims of national superiority through racial purity. Paraphrasing Hill during an interview in 1941, the *Milwaukee Journal* put it most succinctly: "Wisconsin is living proof that democracy can and does work. It is a concentrated segment of democracy, in which people of all heritages work together for the welfare of all. It refutes totalitarian claims of racial superiority." The "state's greatest asset in any emergency," Hill reported, was "its people . . . who cling consciously to the folk customs of their ancestors."[33] As a way to demonstrate the diversity of the population and the way in which "each nationality has contributed something of its heritage to the cultural and spiritual amalgam," the *Journal* reproduced a full-page, simplified color version of the "Ethnic Stocks" map (fig. 5.5).

Even more directly than the *Milwaukee Journal*, the University's "Wisconsin's Changing Population" report made strategic use of ethnic diversity in the propaganda war with Nazi Germany. In his introduction to the report, UW President C. A. Dykstra declared that "at this time when totalitarian demagogues are using their skilled experts of propaganda to depreciate and to destroy all the ideals and values of democratic population policy, it is important that the University publish this statement concerning our population assets and resources."[34] The report was unambiguous about the content of those population assets and what set Wisconsin—and, by implication, the United States—apart from the Nazi regime: a diverse and heterogeneous population mixture that, despite its diversity, recognizes a set of harmonious values and ideals. George Hill expounded this rooted cosmopolitanism with unerring precision. "Contemporary Wisconsin," he said, "is made up of a diversity of cultures. It consists of old stock Americans, of first and second generation Americans from immigrant stocks, and of newly-transformed Americans only recently arrived from their ancestral homelands. It is in this situation in which the temperaments, aptitudes, prejudices, superiorities—all of which stem out of the values and traditions of culture—are fused into the dynamic reality which *is* Wisconsin. Understood and appreciated, these traits form the bulwarks of democracy."[35] Significantly, these "values and traditions of culture" have not been lost through straight-line and one-way assimilation processes, but can be found today on the landscape, in place.

Just as Frederick Jackson Turner had explained the nation's character by the frontier experience, fifty years later George Hill deciphered its constitution in the processes and patterns of cultural pluralism. Neither immigrants nor the people they encounter remained unaffected by cultural contact, and it is precisely such a mixture that created the state's *unified* culture. Hill believed that Wisconsin's harmoniously blended cultural heterogeneity stood in marked contrast to not only

Figure 5.5. "All men are created equal." This full color map from the *Milwaukee Journal* is based on George Hill's "Ethnic Stocks" map (fig. 5.4). Note the use of colored stars for the key and idealized folk figures standing on a United States map. (Courtesy of University of Wisconsin Archives)

Nazi Germany, but "other areas of the nation." Wisconsin, the map demonstrates, "is more than a melting pot, where all constituents have been reduced to one element, for in such a situation the roots of democracy could never have taken hold." What separates the United States from Nazi Germany, and Wisconsin from unnamed "other" [read, Southern] states, is precisely the former's inability to work as a melting pot.[36]

For all the emphasis on diversity and heterogeneity, Hill never advocated strict separatism. Indeed, he had a personal distaste for the term "cultural island," instead preferring "culture type," which implied a connection to a whole. In a revealing 1940 lecture, Hill made pains to note that in the Nationality Project, "we are studying [Wisconsin's ethnic groups] not as cultural islands, but as parts of the mainland of which they are a connected unit."[37] This cosmopolitanism found a further voice in the 1941 *Milwaukee Journal* article in which he summarized his view that "there is no such thing as racial purity or cultural homogeneity, either here or on the other side of the water. But we do have homogeneous ideals and values on which we can anchor our democracy."[38] The appreciation of cultural difference might be the essence of democracy, but that difference was absorbed by unified political values. Hill's pluralism was sufficiently rooted in a vision of unity and commonality, in a cosmopolitanism as parochial as it was universal.

PROVINCIAL COSMOPOLITANISM: FRED HOLMES'S *OLD WORLD WISCONSIN*

Concurrent with George Hill's ethnic mapping project, an amateur historian was planning his own grand interpretation of Wisconsin's ethnic geography. Fred Holmes may not have been trained in the social science favored by Hill, but in many respects their ambitions followed parallel trajectories. Both bought to public light the factor that seemed to summarize Wisconsin's most fundamental attribute: the diverse and seemingly persistent nationality backgrounds of the state's residents. And both found a harmonious blend of distinct cultures within that diversity. But, where Hill's project came to only partial fruition, Holmes's study *Old World Wisconsin* proved remarkably successful, literally rewriting the state's geographical history.

"My Own Wanderings Through Europe Have Been Taken by the Vicarious Method"

Frederick Lionel Holmes made his living as a reporter and lawyer, but exploring and writing about his native state were his real loves.[39] Born

161

of Yankee stock on a farm in Winnebago County, Wisconsin, in 1883, at twenty-three he founded the Holmes News Service, a wire agency that serviced Wisconsin and national metropolitan newspapers until 1927. In that year Holmes shifted professions and was admitted to the bar, remaining a lawyer until his death in 1946. During the final decade of his life, he increasingly turned his attention away from his law practice and toward what he called the "side roads and excursions into Wisconsin's past." These excursions took both physical and metaphorical form as Holmes—by his own calculation—drove nearly every road of Wisconsin and delved into the state's history by reading, interviewing knowledgeable people, and writing up his findings. It was also at this time that he served as a curator at the State Historical Society of Wisconsin. This third career, of professional traveler and amateur historian, found its origins during his youth, when he wrote his first book. Although *Regulation of Railroads and Public Utilities in Wisconsin* (1915) never really caught the public's notice nor provided him an outlet for his burgeoning Wisconsin-centered topophilia, twenty years later three well-received books furnished Holmes with the material and a style that combined travel narrative and amateur historical scholarship. Conceived as a trilogy, each sought to convey some of the "unsung glories of the state."[40]

In the first, *Alluring Wisconsin* (1937), Holmes united historical lore with his deep love of place and travel. His many pilgrimages "relieved the grim and dullness of quotidian life," a condition produced by the grind of modern, urban existence. An escape from modern Wisconsin, *Alluring Wisconsin's* nearly five hundred pages and more than two hundred photographs entreat readers to visit places with such romantic appellations as "sign posts of eternity," "isles of enchantment," "outside modernism," and "Viking survivals." While *Alluring Wisconsin* highlighted romantic *places* at which one might escape modernity, the second book in the trilogy, *Badger Saints and Sinners* (1939), focused on the "colorful" *people* who created these cultural shrines. Although Holmes claims to "undertake no sustained thesis" by lumping together an incongruous cast of thirty-two characters, his upbeat descriptions of even the worst "sinners" suggests a close affinity to the appreciative boosterism and wistful nostalgia of his earlier book.[41]

The third book in Holmes' trilogy was by far his most successful and the one he felt was of greatest importance. *Old World Wisconsin*, Holmes believed, "will grip [people] from the twenty-four nations who make up the Wisconsin melting pot. People who look at this book cannot escape the conclusion that here is a story about Wisconsin never before told."[42] Where the previous books featured distinctly Wisconsin topics, the third concentrated on the "universal appeal" of ancestry,

162

identity, and heritage. Holmes found it remarkable that "no other state has gathered in a melting pot such a diversity of rural and urban foreign groups. . . . To know them from their racial backgrounds through their New World cultures is to understand more clearly the reasons for our hegemony in the family of states." Moreover, those groups could still be found on the landscape as "living history," where "each little transplanted group has its own individuality."[43] This central idea—that the state owed its unique character to its mixture of *still surviving* foreign groups—was Holmes's chief discovery and the message that he wanted to convey to the nation.

Fred Holmes knew at an early stage the book's general themes and how he wished to shape its focus. Like all travel writers, he wished to extend his personal experiences to a readership that could encounter a distant place through his—the travel writer's—eyes. In this case, the distant place was home. Holmes wrote that his own "wanderings though Europe have been taken by the vicarious method of trodding the byways of newer Wisconsin."[44] Simultaneously, the book was also to convey these travels as a "serious" historical treatment. Holmes wanted to "publish the truth only" and illuminate a most important, if previously unacknowledged, aspect of the state's geographical history.[45] As a result, *Old World Wisconsin* does not fit neatly into categories of genre. Not really travel book, nor scholarly history, it makes overtures to both, becoming in the process an excellent example of *heritage.*[46]

"Do Any of the Swiss Farmers Yodel While at Work?": Researching Nostalgia

It was only in 1940 that work on *Old World Wisconsin* began in earnest. Though not trained as a geographer, folklorist, or historian, Holmes did boast extensive proficiency as a reporter and maintained personal contacts throughout the state from his days as a newspaper-man, editor, and lawyer. He made impressive use of these skills as his several dozen trips around Wisconsin enabled Holmes to conduct interviews, take fieldnotes, and collect a vast array of primary documents. From his Madison home, he wrote literally hundreds of letters to old acquaintances and potential informants soliciting information. A closer examination of his methodology is instructive, for it reveals an inquisitive, if romantic mind—two features that most succinctly summarize *Old World Wisconsin*. Here, his research of the Finns of Douglas County and the Swiss of New Glarus is representative.

In both cases, Holmes began his inquiry with letters to prominent members of the community who, it was hoped, could provide not only firsthand knowledge of that community, but advise him about other, knowledgeable informants. In the summer of 1941, he contacted a circuit court judge and old friend who, though not Finnish himself, directed Holmes to Ilmar Kauppinen, the active manager of the Superior Workers Mutual Savings Bank in that city. Kauppinen proved to be an excellent choice as he was very prominent among Finns in northern Wisconsin and quite eager to share his extensive knowledge of their heritage. Kauppinen then put Holmes in contact with knowledgeable informants around the region, all of whom shared a commitment toward their Finnish ancestry.[47] The New Glarus work occurred at roughly the same time and followed an identical procedure. He picked out leaders of the community, in this case, J. J. Figi and Esther Stauffacher, who, in turn, introduced him to the village mayor, the Wilhelm Tell Guild secretary Henry Schmid, and the historian and long-time editor of the *Monroe Times*, Emery Odell.[48] In both cases, Holmes demonstrated his journalist skills at finding the right informants, ones who knew the story he wished to tell.

Once he established contact, he sent questionnaires to a handful of people in each community. Significantly, the questionnaires were not standardized, but were tailored to elicit information that he deemed most important. In actuality they were a list of questions designed to get at the persistence of "old world customs." The New Glarus question list consisted of twenty-seven short questions that reflected a romanticized and essentialized view of ethnic culture. In many cases, Holmes' questions derived from the New Glarus' own recent constructions of identity: "Do any of the Swiss farmers yodel while at work? At Swiss churches do men sit on one side and women other? Do any bands play any airs nationalistic to Switzerland? Are Tuesday and Thursday the only days in which Swiss will be married? Do women always leave the church first? What do you use instead of a bow and arrow [for *Tell*]? Has Lace factory anything to do with Swiss? Do cattle wear bells made in Switzerland?"[49] The Finnish questions were more general and betrayed less familiarity with an essentialized culture. One person was simply asked "what old world customs persist among the Finns?" while Ilmar Kauppinen was given four general questions about Finnish settlement, Old World customs, political beliefs, and festivals.[50] This less invasive method produced better results. Where the lengthy New Glarus question list invoked pat answers and confirmed stereotypes, the Finns offered detailed and vivid descriptions that found their way directly into the final manuscript text. Holmes was more successful in the Swiss community with direct observation and informal visits. He took in the *Wilhelm Tell* play in the fall of 1940, having enjoyed a Swiss dinner at the home of Esther Stauffacher a month earlier. Although he made site visits to Douglas and Bayfield

Counties in the summer of 1941, that chapter relied more heavily on the information supplied to him by knowledgeable correspondents.

Finally, the work on the Finns enjoyed the benefit of skilled copy editing and advice from a leading professional scholar of the region, George Hill. It is uncertain how long and to what extent the two writers of Wisconsin's ethnic heritage knew each other. What is certain is that George Hill worked through Holmes's manuscript and saved the amateur historian from his worst tendencies. Under Hill's guidance, factual errors were corrected and certain topics were strengthened— most notably, Finnish socialism and cooperative movements. Simultaneously, Hill eliminated many of the questionable clichés that shaped other chapters. But where George Hill removed some stereotypes, he advocated other changes designed to put a positive spin on his ancestral group. Fearing, perhaps, that it might cast the wrong light on his people, Hill scratched out Holmes' paragraph on a disreputable Old World–trait, of unmarried "lad and lassies 'bundling' during the long winter nights."[51] This act of censorship is revealing for it highlights the social scientist's own sentimentality and desire for reinterpreting a group often derided for their socialist agitation. Originally "Strange People" in Holmes' text, Finns became "Amazing People" after Hill's editing. He also wanted it known that "only a few . . . may be said to lack pride in their nationality."[52]

While George Hill might have added flattery and deleted a potentially scandalous passage, he also removed a number of factual and interpretative errors, a service that might have served the Swiss chapter well. "Swiss Yodel Cares Away" enjoyed no such expert overview and, perhaps not coincidentally, is riddled with interpretative errors, some of which are historical in nature. They range from viewing the original emigration entirely in terms of increasing population and a dwindling food supply (with no mention of the most important factor: industrialization and its displacement of textile workers) to the dubious view that the settlement was so located because of Green County's resemblance to the Old World. Likewise, the struggles of the first years are turned into "a fortuitous start," giving the false impression that Swiss methods of husbandry were transplanted directly and immediately with emigration. Some errors were contemporary. For instance, although Holmes devotes considerable attention to cheesemaking, nowhere do we read that by the mid-1940s cheese factories had, without exception, ceased operation in the surrounding region. The author then proceeds to make the highly questionable assertion that because "all Swiss cannot be farmers . . . many have become proficient in woodcarving, embroidery, and watchmaking," a statement for which he has the good sense to not find examples. Finally, when discussing these other occupations, he neglects to mention the main employer in town, the Pet Milk condensing plant.[53]

It is impossible to know whether a Swiss George Hill could have saved this chapter from Holmes's worst tendencies; the newspaperman was merely repeating many of the stories and myths told to him by a handful of promotion-minded insiders. At precisely the time that New Glarus reinvented its own "Swissness" through museum work and conspicuous cultural displays, their leaders supplied official culture's chief ambassador with the stories they, themselves, had recently constructed. This dialectal interplay between vernacular and official culture was entirely hidden, as comments from its many reviewers suggest that most readers regarded *Old World Wisconsin* as dispassionate history.[54]

Yet, to judge *Old World Wisconsin* as merely flawed history misses its central message, and betrays a genre error. Or, following David Lowenthal, such a judgment mistakes *heritage* for *history*. Where history remains remote and critical in its view toward the past, heritage thrives on personal immediacy and embraces the past as the building blocks of identity. While neither offers a transparent window to the past, heritage willfully fabricates the past in order to bolster self-esteem and to promote an acceptable vision of the present.[55] Thus, at the same time that the *Oshkosh Northwestern* called *Old World Wisconsin* a "scholarly and thorough study," the newspaper also noted that the book is "a fascinating story, told in an easy conversational manner, so that the reader feels he is being introduced personally to the costumes, characteristics, cuisine, festivals, and lore of the many settlers who [made] their homes in a free America." The book was designed to tell a fascinating story—indeed, the most fascinating story of the state's development by way of travel narrative. Its lie—and its power—derived from heritage masquerading as history.

"Romantic Days Are Fading":
Rejecting Modernity Through Provincial Cosmopolitanism

At its core, *Old World Wisconsin* sought to wrest the essence and soul of the state—its heritage—away from its Yankee political leaders and place it in the hands of the ordinary people who plowed its fields, milked its cows, and staged its folk productions. Highlighting and appreciating the individuality and distinctiveness of the places built by these peoples became the book's raison d'être. Holmes's topophilia bursts through on every page as he recounts becoming "subtly conscious of the centuries of simple living reflected in the daily lives" at the Lake Michigan Luxembourger settlement and of the "nostalgia of

exiled peoples [that] has enriched the world of music and literature."
Summarizing his vision of ethnic place, he concludes that "had it not
been for the invasion of these aggressive European stocks the strong
bedrock of Wisconsin civilization might have been more friable."[56]

This appreciation of ethnic place takes on greater urgency if one
believes, with Holmes, that its existence is threatened. The first chapter
sets the tone for the book with the title "Romantic Days are Fading."
The ethnic places that receive the most tender treatment are those, like
the French village of Somerset, that "alone remain with many Old
World customs unspoiled." The Luxembourger settlement is blessed
because its "way of life" adheres to tradition over "the hurry, confusion, and insecurity found in urban districts." And the Dutch communities along Lake Michigan are favored because, "despite the dreary
efforts of modern civilization to enforce conformity on all peoples, it
has not succeeded."[57] In this way, *Old World Wisconsin*, no less than
Hill's social science, may be seen as a regionalist plea for place distinctiveness, cultural pluralism, and indigenous organicism against the on-
slaught of what Walter Lippmann called the "acids of modernity." An
antimodernist appreciation of ethnic place distinctiveness became
Holmes's safeguard against modernity's tendency to bulldoze difference and tradition.[58]

Holmes' antimodernist reclamation project hinged on his particular vision of cultural pluralism, a vision that I call provincial cosmopolitanism and which bore remarkable similarity to George Hill's. If maintaining distinctiveness was to be admired, there also assumed a
blending between groups—what Holmes calls a "racial admixture"—
whereby everyone would work together to comprise a harmonious
whole. Holmes hoped that nationalities would maintain their customs,
but within a shared political and social framework. Although little interaction occurred between groups in Holmes's text—indeed, each is
treated in random order, almost a separate island unto itself—underlying each chapter is the conviction that the dreams of the many nationalities have "bloomed into a common heritage." More integrationist
than separatist, Holmes's pluralism rested on the belief that ethnic culture could and should be maintained, but only insofar as it contributed
to a composite that was, at its core, welded and unified. Distinctiveness
was admirable; separation, however, was unacceptable. As the work of
Philip Gleason, David Hollinger, and John Higham indicates, Holmes
clearly fit into a growing interwar cosmopolitanism that resituated ethnic Americans within a larger civic culture.[59]

This cosmopolitan vision of a unified, harmonious plurality was
only surpassed by its provinciality. In this way, Holmes differs from
the intellectuals of his time, most notably in New York, who shared a

vision of culture as cosmopolitan. One need only compare Holmes' antimodern cosmopolitanism with the protoseparatist cultural pluralism of Horace Kallen, the early multiculturalism of Randolph Bourne, or the intelligentsia that came after, including Edmund Wilson, David Riesman, and Lionel Trilling.[60] Unlike, say, Bourne, who celebrated the *de*provincializing effect of immigrants on the native-born population, Holmes reveled precisely in the antimodern influence of ethnic life; cosmopolitanism for the Wisconsin historian contained value precisely for its retreat from a homogenizing and modern world. He did not try to make generalizations beyond his state's boundaries, but what he did say about Wisconsin he said with affection, nostalgia, and a posture of authority. Thus, his Badger State parochialism joined with his cosmopolitanism to create a vision of his beloved region's historical and geographical development that grew out of its harmonious ethnic diversity.

Critically, Holmes's provincial cosmopolitanism contained, at its very core, an unresolvable contradiction. If distinctiveness between groups is to be preserved, this would seem to imply the maintenance of the social and geographic distance that, ultimately, Holmes the Yankee found unacceptable. Interaction between—and a fusion of—cultures would lead to the dissolution of the ethnic place that Holmes so revered. Fred Holmes was not alone in finding himself at the center of this inconsistency. While anthropologists regarded the tight cultural cohesion in "primitive groups" as healthy, other social scientists were decrying its byproduct (or cause?)—ethnocentrism. Philip Gleason stated the matter with precision when he asked, "How could diversity be a good thing if the ethnocentrism that was central to preserving ethnic distinctiveness was such a bad thing?"[61] Since the question was never put in such stark terms, an answer was never given. He must have sensed the tension inherent in his pluralism, however. For Holmes, the way out of this contradiction was to thoroughly and completely depoliticize his subject material.

Thus, it is axiomatic that this reappraisal of, and appreciation for, ethnicity came at a time when ethnic groups were becoming less a potential threat. Even the German-speaking people, many of whom could recall the bitter memories of experiences during the Great War, were romanticized. Articulating an implicit theme of unity in diversity, *Old World Wisconsin* stripped controversy and politics from the cultural identity of the various groups on display. Holmes tread closely to creating what Barbara Kirshenblatt-Gimblett has called the "banality of difference," whereby the "proliferation of variation has the neutralizing effect of rendering difference (and conflict) inconsequential."[62] Difference was fine as long as it was acceptable, nonthreatening, and beyond politics. Walter S. Goodland, Wisconsin's sitting governor, recognized and clearly approved of this message when he wrote that although the "Old World nationalities" are maintaining their values, they are losing the "identities," thus guaranteeing greater unity and harmony. The *Milwaukee Journal's* review put the matter accurately when it declared that "there is no feeling, reading the book, that these are 'foreign' people. Rather, you seem to look upon them as part of the big Wisconsin family."[63]

Pluralism was thus contingent upon a larger framework of consensus, a consensus achieved through tourism. This is illustrated in a letter by another *Old World Wisconsin* fan, Charles Gillen, who found that "Dr. Holmes has done an extraordinary thing: he has made it possible for the widely different groups settled in Wisconsin to visit one the other within a few hours; to learn the best of their traditions, customs, artistry, and ethics; and thus to *make each group more appreciative to the others*. I can think of no finer way in to increase the *harmony* that should prevail in America, and to keep America the wonder of civilization and the guarantor of the peace of the world."[64] There is ample evidence to suggest that many other readers took this basic message to heart. Like Charles Gillen and Giradela Wackler, Grace Bloom (a second-generation Swedish American from Osceola) found the book a testimony to both harmony between peoples and an invitation for ethnic tourism. Although she and her husband had traveled the state extensively, she was "frank to admit that you have found greater drama than we could." But now, armed with *Old World Wisconsin*, "the first Sunday morning we can . . . we are going up to the Russian church at Clayton to attend the Greek Orthodox service." These readers agreed with attorney O. A. Stolen of Madison, that Holmes "caught the spirit and the 'flavor' of each nationality" and, by visiting the home districts of these peoples, a greater appreciation of diversity and the state as a whole would result.[65]

Such an audience grasped the book's subtext that Holmes's provincial cosmopolitanism, at its very core, revisualized ethnic place as *tourist place*. His "vicarious method" of traveling Europe in Wisconsin relied on a view of culture that was a tourist vision of culture. Looking for the essences of quaint peoples, ethnics become synonymous with their representations: the Swiss *are* yodelers, Norwegians *are* lutefisk eaters, the Welsh *are* singers of hymns (figs. 5.1 and 5.6). As Jonathan Culler notes, the tourist vision "is interested in everything as a sign of itself." It may be true that some Swiss may yodel, some Norwegians may eat lutefisk, and some Welsh may sing hymns, but the tourist vision of culture essentializes the representation—one is seeing the essence of "Swissness" or "Norwegianness" or "Welshness" through

169

Figure 5.6. Hungarian wheat festival near Milwaukee. This photograph was typical of the illustrations that accompanied *Old World Wisconsin*. Holmes's provincial cosmopolitanism relied on seeing the essences of a people in their representations, a strategy employed by many ethnic communities themselves. (Photo by Arthur M. Vinje and reprinted courtesy of the State Historical Society of Wisconsin, Fred Holmes Collection, neg. WHi [V51] 31)

these practices. The crucial role of human agency and social constuctedness of that culture is effectively hidden from view.[66]

Fred Holmes, Louis Adamic, and the Therapeutic Necessity of Ethnic Pride

While most readers found this approach commendable, at least one reviewer demurred. For Louis Adamic, *Old World Wisconsin*'s provincial cosmopolitanism focused too narrowly on what he disparagingly called Holmes's "tourist's eye." Although he approved of Holmes' emphasis on "Europe's past and present influence" in Wisconsin, and of the author's acknowledgment of both the diversity of these peoples and their ability to work together, Adamic severely criticized the book's "lushness and sentimentality," and said that its author unduly stressed "the more obvious and less lasting evidence of the transplantation" and downplayed more important, but "less visible" aspects of that transplantation.[67] Adamic, however, was vague about what precisely constituted those more important, but less visible attributes. In many respects, I argue, Adamic and Holmes had more in common than the former would admit. For, though their writing styles may have been at odds, both positioned ethnic heritage as a means of overcoming the malaise of modern America.

A Slovenian immigrant, Louis Adamic ranked among the most popular and prolific American writers of the 1930s and 1940s. His many books, articles, and speeches grappled with the ethnic question in American life—with an intensity and commitment matched by few. Although a full assessment of Adamic's far-reaching contribution to American thought on race and ethnicity has yet to be made, his impact was significant. Popular discourse about ethnic identity during this period relied heavily on his work as both elites and nonelites alike reflected on the place of non-Anglos in American society. One scholar has referred to Adamic as the ethnic "pied piper" of his generation, and the editors of the 1980 *Harvard Encyclopedia of American Ethnic Groups* conceded that their monumental project had its origins with Adamic.[68]

Like Fred Holmes, but on a nationwide scale, Adamic sought to rewrite American history as ethnic history. As Holmes aimed to unite Wisconsin's Yankee and "Old World" pasts, so did Adamic wish to unite "Plymouth Rock and Ellis Island." For Adamic, a proper under-standing of America's past as well as its contemporary makeup re-quired an "intellectual-emotional synthesis of old and new America; of the Mayflower and the steerage . . . of the Liberty Bell and the Statue of Liberty. The Old American Dream needs to be interlaced with the immigrant emotions as they saw the Statue of Liberty. The two must be made one story."[69] But where Holmes—the Wisconsin Yankee—sought this reconciliation as an antimodernist move to promote accept-able difference, Adamic—the Slovenian-American—operated on a more personal and psychological level. The resolution of emotional distress, rather than the recovery of romantic "scenes from the past," defined his mission. In an especially influential article written for the November 1934 issue of *Harper's* magazine, Adamic outlined the dis-tress that he detected during his visits with some of the "Thirty Million New Americans." Written after an extensive tour of ethnic communi-ties in New York, Michigan, Minnesota, and Wisconsin, where the pop-ulation is "preponderantly 'foreign,'" Adamic tried to explain the emo-tional malaise of the second- and third-generation Americans he met. "The chief and most important fact . . . about the new Americans," he wrote, "is that the majority of them are oppressed by feelings of

inferiority." Where Holmes saw "quaint" Cornish, "quiet" Luxemburgers, "thrifty" Belgians, and "strong, virile" Norwegians, Adamic encountered people who "cannot look one in the eye. They are shy. Their
limp handshakes gave me creepy feelings all the way from New York
to the Iron Range of Minnesota." The problem, it seemed, was that
the immigrant generation was "too inarticulate" to transmit to their
children "a consciousness . . . of their being part of any sort of continuity in human or historic experience." The solution to the dilemma was
a typically *therapeutic* one, namely building a sense of belonging by
passing along "a knowledge of, and pride in, their own heritage."[70]

As far apart as these two writers may seem at first, and as Adamic
would have wished, a remarkably similar vision of pluralism and unity
undergirded their thinking. Although Adamic insisted on the importance of ethnic studies and the necessity of ethnic pride, "he did not
envision, nor could he accept, an America of separate cultural islands."[71] Rather, and very much like Holmes and George Hill, Adamic
called for an appreciation of each group's contribution, for tolerance
of diversity en route to a unified American republic. This cultural integrationist perspective was spelled out perhaps most succinctly in the
1939 statement of purposes of *Common Ground*, a journal edited by
Adamic for one and one-half years. With an editorial board comprised
of such notables as Van Wyck Brooks, Mary Ellen Chase, Langston
Hughes, and Thomas Mann, Adamic's journal was published by the
Common Council for American Unity. The Council's purpose, as outlined by Adamic, was "to help create among the American people the
unity and mutual understanding resulting from a common citizenship,
a common belief in democracy and the ideals of liberty, the placing of
the common good before the interests of any group, and the acceptance, in fact as well as in law, of all citizens, whatever their national
or racial origins, as equal partners in American society."[72] Adamic's
cultural integrationism, Holmes's provincial cosmopolitanism, and
George Hill's social science shared the identical ambition of promoting
unity in diversity. It was Adamic, not Holmes, who wrote that "the
pattern of America is all of a piece; it is a blend of cultures from many
lands, woven of threads from many corners of the world. Diversity
itself is the pattern, is the stuff and color of the fabric."[73]

The major divergence between the two stemmed more from personal background and temperament than serious philosophical differences. Where Holmes wrote with an upbeat and unwavering optimism, Adamic tended to concentrate on the difficulties of
adjustment.[74] However, both shared a tendency for attributing group
characteristics to individuals where the leap to stereotyping was a very
short hop. With words that might well have come from *Old World Wis-*

consin, Adamic wrote that same year of "spice" added from "the flavor of cynicism and humanitarianism from the Jews, sex and sophistication from the French, and sentimentality and love of comfort from the old fashioned Germans, and you have a rough outline of essential Americanism" (fig. 5.7).[75] Both were infected with the filiopietism and the search for contributions or immigrant "gifts" that underscored much ethnic writing during these decades. And both regretted that, as Adamic put it, "there has been entirely too much melting away and shattering of the cultural values of the new groups," while at the same time noted with approval that "there has been more getting together among Americans than ever before." The key, of course, rested in the view that unity could be achieved through diversity, through a rooted or provincial cosmopolitanism.[76]

Herbert Kubly, for one, immediately recognized such shared philosophical terrain. Writing for Adamic's journal *Common Ground*, the New Glarus native and future National Book Award winner describes a return visit to his home village. In words evocative of Holmes's *Old World Wisconsin* or Adamic's *A Nation of Nations*, Kubly finds that "New Glarus has rediscovered itself as I rediscovered it." The people of the village, he notes, "have largely discarded the superficial values of the postwar years when every manifestation of their Swiss heritage was frowned upon. Those years did not bring them the America they sought." Now, as evidenced in the new Swiss museum, frequent yodeling concerts, new "authentic Swiss architecture," and the annual *Wilhelm Tell* play, "they are recapturing the uniqueness of their pasts and in it are finding the spirit of the New World." Thus are New Glarners proudly maintaining cultural diversity all the while that they contribute to a shared and united vision of America; they are provincially cosmopolitan.[77]

Despite their common vision, Adamic completely missed the redemptive nature of Holmes's ethnic tourism. Traveling to the different ethnic communities was undertaken not merely for recreation, but to discover precisely the diversity that Adamic so championed. James Gray, a reviewer for the *St. Paul Dispatch*, recognized the shared vision immediately. He believed that "Louis Adamic would whole heartedly agree with Mr. Holmes' attitude." Writing in the shadows of Hitlerism, Gray asserted that "we live in a moment when local antagonisms against minorities and national feeling against certain groups might easily result from the tensions of this war. It would be pleasant to think that communities of Wisconsin and also like our own might be able to offer hints of a more profitable way of receiving the descendants of foreign groups wholeheartedly into our way of life. Any method of producing a real spirit of fraternity is a good one. But perhaps the

Figure 5.7. "The sentimentality and love of comfort among old fashioned Germans." Louis Adamic, no less than Fred Holmes, conceived ethnic identity in broad, cultural terms. This photograph, *The Bavarian*, was one of many that appeared in his journal, *Common Ground*, at the beginning of the Second World War. (Reprinted with permission from *Common Ground* [1941] 72; courtesy of the Immigration and Refugee Services of America)

very best is that of Fred Holmes' exuberant receptivity."[78] No less therapeutic than Adamic's cultural integrationism, this "spirit of fraternity" was best achieved by visiting those provincially cosmopolitan communities.

Establishing a Paradigm:
Assessing the Impact of *Old World Wisconsin*

Although it is never easy to judge precisely the impact of any one book, there is ample evidence to suggest that Fred Holmes's *Old World Wisconsin* defined for many Wisconsinites the historical development of their state's geography and essential character. In many ways, "Wisconsin culture" has come to mean "old world culture," with the most prominent example being the creation, in 1976, of the state's most significant and ambitious museum, Old World Wisconsin, a historical village in Eagle. With its town hall named "Harmony" and surrounded by a constellation of Euro-American vernacular buildings, Old World Wisconsin, the museum, bears an unmistakable resemblance to its namesake, *Old World Wisconsin*, the book.[79] As Holmes's book was a "new kind of Wisconsin book," so is the Old World Wisconsin outdoor ethnic museum the first of its kind: the "only multinational, multicultural 'living museum' in existence."[80]

Well before the establishment of a museum, however, *Old World Wisconsin*, the book, was exerting a powerful influence; it became an integral part of the state's high school history curriculum, while sales were sufficiently good to merit a second printing within the first year (fig. 5.8).[81] Letters and book reviews made it clear that the book, as Maude Maxwell Munroe of Baraboo put it, "struck a chord" with many readers across Wisconsin.[82] While others, like many of the Lake Michigan Luxembourgers, took offense at *Old World Wisconsin*'s stress on tradition while ignoring "the fact that as a race of European people they soon lost the casual clannish identity he ignores."[83]

Most, however, agreed with the *Eau Claire Leader* that "whatever your nationality before becoming American, you will find a glow of pride in reading *Old World Wisconsin*." In words that Louis Adamic might have wished to read about his own books, the newspaper suggested that ". . . in these days of racial strife and antagonisms against minorities [it helps] to read Mr. Holmes' refreshing book and see how peoples from many lands have lived in peace and harmony while retaining their pride in their own old world backgrounds."[84] In short, to cite the Wisconsin Arts Board as it prepared to ready the state for its 1998 sesquicentennial, "Fred Holmes' *Old World Wisconsin* established

Figure 5.8. Display of *Old World Wisconsin*. Fred Holmes's *Old World Wisconsin*, with sales strong enough to merit a second printing during its first year of publication, popularized the view of Wisconsin as the Old World state. Holmes wrote to his publisher that he especially liked this display at Brown's Bookstore in Madison. (Courtesy of State Historical Society of Wisconsin, Fred Holmes Collection, neg. WHi [X3] 51275)

a paradigm for understanding Wisconsin's cultural geography as a mosaic of evolving but persisting ethnic cultures."[85]

CONCLUSION

Fred Holmes and George Hill, coming from as different occupational and ethnic backgrounds as they did, told remarkably similar stories about the fundamental composition of Wisconsin, stories that taught the state's residents to appreciate ethnic diversity and difference. Both *Old World Wisconsin* and the 1942 "Ethnic Stocks" map dramatically portrayed that diversity in media that were both highly readable and convincing. Such cultural productions served to promote the view that assimilation had not occurred, and that Wisconsin was all the better for it. Of course, this is telling only one side of their shared vision. As they sought to convey messages of the appreciation of difference, both writers also sought to absorb those differences within a larger, integrated culture. Wisconsin may not be a melting pot, but, as Hill put it, the state was a "cooperatively blended idea out of many cultures."[86] Diversity, in other words, served as the basis for unity.

This provincial cosmopolitanism was a message as powerful as it was well received. Four years after the publication of *Old World Wisconsin*, the state celebrated its centennial with a flourish of parades, pageants, and displays designed specifically to reinforce this very theme.

The State Historical Society, in particular, took the lead in promoting the vision of Wisconsin as "Old World." It commissioned Ethel Theodora Rockwell to stage "Children of Old-World Wisconsin," a pageant written for the production of elementary school children and presented in over fifty schools throughout the state that year, as well as "A Century of Progress, Cavalcade of Wisconsin." The climax of the 1948 celebration, however, occurred when 125,000 people from across the state descended upon the capital to observe and participate in a day of speeches, parades, and performances. Featuring Swiss yodelers from New Glarus, Dutch "Klompen" dancers, Italian singers, and many more such performers, Statehood Day dramatically reinforced the provincial cosmopolitanism ideal of unity in diversity.[87]

The 1948 centennial brought to fruition the ideas that George Hill and Fred Holmes strove to delineate. It marked a turning point in which ethnic diversity found a legitimate, if limited, place in defining the state's culture. The limitation, if one may call it that, was based on an integration of shared beliefs and culture that undergirded the very idea of difference. Ethnic cultures should be sampled and enjoyed, for ethnic awareness would bring both self-esteem and tolerance for others. Awareness of cultural differences could enhance another value as well. Fred Holmes recognized better than anyone the travel possibilities inherent in ethnic difference, a theme that the New Glarus Swiss were just beginning to test. Nearly twenty years later the theme developed into a full-blown expression of place as ethnic tourism hit its stride in the mid-1960s, the focus of the next chapter.

But George Hill's ethnic mapping and, especially, Fred Holmes's popular writing did more than merely foreshadow ethnic tourism in places like New Glarus. These two state-level cultural productions were both a reflection of, and a context for, memory work at the local level. In New Glarus, the performance of the 1935 historical pageant, the creation of a permanent site of vernacular memory in the Swiss Historical Village, and the invention of the *Tell* tradition all occurred at a time when Hill and Holmes were praising that very work. Indeed, by suppling a selective view of their own community, leaders in New Glarus actively shaped their own representation in *Old World Wisconsin*. The legitimation bestowed upon the ethnic community from these respected writers is of an importance that can hardly be overestimated. One needs only to recall the vast difference between the 1918 Loyalty Legion's "Sedition Map" (fig. 3.1) and the popularization of Hill's "Ethnic Stocks" map (fig. 5.5) to realize the sea change underway. Communities like New Glarus understood that transformation well and, with the assistance of cultural agents like George Hill and Fred Holmes, contributed to it.

PART III

Making the Transition to Tourism, 1962–1995

6

"Swisscapes" on Main Street

Landscape, Ethnic Tourism, and the Commodification of Place

In 1964, New Glarus native Herbert Kubly described a recent visit to "the village of Wilhelm Tells" where he took in a performance of the annual community pageant. The fourth-generation Swiss American had heard from friends that the village had changed considerably since he left twenty years earlier; he felt the need to see it again for himself. A self-described *Wunderlicher*, or "odd one," Kubly preferred reading books to milking cows, a disposition that pained his successful farmer father, but one that led the young Herbert first to the university in nearby Madison, then to an international career as a journalist and fiction writer, and finally to New York, where the National Book Award author was now teaching literature at Columbia University.

Several years earlier, the *Wilhelm Tell* drama had moved its outdoor stage to Kubly's family farm, bringing with it considerable pride for Herbert's father, who had participated in the play since its first staging under Edwin Barlow in 1938. Typical of many area farm families, Herbert's mother and sister were equally involved in *Tell,* with the elder Kubly working long hours on costumes and the younger serving as both director and performer for many years. Although he had attended the drama several times before and had written articles about it for both *Time* magazine and *Common Ground* in the 1940s, the Wunderlicher had yet to see *Wilhelm Tell* performed at his family's recently renamed "Wilhelm Tell Farm."[1]

The performance, as it turned out, attracted record numbers and went off without a hitch. It remained pretty much as Kubly remembered it from years before, with the pageant's dramatic speeches, its colorful costumes, and rigid scene blocking. Even at its new location, the tradition seemed firmly ensconced in the midwestern town. What gave Kubly cause for reflection, however, was the metamorphosis in the community itself. As he rode the bus south and west from Madison into the rolling Green County landscape, the New Yorker appreciated the unchanging stability of the herds of Swiss Brown cows and Holsteins knee deep in clover and alfalfa. It was upon reaching the village, though, that things seemed different. Gone was the familiar drab, if neat, veneer that he had come to associate with home. In its place was a freshness and visible boosterism that seemed as out of place as it was new. Kubly described his impression of New Glarus, as seen from the Greyhound bus, in this way:

The town's familiar profile—a slender church spire rising from the hills—came into view. A large billboard painted with Wilhelm Tell and his son against a backboard of mountainous and blue skies read, "Don't miss the Swiss! Visit New Glarus. Schiller's *Wilhelm Tell* annually." The sleepy village I remembered had no need for street names, but now brand-new street signs ... were mounted under brightly colored shields of the twenty-two Swiss cantons. Similar medallions were mounted on the facade of the local movie [theater] and on a new "Alpine Restaurant." A hand-carved bulletin board in the churchyard looked like an Alpine prayer station. It was Saturday afternoon and the streets were crowded and bustling as a fair.[2]

The outstanding yodeling concert that evening came as much a surprise as did the lack of available parking space. Though yodeling had been a part of the community for as long as Kubly could remember, never had he heard more skilled renditions of tunes that had been part of his youth. Singing like "a bewitched flute," the star of the evening, oddly enough, was a professional yodeler from Switzerland hired by the community specifically for such occasions and to manage the village's largest restaurant.

Finally, there was the original Chalet, "the most elegant house in New Glarus." Young Herbert had been a frequent guest at Edwin Barlow's Chalet of the Golden Fleece, taking part in many of the convivial parties that ran well into the wee hours of the morning. Shortly before the world traveler's death, Barlow had presented the Chalet and its contents to the village as a museum. Though clearly a most generous gift, the effect on the visiting Kubly was chilling: "Inside, the rooms, which I remembered as settings for lighthearted revelries, were melancholy and funereal. Ropes and glass protected the furniture, paintings, books, and china with which I once had intimately lived. The chalet, it struck me sadly, was a mausoleum of my youth."[3] In large measure, Herbert Kubly's instincts proved correct. By 1964, the New Glarus of his youth seemed as far away as New York must have during his Wisconsin visit. One thing had remained constant however: the search for

Figure 6.1. "What we were seeking was an identity." Herb Kubly appears here "in Native Dress" with his mother at an early performance of *Wilhelm Tell*, ca. 1940. Note the photojournalist's camera. (Courtesy of Mrs. Herbert Kubly. Reprinted, with permission, from Herbert Kubly, *At Large* [New York: Doubleday, 1964], plate 4)

heritage. "Whatever we were," Kubly concluded, "whether it was Swiss trying to be American or Americans trying to be Swiss, what we have all been seeking for a century is an identity" (fig. 6.1).[4]

Commodifying Place: Making a Heart and Soul into a Lifeblood

Herbert Kubly's 1964 description of his return to the "village of Wilhelm Tell" provides a useful benchmark for a discussion of the central issues facing such communities over the past thirty years. The new, Swiss-style landscapes, the large crowds demanding more parking space, the professional quality musical performances, and the "museumization" of the Chalet all point to a new direction in Kubly's hometown. By 1964 New Glarus was developing into an ethnic theme town where the identity for which Kubly and his fellow New Glarners had

183

been searching was becoming a commodity.[5] Previous chapters have shown that as early as 1853, eight years after the arrival of the first colonists, the Swiss commemorated their ancestral heritage and attracted a crowd in the process. These commemorations were to continue with decennial anniversaries of the village's founding, with yearly Volksfeste and, since 1938, annual presentations of Schiller's *Wilhelm Tell*. By the early 1960s cultural displays were part of the village's cultural heart and soul. It was only at this time, however, that New Glarus' heart and soul became its economic lifeblood.

As early as the 1930s, an increasingly self-conscious ethnic identity, tied to a burgeoning tourist trade and legitimated through the official culture of the state, brought forth the newly minted ethnic landscapes that I shall refer to as "Swisscapes" (fig. 6.2). My chief interest lies in the efforts to create a recognizably themed place, in landscapes deliberately contrived to appeal to the outsider. These places, mockingly and delightfully coined by J. B. Jackson as "other-directed," offer exceptional windows into the intentionality of landscape production. And, as their contrivance intersects with identity and heritage, other-directed places cannot help but to shimmer with contestation and conflict.[6]

But conspicuously constructed landscapes may be other-directed

Figure 6.2. "Swisscapes" on Main Street. Canton shields, Swiss-style gables, and potted geraniums rework Main Street into "Wisconsin's most picturesque town." (Photograph by author)

184

in a second way. Not only does a New Glarus, or many a Chinatown, attempt to lure the gaze of an outside "other," it attempts to *represent* itself as an "other." Such a place is an interestingly and uniquely (post)-modern example of what Pierre van den Berghe calls "ethnic tourism." Ethnic tourism differs from other kinds of tourism in that the product being offered is an experience of the authentic, unspoiled, and exotic culture of the "natives" as well as an escape from the alienation of modern industrial society.[7] Such relations have been most frequently described in very negative terms for their tendency to create a situation rather like a human zoo, to facilitate heritage's devolution into "drive-through history." Dean MacCannell, for one, is miffed that "when an ethnic group begins to sell itself, or is forced to sell itself as an ethnic attraction, it ceases to evolve *naturally*." The "authenticity" offered to tourists is often staged and packaged specifically for tourist consumption. Similarly, tourist-native interaction is frequently between haves and have-nots, with all the inequalities of wealth, status, and power that such asymmetrical relationships imply.[8]

These negative aspects are very much a part of ethnic tourism, and can be found as easily in New Glarus as San Cristóbal. Yet, there is ample evidence to suggest that the complete story is a great deal more complex than the above critique would suggest. Authenticity, like tradition and ethnic identity, is far from an either/or proposition. Sentiments toward authenticity—so critical to other-directed places—are a matter of status discrimination and, as a corollary, a matter of power. Not a property inherent in an object—fixed in time and space only to be lost—authenticity, Edward Bruner argues, is best viewed "as a struggle, a social process, in which competing interests argue for their own interpretation of history." Or, as Arjun Appadurai notes, authenticity today is part and parcel of a political economy of taste; whoever enjoys the right to authenticate—to define what's really real—wields considerable power indeed. Likewise, externally imposed views of authenticity from representatives of official culture may be appropriated and subverted by less powerful groups. The methodological implication is the rise of intentionality and human agency, as competing definitions of what is "authentic" are negotiated, contested, and up for grabs.[9]

The "touristification" or commodification of New Glarus has its roots in the 1930s and 1940s, but accelerates with the closure of the community's most significant employer, the Pet Milk condensery in 1962, just two years before Herbert Kubly's memorable description. Today, ethnic tourism is the community's chief industry, a fact that is embraced by many, disliked by some, but acknowledged by all.[10] This chapter, then, examines the role of landscape in shaping that industry

185

and, more generally, the social-economic constellation of the community. At the same time, I highlight the way in which landscape alternates as medium of group identity and lightening rod for controversy. I aim to peel back the mask of innocence surrounding New Glarus's Swisscapes and interpret the tremendous changes of which, beginning in the early 1960s, Herbert Kubly caught a momentary glimpse.

CONSPICUOUS CONSTRUCTION: ETHNIC ARCHITECTURE IN AMERICA'S LITTLE SWITZERLAND

"One May Not Compare These Streets with Those of Our Lovely Swiss-land": Landscape before Swisscape

Although the town has claimed Swisscapes as part of its identity for decades, early sketches and travel accounts of New Glarus indicate that, with few exceptions, the Swiss were quick to adopt American building traditions.[11] One such traveler, apparently the first to visit from "die alte Heimat" (the old homeland) in 1894 reported that, due to the intractable grid, American housing styles, and muddy walk ways, the landscape took on a decidedly "New World atmosphere." Daniel Dürst found that, despite the village's name, "one may not compare this scene and streets with those of our lovely Swiss-land."[12] Another traveler thirty years later found that the village, like its train station, "is typical, like a thousand that you see—each one like the other on the transcontinental line in the U.S." (fig. 6.3).[13]

The village's plan, platted by Swiss immigrant W. Blumer, had an American aura from the beginning.[14] The streets, which cut at right angles and lay like a rigid grid over the gentle incline of the hillside, joined with the frame and brick houses to create the appearance of an unremarkable American village (fig. 6.4). Most typical of all, however, were the village's two main streets, both of which received more inspiration from nearby Midwestern towns than from Elm, Schwanden, or any other village in Canton Glarus (fig. 6.5).

There is some evidence that the very earliest years witnessed the building of structures with old world antecedents. John Luchsinger, for instance, noted in 1879 that the village had a "somewhat unamerican [sic] appearance" and six years earlier Emmanuel Dettweiler from Basel noted that some of the buildings were constructed "in Swiss style."[15] Luchsinger and Dettweiler were noticing, above all, the village's second church, the so-called stone church, built in 1858 by a local immigrant from Glarus, John Becker.[16] Correctly described as an "architectural curiosity in America," the onion-shaped dome and white mortar indeed gave the stone church visual distinction (fig. 2.14).

Figure 6.3. "One may not compare this scene and streets with those of our lovely Swiss-land." An utter lack of the foreign or distinctive characterizes the early New Glarus landscape. (Courtesy of Elda Schiesser Collection)

Figure 6.4. *New Glarus in the early years,* ca. 1860. This early painting depicts the grid pattern and nondescript building styles of nineteenth-century New Glarus. Note the sole exception: the onion-domed Swiss Reformed Church in the lower right. (Courtesy of New Glarus Historical Society)

Figure 6.5. West side of Ennenda Street. Only the name of the street—chosen in honor of a Canton Glarus village—attests to the ethnic character of the Swiss community in the turn-of-the-century landscape. (Courtesy of New Glarus Historical Society)

Likewise, a *Schützenpark* (shooting park) added in the 1880s conferred an unmistakable Swiss character. Finally, the village plan of 1873 shows for the first time that streets were named after the home villages of Canton Glarus: Schwanden, Engi, Ennenda, and Dießbach. On the whole, however, these landscapes were subtle manifestations of an insular, guarded community that felt little reason to conspicuously construct its identity through landscape. Swiss ethnic identity remained, if you will, backstage: that is, in the church, local taverns, businesses, and homes. It determined who one married, with whom to conduct business, and where one should worship. More conspicuous "front stage" displays were evident, too, but during the "time out of time" of festive occasions.[17] For outward displays of "Swissness" on the landscape, we need to wait until the Great Depression and the work of the Swiss immigrant Jacob Rieder.

Swisscapes, Phase I: "A Chalet to Breathe the Atmosphere of the Homeland," 1937 and 1947

Jacob Rieder was born in Chur, in the canton of Graubunden in 1888.[18] Displaying an early interest in architecture, he entered St. Gallen Polytechnic in 1910 to study traditional Swiss building styles and techniques (fig. 6.6). His studies there led him to produce a number of detailed drawings and plans based on folk styles, notably those of Canton Bern. Rieder brought these drawings with him to the United States when he emigrated in 1913 and they remained, according to one longtime friend, "the center of his life."[19] He continued making plans of Swiss-style homes, eventually winning a Juror's Prize from the Yale University School of Architecture for plans submitted at the 1933 Chicago Century of Progress Exposition (fig. 6.7). Upon his arrival in New Glarus in 1911, he began a career as a smalltown builder who completed dozens of local residences in both the village and the surrounding country; most are nondescript and blend into the American patina of the landscape. Two notable exceptions exist, however, that altered not only that American veneer, but also the life history of the community.

The first building was commissioned by Edwin Barlow. The eccentric insider/outsider's unique social position of being born into the

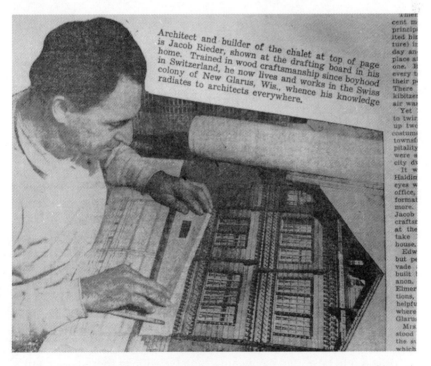

Architect and builder of the chalet at top of page is Jacob Rieder, shown at the drafting board in his home. Trained in wood craftsmanship since boyhood in Switzerland, he now lives and works in the Swiss colony of New Glarus, Wis., whence his knowledge radiates to architects everywhere.

Figure 6.6. Jacob Rieder at work. With extensive training in the design and building of folk housing at the St. Gallen Polytechnical University, Jacob Rieder was in a unique position to bring traditional Swiss chalets to America's Little Switzerland. (Courtesy of New Glarus Historical Society)

Figure 6.7. Jacob Rieder's chalet plan. This watercolor plan of a Swiss chalet was one of several that Rieder continued to make in the United States, and which won him distinction at the 1933 Century of Progress exposition in Chicago. (Courtesy of New Glarus Historical Society)

community and yet removed, combined with years of experience living in Switzerland, enabled Barlow to conceive of a building, the likes of which New Glarus had never seen before: a replica of a Swiss chalet dedicated to the memory of the village's early pioneers. The idea apparently came to Barlow in 1935 while on Long Island. There he read of the community's ninetieth anniversary historical pageant and became convinced that the filiopietism displayed on stage could be constructed materially on the ground. After congratulating the town on the successful celebration, the "former Green County resident" rhetorically asked if anyone "[could] think of a more fitting tribute to be paid to the early founders, than commemorating the first century of their adventure with the dedication of a Swiss Museum, to perpetuate their memory? In such a Swiss Chalet breathing the atmosphere of the land which they left to seek a home, and fortune across the sea, could be kept the recorded virtues of the past on into the future, with existing records might be added a genealogical tabulation of their descendants, and last but not least, the history of your Village."[20] This Swiss Museum would not be built as just any old museum, but as a "replica of an old Swiss chalet with its roof in reproduction of a shingled one held

190

down by stones." Such a structure, Barlow felt certain, "would prove unique in itself, but [I] am certain [it] would soon achieve quite an assortment of interesting things." As a collector of "valuable Swiss items," Barlow offered to help supply the Swiss Museum, but it was up to the village to build the chalet. "Should the people of your village have the desire to undertake such a Memorial to be handed down to posterity as an example of Swiss tradition and custom," Barlow concluded with the utter certainty that became his trademark, "that fine Patriotism will create a monument unique in the State if not in the country."[21]

Barlow's proposed Swiss museum was met with the customary indifference that new ideas injected from outside the community usually received. It *was* rather strange, after all, and certainly at odds with the more locally derived sense of heritage felt by key community members such as Esther Stauffacher, J. J. Figi, and Gilbert Ott. *Their* museum, it will be remembered, took the form of a log cabin patterned after the kind built by the immigrants in the New World. No chalet-type structures had ever been built in the community, nor should there have been reason to construct an alpine dwelling in the gently undulating, southern Wisconsin countryside. Barlow, of course, would not be deterred by mere indifference, and two years later decided to build the chalet by himself, as a home for his aunt.

This was the opportunity that Rieder had long waited for: he quickly busied himself with the construction of Edwin Barlow's so-called Chalet of the Golden Fleece (fig. 6.8). Using designs that he had been tinkering with since his school days at St. Gallen, Rieder constructed a building that conformed in the most minute details to his vision of the typical Bernese Alpine chalet (fig. 6.9).[22] From his own ornately carved woodwork both inside and outside to the elaborate *Deutsche Sprüche*, or aphorisms, that adorn the exterior, Rieder's chalet became the visual focal point for the community's reawakening ethnic identity. It also became a social focal point, for it was at meetings and social occasions in the Chalet that Edwin Barlow and his associates held the first meetings in preparation for the production of *Wilhelm Tell*. That the Chalet of the Golden Fleece would have been as out of place in Canton Glarus as it was in New Glarus mattered little for the many guests during these early years.[23] As one informant put it, himself a frequent guest as a young man, "It was a great place to get together. You could only go by invitation, and Barlow was pretty snooty about who he invited over. He was a pretty strange guy, but he could sure throw parties."[24] The world traveler donated the Chalet and its rather bizarre collections to the village in 1955. It has been open to the public as a museum ever since.

191

Figure 6.8. Chalet of the Golden Fleece. Edwin Barlow's tribute to Swiss culture and his gift to the community, the chalet was built by Swiss architect Jacob Rieder in 1937. The "authentic replica" features German aphorisms, rocks holding down roof shingles, and intricately carved woodwork. (Photograph by author)

Figure 6.9. Close-up of Jacob Rieder Bernese chalet design. This undated watercolor served as a model for Rieder's eventual Chalet of the Golden Fleece. (Courtesy of New Glarus Historical Society)

Figure 6.10. Ernest Thierstein, ca. 1940. Ernest Thierstein, cheesemaker/tavernkeeper, commissioned Rieder to build a house that reminded him of his Swiss home. He is pictured here in good spirits during a Volksfest. (Courtesy of Elda Schiesser Collection)

Rieder's second commission to build a chalet came ten years later from a Swiss immigrant named Ernest Thierstein (fig. 6.10). Thierstein, a key member and leader of New Glarus's various Swiss voluntary associations, remains legendary within the community for his thirty-four year portrayal of Walter Fürst in the *Wilhelm Tell* drama, and for his yodeling prowess (fig. 4.12).[25] Upon his retirement from a long life

Figure 6.11. Emmentaler chalet. Jacob Rieder's second chalet, modeled after structures of the Emmental region in Canton Bern, was commissioned by a Swiss native of that region. (Photograph by author)

of farm labor and tavern keeping, he asked Rieder to design a home reminiscent of his native Swiss Emmental Heimat.[26]

The Emmentaler Chalet, as it is now known, is even more grand in scale and elaborate in details than the Chalet of the Golden Fleece (fig. 6.11). Rieder himself did most of the carpentry work, taking great pride that no power tools were used in its construction.[27] Indeed, the house may be considered a piece of *Gesamtkunstwerk*—a complete artwork—as Rieder executed the cabinetwork, furniture, and interior paneling, all in the name of Bernese-style folk building. On the exterior, the cedar-shake roof slopes outward from the distinctive beveled peak, the distinguishing feature in traditional Emmental structures. Where the Chalet of the Golden Fleece bears a traditional Swiss building style merely because of its generalized popularity, Thierstein's Emmentaler Chalet resonated with a man who longed to return to a homeland.

The house functioned as a gathering point for those visiting from that homeland as well as for a tight circle of Thierstein's fellow immigrants. The Männerchor and the New Glarus Yodel Club met here on a weekly basis for practice and endless rounds of *yass*, the favorite card game of the Swiss. Similarly, Thierstein, the former tavern keeper, followed his practice of boarding Swiss immigrants during their first

194

few months in the new world. His Emmentaler Chalet became a way station for the dozens of men and women who emigrated to Green County in the years that followed the Second World War. Like Barlow's Chalet of the Golden Fleece, the Emmentaler Chalet conspicuously constructed ethnic place as a way to shore up receding group memory.

Swisscapes, Phase II: "A Way Through this Black Period," 1962 to the Mid-1970s

The next period of conspicuous construction can be given a very precise beginning date: February 15, 1962. This was the day that the community's only major industry and chief employer—the Pet Milk condensing plant—suddenly shut down its operations and fired more than 140 full- and part-time employees. Directly affecting nearly 38 percent of New Glarus's working population, an impending crisis seemed at hand.[28] The local paper assessed the situation directly, saying, "The news to close the New Glarus plant came as a complete surprise and was met with a feeling of despair as the news spread over the entire community. What had been a source of economic stability for the community for 51 years is being removed. To the more than 80 families who depend on the production of Pet Milk for their livelihood, this came as a blow which was felt on the streets, in the schools, the homes and in every business place in the community."[29] Of course, it was not only the eighty families who suffered from the plant's closure. Grocery stores and other businesses lost trade when families had to cut back on their expenditures. Farmers came to town less frequently since they now sold their milk to other buyers; and many families moved away, looking for greener pastures. In many ways, the village had been something of a company town for the previous half century, as more than one-third of the working population at any one time relied on Pet Milk for their income. As dramatically as industrialization had transformed the region in the early part of the century, bringing an end to small-scale cheesemaking, so now did its inverse twin threaten to alter the socioeconomic structure of the community.[30]

Meetings were quickly scheduled among the village elite to discuss the community's future. More than two hundred people attended a meeting at the local elementary school sponsored by the newly formed Industrial Development Corporation. While attracting new industry headed everyone's list of priorities, it was becoming clear that the community could not count on that alternative. More important was the recognition that, as mayor Waldo Freitag put it at the time, "the tourist business could become an important factor in recovery for the community." This move to ethnic or heritage tourism, Don Mitchell notes (for

Johnstown, Pennsylvania), is a typical one for communities undergoing economic stress associated with deindustrialization.[31] One restaurant owner—a Swiss immigrant—suggested less than a decade earlier that "tourists will be attracted to New Glarus as the village becomes more and more a Swiss town."[32] Such ideas were easy to ignore during a self-confident period of gainful employment, however; only with economic anxiety did tourism and landscape become important.

Thus, in 1962 the village elite brought several experts on Wisconsin tourism to the community, including marketing agents from the Wisconsin Dells, rural sociologists from the University of Wisconsin Extension, and the Wisconsin Chamber of Commerce. Predictably, each suggested "the tourist angle" as a way through "this black period." The move toward tourism seemed to be a natural one for a community that had become so well known for its productions of *Wilhelm Tell*. Attendance at the festival had been increasing steadily, and more and more people were visiting the two village museums. Yet, in many ways the village didn't *seem* as Swiss as its promoters thought it should. It was not enough, as one resident put it, just to be Swiss: "We had to start *looking* Swiss."[33]

Several recommendations by the newly formed Community Betterment Group pointed in the direction that these business leaders felt New Glarus should move to "provide for greater tourist appeal." First, the Betterment Group suggested that residents dress in Swiss costumes for all special events for the summer, an appeal that it knew would be met with resistance. Next, a new emblem "typical of the community and its Swiss background" was developed, with the emphasis on Wilhelm Tell. Then, there was the matter of neatness and the "appropriate" presentation of place. After a rash of vandalism that spring, most notably the smashing of windows at schools and homes throughout the community, business leaders pleaded with residents to exercise control over their children for the betterment of all. Parents were implored to "educate children in New Glarus to take seriously the neat appearance of the community and to challenge [other] parents and school officials to stress the importance in neatness to community interests in attracting tourists."[34]

At its foundation, the move toward the so-called tourist angle hinged on what one report at the time called a "conscious influence over the appearance" of the village. Describing the entrance to the village, but speaking for the community as a whole, hotelier Robbie Schneider related to a reporter that "we believe that we could attract more tourists into town if we fixed [it] up."[35] Fixing it up required rethinking decades-long patterns of building. During the "black period" this reassessment was summarized by one of the visiting rural sociologists in the following way: "Responsible, forward-looking citizens must become concerned with the possibilities of a headlong progression towards visual disharmony and to provide protection for historical significance and scenic beauty. The people of New Glarus must form a *conscious influence over the appearance of their community* if they are to realize the rewards of emphasizing the beauty found in the unique Swiss architecture."[36]

It would appear that local residents and business agreed with such sentiments, for as a result of these meetings and discussions, New Glarus experienced a flourish of Swiss-inspired renovations and new building. That summer, residents watched as two new businesses were constructed in loosely chalet fashion, three nineteenth-century commercial structures received facelifts, the ten-year-old theater became "Swissified," and canton and village shields were placed on street posts and other buildings that could not afford complete renovation. No wonder, then, that Herbert Kubly barely recognized the town two years later. By 1970, Madison's *Capital Times* newspaper could report that the "the business district . . . is achieving an all Swiss-look."[37]

That look varied considerably from the ornately detailed appearance of Rieder's chalets. The rapidity with which both residential and commercial structures became suddenly "Swiss" is evident in the manner in which they were built. In most cases, little more than an overhanging balcony was added to a preexisting structure (fig. 6.12). Or, if a new building was erected, it made only slight gestures toward a stereotypical interpretation of Swiss building style. This should not be surprising given that, unlike Rieder, very few of the builders during this period of conspicuous construction had any formal training in Swiss building traditions (fig. 6.13).

In large measure, the audience for the architectural performance had shifted at this time as well. Where Rieder created Swiss-style buildings for his Swiss neighbors, the chalets of the early 1960s were intended largely for passersby, for an admission-paying audience en route to a festival or a gift shop. True, unlike other conspicuously ethnic communities like Pella, Iowa, or Holland, Michigan, New Glarus themed its downtown without zoning regulations.[38] But, the purely gestural nature of the landscapes suggests that New Glarners held a pretty low opinion of the tourists. There is some evidence, however, that in many (but not all) instances, these newly built, outwardly directed landscapes stemmed from and reinforced an inwardly focused sense of self and identity. Don Ott's chalet offers one such example.

Don Ott is a fourth-generation Swiss American who proudly traces

Figure 6.12. Phase II Swisscape: Strickler's Sausage Store. With its overhanging balcony, "Strickler's Chalet" was one of the first "restorations" during the period surrounding Pet Milk's closure. (Photograph by author)

Figure 6.13. Chalet-style bank. A simplified chalet-type structure of the middle period, the Bank of New Glarus (1970) replaced a neoclassical structure. (Photograph by author)

his ancestry back to the original colonization.[39] Like his father before him, Ott worked in the local lumber yard where he supervised much of the building in town, including the 1960s commercial construction. He also found time in 1963 to build his own Swiss-style chalet. The Chalet Ott followed no formal blueprints, only the most general plans. He describes his building method this way: "We built it bit by bit with wood from the lumber yard. We'd work out problems as we went because all the ideas were in my head. I didn't have much money so we had to improvise a lot." Inspiration came, not directly from Switzerland (he had not yet been there), but from Rieder's two chalets, where he had spent a great deal of time with those he calls the "old-time Swiss." Ott acknowledges that in many respects his chalet isn't "technically correct." For one thing, (due to the sort of lumber he could most easily afford) the interior paneling was hung horizontally instead of vertically, the way Ott says "they really should be." Similarly, the panels are of different lengths, and when viewed from across the room appear to be unevenly placed. But, he adds quickly, "if someone doesn't like [it], tough," for both he and his wife Pauline assert that their house is part of who they are. Reflecting this attachment to place, Chalet Ott became in the 1970s through the 1980s what Thierstein's chalet was during an earlier era: a central meeting point for the Swiss community and those visiting from Europe (fig. 6.14).

Don Ott's chalet—and numerous other houses and businesses like it—is an excellent instance of the emergent quality of authenticity and its negotiability. As annual performances of *Wilhelm Tell* enabled third- and fourth-generation Swiss Americans to discover and re-create their own ethnic identity, so did the material "performance" of creating an artifact—in this case a chalet—arise from, and in turn recast, a sense of ethnic identity. Accordingly, rather than seeing these landscapes as poor imitators of a somehow more authentic architecture elsewhere, I contend that they be considered "authentically," if also conspicuously, Swiss American. Many are vernacular, if we understand that term, as J. B. Jackson does, to be "identified with local custom, pragmatic adaptations to circumstances, and unpredictable mobility."[40] A far cry from the formalism and stability of elite or official landscapes, vernacular landscapes are usually built step by step, with materials close at hand, and are responsive to locally perceived pressures and inspiration. In the tangled interplay between market, creators, and consumers, artisans like Ott appropriate an externally imposed notion of what is genuine.[41]

Figure 6.14. New Glarus vernacular: Don Ott and his chalet. Based on knowledge of local Swiss building traditions and built bit by bit with materials from his family's lumber yard, Chalet Ott exemplifies a New Glarus vernacular. (Photograph by author)

Swisscapes, Phase III: "Building Based on Authenticity and Tradition," The Mid-1970s to 1995

The final stage of conspicuously constructed Swisscapes may be considered an extension of the sort of residential and commercial building found in the 1960s and early 1970s, but with a clear departure in both style and intent. Since the mid-1970s and continuing today, ever greater attention has been paid to notions of authenticity. Financed in many cases by Swiss capital, Swiss-trained architects and craftspeople have been commissioned to build landscapes that more accurately reflect Swiss folk building traditions. As one Swiss immigrant to New Glarus put it: "If you want to build a chalet-type building, then build an authentic Swiss chalet."[42] Significantly, this period of building was accomplished, like the first and unlike the second, by trained architects. Local builders have continued to erect Swiss-influenced buildings, but more often architects with training in, and comprehensive knowledge of, Swiss building traditions have come to the fore.

Most important was a Madison architect named Stuart W. Gallaher who, though not Swiss himself, received extensive training in Swiss building. Gallaher, best known for his design of the Olbrich Botanical Gardens' conservatory in Madison, was sent to Switzerland with the backing of New Glarus capital to receive that training. On more than a half dozen trips with his New Glarus clients and more on his own, Gallaher surveyed the landscape, collected dozens of books on Swiss building, and met with Swiss architects. As his chief patron put it, "Stu wanted his buildings here to be as authentically Swiss as possible in every detail."[43] For one structure, Gallaher went to Appenzell specifically to photograph houses in that canton for ideas.

More typical, however, are the numerous "generic chalets" such as the Chalet Landhaus, Gallaher's most significant "Swiss" structure (fig. 6.15). In a manner that has become the standard for postmodern architecture, Gallaher cobbled together motifs, detailing, and linework from different regions to produce what his client, Hans Lenzlinger, calls a "generic chalet." Lenzlinger, the owner of numerous business interests in the community and himself a Swiss immigrant relayed that, "you can't really call [the Landhaus] any one particular style; it's made up of different features taken from different building types across Switzerland."[44] This regional pastiche, joined with the imposing scale (the two cojoined buildings have sixty-seven guest rooms), separate these architecturally designed chalets of the 1980s from Rieder's two chalets of the 1930s and 1940s. Those, it will be remembered, were designed to reflect the building of a specific locality in the regionally diverse Switzerland. Hans Lenzlinger recalls that just after the Landhaus was

Figure 6.15. Chalet Landhaus, close-up. The intricately detailed woodwork is representative of the heightened sense of authenticity in recent building of a "generic chalet." (Photograph by author)

opened, "Swiss architects came here and they were appalled. Not at the styling or details, but at the scale. They couldn't believe that you would build such a huge chalet. And with AC [air conditioning] in the windows! They really didn't like that. But you have to consider the real money involved. There is no way that a small-scale chalet would pay for itself. We wanted a hotel large enough to handle bus and group tours. This really demands a very large size building."[45] Lenzlinger candidly spoke to the double coding so often embodied in postmodernism. On the one hand, buildings such as the Landhaus express a traditionalism that is slow moving, "full of clichés, and rooted in family life." Conversely, such buildings are also deeply rooted in a rapidly changing society, with "new functional tasks, new materials, new technologies and ideologies" (fig. 6.16).[46]

Gallaher's chalets, despite their scale and eclecticism, successfully evoke that most precious quality of postmodern design: ambiance. The Chalet Landhaus, in particular, effectively re-creates the atmosphere of the old country. It is a building, its chief woodworker—himself a Swiss immigrant—believes is "based on authenticity and tradition."[47] Largely through detailed interior work, "the mood [of the place] makes you feel like you're back in Switzerland," one New Glarus resident relates. Working for several years now in the housekeeping staff, but

202

Figure 6.16. Photographing the Chalet Landhaus. Swiss-looking landscapes are central to the tourist gaze. Here the Chalet Landhaus motel serves as the backdrop. (Photograph by author)

having also spent many long evenings with friends in Switzerland, she would seem to be in a position to judge such a subjective quality: "When I am working there and see the entire dining room filled with people speaking Swiss and carrying on, it seems to me that I am not in Wisconsin at all."[48]

The Chalet Landhaus departs from the earlier Swiss building traditions of, say, Don Ott in a couple of ways. First, it and other more recent buildings are based on substantial research. Whereas Ott's construction style is bit by bit and based on what he has seen and known in New Glarus, the more contemporary design is based on extensive library research and visits to Switzerland for inspiration. Second, this research and building is based on considerably more capital than was at hand in the early 1960s. A decade after the collapse of the local economy through Pet Milk's departure, New Glarus witnessed sizable economic growth and capital concentration in the hands of a few local businesses. These businesses, significantly, could *afford* to create a landscape based more on research and less on local knowledge. This change was necessary, for tourism in New Glarus is based on the perception of it being an authentic Swiss village. Indeed, by 1993 more than 90 percent of all tourists felt that the town's "authenticity" was a major factor in their deciding to visit New Glarus.[49] The result is a

Swiss American landscape more consciously informed by Old World building traditions than its founders could have imagined possible 150 years ago.

THE LIMITS TO LANDSCAPE: CONTESTING SWISSCAPES

Until the early 1960s, Rieder's Swiss-inspired landscapes fostered ethnic awareness but offered little in the way of other-directedness, as a prosperous industrial base provided economic stability for the community. The Pet Milk closure changed all that, and community members turned to landscape as a way to grapple with sudden and negative change. In some cases, dwellings and commercial structures dovetailed with a strong sense of collective identity and public memory; in others, the very notion of landscape-as-clue-to-culture was questioned. The rapidity with which landscape became Swisscape raised objections within the community over the new direction it was being led. Beginning in the mid-1960s and escalating over the next several decades, discordant voices questioned the wave of other-directedness. Two projects, in particular, pushed the limits of conspicuous construction in a complex, working community.

Bus Station or Chalet? The Hall of History

The idea that the Swiss Historical Village museum should contain a separate building dedicated to artifacts and exhibits from the homeland dates back to the earliest years of the open-air museum. Though the plans were summarily dropped for rather obvious logistical problems, those committed to the concrete preservation of their ancestral past hoped that, with time, a building might be erected that would be "partly furnished by Glarus, Switzerland as well as with articles that came from there."[50] With the high cost of trans-Atlantic travel and in the days before satellites and television, very little contact, except for personal letters between relatives, existed between Glarus and New Glarus in 1939. Moreover, the struggle to erect small, pioneer-focused structures demanded more resources than the early organizers imagined and quickly took precedence over any explicit "history building."[51]

During the early 1960s, when conspicuous construction was literally changing the face of the community, increasing interaction occurred between the two Glaruses. Triggered largely by a successful Yodel Club "pilgrimage" to Switzerland in 1960 in which 158 Swiss Americans traveled to the old country, long-forgotten relations were renewed between individuals and official agencies across the Atlantic. Though many in the tour group were Swiss immigrants themselves, many more were third- and fourth-generation Swiss Americans for whom this was the first tangible contact with the relatives and places of their family's shared memory. Yodeling on the runway at Zürich's international airport and walking through the urban streets of Glarus, Bern, and Zürich in folkloric costumes, the Yodel Club and their followers created something of a sensation on the order of a Swiss Rip van Winkle. To their Old World hosts, the American Swiss yodelers seemed frozen in time and a quaint reminder of how things once had been in a now modern Switzerland where automobiles outnumbered cows (fig. 6.17).[52]

As a result of this new connection, *"Swiss* Swiss" (contrasted with *New Glarus* Swiss) involvement in the American village's festive culture increased dramatically. In 1965, Canton Glarus donated display materials for a historical exhibit at the village hall in observation of the 120th anniversary and, that same summer, contributed one thousand dollars toward the creation of a permanent exhibition hall. This money was to be used as the groundwork for a so-called "Hall of History," in which the memory of the old country could be perpetuated and ties between the two Glaruses strengthened.[53]

With this seed money and some locally raised funds in hand, curators at the Swiss Historical Village began the process of deciding the shape, form, and location of the new history building. The open-air museum had grown substantially since its creation in the late 1930s and by 1965 included replicas of seven pioneer-style buildings, including a log church, cheese factory, and a general store. The historic village had, as the recently formed (Glarus, Switzerland based) Swiss Friends of New Glarus noted, "an almost prehistoric air" where the atmosphere was designed for *frontier* or *pioneer* time travel. Like most pioneer museums, the pre-1965 core of the Swiss Historical Village returned visitors to their roots—not the roots of a remote Old World origin, but to a generic Turnerian frontier. All the buildings (with the exception of the first community house and log church) stressed the explicitly nonethnic days of the pioneer, a message that became the historic village's predominant theme. The setting had become one to which *any* Euro-American, raised within the mythology of the pioneer and frontier past, could hopefully relate.[54]

The history hall was to be different. The intent here was to connect explicitly with the *Swiss* homeland, not the pioneer past. As such, it was given a central location—in both the geographic and hierarchical sense—within the museum complex. In order to reflect this intention,

205

Figure 6.17. Yodeling on the runway. During their 1960 tour of Switzerland, the New Glarus Yodel Club and their entourage of more than 150 New Glarners visited the "homeland," an event that triggered renewed contact between Glarus and New Glarus. (Courtesy of New Glarus Historical Society)

the original designs for the Hall of History took the shape of a chalet-type structure. Local builder Wayne Duerst, an American of Swiss heritage, drafted plans for a chalet that would serve as a focus of the Swiss background of the area's residents. With so much local chalet building on the part of businesses and residents, such a structure fit nicely into the architectural context of the mid-1960s (fig. 6.18).

Not everyone saw things this way, however. Paul Grossenbacher, for one, felt that the chalet-type structure would be inappropriate to the setting and "did not like it at all." A Swiss immigrant from Canton Bern and a prominent member of the historical society's board, Grossenbacher noted that the design "caused great concern and dissension among the Board of Directors" because it was "not an authentic Swiss chalet." Others felt similarly, but the largely American-born board members found the design to be "real Swiss."[55] Dismayed at the design, Grossenbacher forwarded copies to the Embassy of Switzerland in Washington, D.C., and to Jakob Winteler, the archivist of Canton Glarus and the one most responsible for obtaining the 1965 exhibition and monetary gifts.[56]

Swiss reactions to the chalet-type design were even more disapproving than Grossenbacher's. Upon seeing the plans, Winteler and his colleagues "were honestly shocked, because the way we look at it,

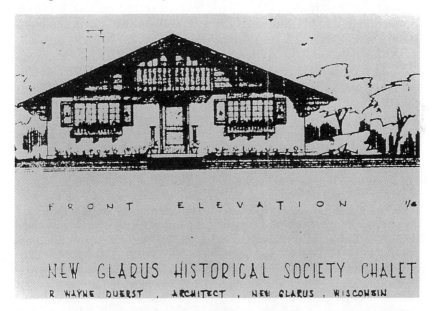

Figure 6.18. Chalet design for the Hall of History, ca. 1966. The locally rendered plans for the historic village's Hall of History, by Wayne Duerst, were originally of a chalet-type structure. (Courtesy of New Glarus Historical Society)

what is planned is not a happy solution." After reminding Grossenbacher that he was the one who worked tirelessly for the project's funding, the archivist outlined his objections in the most detailed and
straightforward manner. The American architect, he wrote, "with great
care made a plan for a beautiful chalet, but the style is not Swiss much
less Glarus style. Houses like this are built in Austria, in Tirol." Unlike
the Chalet of the Golden Fleece, "a chalet with the correct proportions
[that] could just as well stand somewhere in the Canton of Bern," the
museum's proposed chalet would be an embarrassment. He reminded
Grossenbacher that the earliest known pictures of the village do not
betray "one single Chalet." Winteler then concluded with a fervent plea
to reconsider the building's design, saying, "With an open air museum
you surely want to create a *true* picture from the first decades of New
Glarus. Therefore please! No foreign object with an Austrian Chalet. I
plead with you from the bottom of my heart . . . postpone the building.
I think of the well meant efforts to give New Glarus a Swiss look by
building Swiss fronts to business places and private homes. I think it
is terrible. But those are but trifles and they don't count nearly as much
as a complete false construction of a Hall of History."[57] The archivist,
of course, was preaching to the already converted. But his suggestion,
seconded by Lukas F. Burkhardt, the cultural counselor of the Swiss
embassy in Washington, D.C., that a Swiss architect become involved
was accepted by the historical society.[58] A well-known architect named
Jakob Zweifel was then brought into the picture.

At the time one of his country's most prominent architects, Zweifel's work centered largely in his hometown of Zürich.[59] His design
departed radically from the chalet-type structure of Duerst and from
the existing pioneer replicas. The problem of creating a structure that
linked both sides of the emigration should not be solved, he felt, with a
historicist framework, but with a modern, internationalist perspective.
Zweifel outlined his vision in 1966 this way:

The studies until now have shown that it will hardly be possible to build a
historical building for the exhibit hall. The imitation of an old building style
for a new building would produce an unsatisfactory result. Real, historical
documents should not be placed in an architectural falsification [*in einem bau-
lichen Falsifikat*]; a construction that pleases neither the professional historian,
nor the master builder, nor the requirements for an exhibition cannot be suitable. Styles of the pioneer period are displayed in the buildings of the Swiss
Historical Village. The Hall of History, on the other hand and in clear contrast
to these buildings, should be built in a *construction style of our own time.* . . . Just
as the resources and skills of the pioneers have been displayed in order to
correspond with the necessities of a school, a blacksmith shop, etc., so should
we display our construction skills.[60]

Figure 6.19. Plans for a modern Hall of History, 1966. Swiss architect Jacob Zweifel envisioned a Hall of History "built in the construction style of our own time." Authenticity was to be found not in the chalet-type structure, but in the match between form and function. (Courtesy of New Glarus Historical Society)

Authenticity, for the modernist architect, was to be found in a trueness to materials, in the correspondence to today's architectural trends, and in the essential match between form and function (fig. 6.19).[61]

Three years later, after consultations with the leaders of the Glarner Heimatschutz (the canton's cultural and historic preservation agency) in Switzerland, Zweifel's Hall of History was built. The Glarus-based Swiss Friends of New Glarus raised more than 100,000 Swiss francs for the installation of the permanent and temporary exhibits.[62] The modern, open construction achieves the Swiss architect's vision of a jarring contrast to the homey, smaller structures that surround it (fig. 6.20). Standing in the middle of the historic village, the Hall of History—with its strong horizontal lines, redwood paneled interior, and continuous band of glass panes for a roof—reminds the visitor that she is not back in the nineteenth century, but at a historical re-creation in the twentieth. The immediate local reaction, as one might predict, was less than favorable. Paul Grossenbacher recalls that, though he admired the modern building, "many did not like it—some called it a bus station."[63]

Not only is the contrast a visual one, but it is also one of cultural differences, differences that occur between vernacular and official culture, between nostalgia and reason, and the New and Old Worlds. Most immediately, the Hall of History conflict points to the very differ-

Figure 6.20. The Hall of History. Surrounded by replicas of the pioneer period, the Hall of History forms the nucleus of the Swiss Historical Village. The open, modernist structure, once disparaged as a "bus station," has also been called "the only true Swiss building in New Glarus." (Courtesy of New Glarus Historical Society)

ent agendas between heritage and history. Whereas the other historic village buildings were constructed by local people with plans drawn up by local carpenters, the Hall of History—"the only true Swiss building in New Glarus," as one Glarus, Switzerland, magazine recently put it—was designed by a well-known international architect with training in the latest modern styles.[64] The culture that Zweifel represented, an official culture, had only disdain for the heritage-based nostalgia of the local or vernacular culture as represented by the small-scale structures that surround the Hall. The New World wanted to look old and the Old World wanted to look new. In the end, newness won out, though ambivalence remains.

A Tale of Two Swiss Villages: "Swissness" Gone Too Far

One of the dominant players in the redevelopment of the community following Pet Milk's 1962 closure was a young Swiss immigrant named Robert (Robbie) Schneider. After answering an advertisement for a professional yodeler placed in several Swiss newspapers by the New Glarus Yodel Club, Schneider immigrated to the Swiss colony in 1956 (fig. 6.21). His credentials to lead the Yodel Club were impressive as the

210

Figure 6.21. Flag thrower, yodeler, and businessman Robbie Schneider. Shown here standing in front of the Emmentaler Chalet, Robbie Schneider brought not only expert yodeling skills, but also tourist industry experience, from his native Switzerland. (Courtesy of Elda Schiesser Collection)

prize winner of several international folklore competitions. His flair for Swiss folkloric productions quickly won him a prominent place in his adopted home as he became the central attraction for performances in New Glarus, as well as for their increasing number of traveling engagements (fig. 6.22). It was Schneider, for instance, who organized the 1960 Yodel Club Swiss homeland tour that rekindled personal and institu-

Figure 6.22. The yodel club at the time of their Switzerland tour, ca. 1960. Robbie Schneider, *second from the right in the middle row,* immigrated to the Wisconsin community after answering an advertisement placed in Swiss newspapers by the New Glarus Yodel Club. (Courtesy of New Glarus Historical Society)

tional ties between Glarus and New Glarus. It was also Schneider who decided that the yodelers were "cheapening whatever they do by not charging," and so introduced monetary exchange into the Yodel Club's repertoire.[65]

At the time that Herbert Kubly described Schneider's "bewitched flute" yodels after his visit in 1964, the young performer had recently become a businessman as well. In 1960, after he organized and led the trip to Switzerland, he convinced several Yodel Club members to extend him enough credit for a down payment on the historic, but dilapidated, New Glarus Hotel. The first item of business was to make it more "Swiss" by serving Swiss specialties, offering regular Swiss entertainment, and two years later, during the "black period," making over the Hotel's facade to create "a Swiss face." Soon, a gift shop and tourist information booth were added and the restaurant was remodeled in Swiss style and renamed "Robbie's Yodel Club."[66]

Schneider, an engineering graduate of the University of Zürich, picked up the tourism business quickly. In short order, the 1853 New Glarus Hotel—complete with its new Swiss facade—burst with renewed excitement as it quickly became the new center of Swiss tourist culture. Schneider was the first to bring in organized bus tours, thereby stretching out the tourist season beyond the few, separate weekends that clustered around festivals. He became heavily involved in state tourism efforts, being appointed to the Travel and Vacation Committee of the Wisconsin State Chamber of Commerce, to the information committee of the Department of Natural Resources, and as chairman of the Green County Tourism Committee.[67]

As one of the first to recognize the importance of landscape in attracting visitors, he proposed the wholesale reworking of the downtown into a miniature Swiss village. The plan, known as Project Edelweiss and modeled after what Schneider saw in the Danish community of Solvang, California, called for Swiss architecture beyond the haphazard and "technically incorrect" building that characterized the second phase of conspicuous construction. It also called for the sale of ethnic goods in shops and the attraction of Old World artisans to produce salable crafts on site. Other business leaders rejected the plan—some from fear of Schneider's growing importance in the town—and the community proceeded to "Swissify" its appearance, building by building, owner by owner, with no general oversight or control over the refashioned landscape.[68]

Schneider, not one to be deterred by mere lack of support from his fellow businessmen, took his idea of a miniature Swiss village outside the village limits. There, on 165 acres of prime dairy land that he had recently purchased, Schneider proposed to build his own Swiss theme park. The park, to be known as the "Swiss Village," seemed to offer the solution for which the performer/businessman had been searching. During the late 1960s and early 1970s, when Schneider began to seriously consider the idea of a theme park, attendance figures for the festivals and museums were going through the roof. In 1968 alone, more than eight thousand people descended upon New Glarus for performances of *Wilhelm Tell*, and almost two thousand came for the new "Heidi" festival, invented only three years earlier. Moreover, visitation to both museums soared past the 20,000 mark and the downtown streets were becoming so crowded that community residents began leaving the village in droves during busy weekends.[69] With the same assurance that he introduced paid performances for the Yodel Club, Schneider began work on his theme park.

In 1974 Schneider released the first plans for his Swiss Village, the result of "data gathered during a five year research program." All told, the experts from twenty different local, state, national, and international agencies helped the increasingly well-connected Schneider to survey the market potential, site characteristics, and locational advantages of his theme park, at the impressive cost of $36,000.[70] The University of Wisconsin Extension, in particular, became closely involved, producing three different studies that all came to the same conclusion: "the Village will be the only authentic Swiss resort complex in the United States."[71]

The use of the word "authentic" here is quite interesting, given the form and content of the theme park. The project was to be completed in two phases on seventy acres of Schneider's 165-acre property. The scale and magnitude of the Swiss Village theme park—estimated to require a six-million-dollar investment—were impressive and worth enumerating. The theme park was to include a 76-unit luxury motel, replete with a 300-seat restaurant and 250-seat cocktail lounge in the same building that would also house three separate meeting rooms for banquets and conventions, an indoor swimming pool, a beauty salon, floral shop, travel agency, sporting goods shops and business offices; a Swiss cheese factory (recall that the last local cheese factory closed its doors some thirty years earlier) with a visitor's viewing area and retail store, designed to produce roughly 4,000 pounds of Swiss and other varieties daily; craft and speciality shops where Swiss artisans, bakers, and butchers would work on site (including a Swiss pastry shop, a meat and delicatessen shop, a hand-embroidery shop, a wood carving and wood sculpture studio, and an art shop where Swiss scissor-cuttings and paintings would be made and sold); a museum constructed in the form of a cross dedicated to "Swiss folklore and history" and [taking a cue from Disney himself] centered on "a two-story replica of the Matterhorn"; a "medieval Swiss water tower"; an ice rink and four tennis courts; a cable car lift; and "America's only crossbow shooting range" (fig. 6.23).[72]

The sheer audacity of the Swiss Village theme park staggers the imagination. How, by any stretch of the imagination, could the theme park with a "two-story replica of the Matterhorn" as its centerpiece be viewed as "authentic?" This was not a trivial question, and one that tested the creative energies of Schneider. In the end, "authenticity" hinged, first, on the entrepreneur's strategic use of the concept's malleability, and second, on a view of landscape that demanded precise simulation.

Likening his proposed Swiss Village to Colonial Williamsburg as its "only counterpart in the nation," Schneider stressed that what set his theme park aside would be its "emphasi[s] on the arts, foods, and traditions of *only one nation*." Where Disneyland might confuse people by jumbling a New Orleans ambiance with that of an African jungle cruise, the Swiss Village would have none of that. The managerial staff would be comprised only of Swiss natives recruited from that country "to serve in various capacities at the new resort." The attractions—

Figure 6.23. Plans for the Swiss Village theme park, 1974. Making use of "authentic Swiss architectural styles" depicted here, and with a 76-unit luxury motel, a 300-seat restaurant, Swiss folk craft shops, a Swiss history museum, a medieval Swiss water tower, and "America's only crossbow shooting range," the Swiss Village tested the limits of commodified ethnic heritage. (Reprinted from the *New Glarus Post*, 26 July 1974, courtesy of New Glarus Historical Society)

cheesemaking, crossbow shooting, the Matterhorn—would conform to stereotypical notions of Swissness. Most importantly, however, the architecture of the entire resort complex would reflect an integrated, shared style. A far cry from the unplanned, haphazard approach to Swisscapes in the community itself, the 25 buildings of the theme park would all be built in the style common to the Simmental region of Canton Bern. Architects trained in Switzerland produced lovely plans of the "authentic Swiss architecture" from the region whose most well-known village, Gstaad, served as the model.[73]

Where for the Swiss modernist architect Zweifel authenticity was to be found in a trueness to materials and an honesty about age, for Schneider, authenticity corresponded to the closeness of fit between the theme and the landscape of his Swiss Village. By purging signifiers of anything non-Swiss, Schneider hoped that a theme park could become, as it were, real. In a way that has become emblematic of our age, his Swiss Village sought to create—through fantasy—a context of "authentic Swiss life."[74] Schneider was correct to be so concerned with the perception of the theme park, for without a sense of authenticity, it would offer little that was different from other tourist destinations and thereby lose its critical appeal.

More immediately, however, it was not tourists, but the people of the other Swiss village down the road, New Glarus, who voiced concern over the scheme. The feeling among many at the time, though largely unarticulated, was that capitalizing on the past had gone far enough. One longtime New Glarus resident, Millard Tschudy, summed up his opposition to the "Swiss Village" this way: "It's too cut-and-dried, too dollar-and-centsy. They're going to have crafts people from Europe, and what we're interested in is preserving the things that were grounded here. What they'll have may be authentic Swiss, but it won't be authentically New Glarus Swiss."[75] The question of authenticity, of course, was a tricky one, as Schneider's architecture would be more true to Swiss design than the facades constructed during the 1960s.

At the core of Tschudy's remark is the issue of whose authenticity is at stake and who has the power to define what is authentic. Many felt that control of their Swiss identity was slipping away and resented a relative newcomer (even if he was from Switzerland) cashing in on *their* town's heritage. At a heated village board meeting in the spring of 1974, one of the most vocal opponents charged that "even the name 'Swiss Village, Inc.' was a name people used for New Glarus." Foreshadowing the wave of protest unleashed in the wake of Disney's proposed heritage theme park near Washington, D.C.—called, appropriately enough, "Disney's America"—Schneider's Swiss Village ran the risk of sanitizing local history beyond recognition. One of the university consultants most directly engaged in the theme park concluded that "in the end, Robbie's vision did not fit with the conservative Switzers who, though they may love to yodel, can't stand seeing someone else profit from it."[76]

After four years of endless haggling over his Swiss Village, after leasing out his New Glarus Hotel to raise money for the federal loan guarantee, and after nearly twenty years of living in New Glarus, Robbie Schneider—the master yodeler cum businessman—pulled up stakes and moved to Colorado.[77] Unrepentant, Schneider scoffed at the hint that he was commercializing on the heritage of New Glarus. "Eighteen years ago, I came here to teach these people songs they're singing today," he relayed to *New Yorker* journalist Calvin Trillin during the heat of his battle. "Ninety years ago, there was no yodeling club. If they hadn't been proud to show all America what they are doing, they wouldn't put on a play for charge; people would still have to go to Switzerland to see 'Wilhelm Tell.' They wouldn't have brought me here. They'd still be sitting around a tavern yodeling for a bottle of wine."[78]

Schneider was right, of course. Marketing and selling the cultural displays that defined a very different sense of Swiss ethnic place one hundred years ago had, by 1974, become the economic foundation of a deindustrialized Midwestern community struggling to survive. In a village-wide survey compiled in 1977, three years later and one year before the eventual demise of the Swiss Village theme park, more than half the community residents responded favorably to a question about their opinion of New Glarus "as a 'Swiss tourist village,'" recognizing the jobs that tourism generated. Yet, that same survey found that more than a third of those same residents did *not* support the expansion of more tourist shops and facilities even if they would create more jobs.[79] These findings reflect less an opposition to ethnic tourism per se than to its tight control in the hands of one person. Before Schneider's proposed Swiss Village, Swissness in the community was largely individual and artisan in nature—a point demonstrated in the fact that the community had both willingly built Swiss-inspired store fronts and already resisted the adoption of Swissifying zoning laws that would codify and mandate such construction. One person's expression of authenticity and Swissness, no matter how closely patterned on the "real thing," stood the chance of becoming hegemonic, both culturally *and* economically.

CONCLUSION

Landscape, because of its visibility, tends to mask or naturalize social relations. This is to say that what we see on the landscape—the workers' homes of a planned industrial community, the redevelopment of a central business district, or the conspicuously ethnic-styled buildings of a small town—stem from the social, economic, and political ideologies of their creators and from individual creativity and inventiveness. These relations are normally hidden from view, giving the landscape a quality of naturalness, as if its construction were somehow inevitable. Such inevitability is hardly characteristic of New Glarus's Swisscapes, despite the understandable claim by some residents that "the downtown is Swiss because we're Swiss." Such a statement—commonsensible though it may seem—conceals the work, contingency, and frequent contestation involved in landscape production. Wisps of nostalgia in the 1930s and 1940s, and economic demand for tourist revenues in the 1960s and 1970s, combined with individual initiative and led to the creation of a distinctive landscape. Specific individuals active at critical junctures can have an enormous impact in the process of ethnic place invention. Never, in other words, were Swisscapes part of some mysterious and uncontextualized zeitgeist, nor did they appear disconnected from the cultural logic of the community's political economy.

An important corollary is that such "texts" ripple with an inherent instability. Interpretations of landscape hinge, in part, on the perspective from which one views it, and here is precisely the point where conflict can arise. The immense success of New Glarus in attracting the attention (and tourists) that it has over the years has rested largely on the ability of community leaders to neutralize the dissension that inevitably arises in small town life. In two instances, however, tensions surfaced that could not be squelched. This is especially critical when it is the image of the place—and not its manufactured products—that are offered for sale. In the first, the Swiss Historical Village's proposed Hall of History, the conflict was resolved amiably, though not without tension. In the second, the creation of a Swiss theme park, the conflict threatened the community to its core and eventually led to the departure of its most successful businessman. Each is an outstanding example of the stress that quakes underneath even the most seemingly benign of landscapes.

As it makes it way from a manufacturing and agricultural community to one increasingly reliant on tourism, New Glarus finds itself at a crossroads. Unlike "Frontierland" at Disneyland, or Colonial Williamsburg, New Glarus is a "real" place; the high school sends its kids to the state basketball tournament, retired Swiss farmers worry about their property taxes, and everyone wonders if there will be parking downtown next weekend. The community relies on, and supports itself by, its image of being an authentic ethnic community untouched by forces of assimilation and homogenization. The question remains,

How long can that last? The reviving and inventing of vernacular traditions, David Harvey correctly notes, becomes more critical as places become differentiated in order to define cultural and symbolic capital.[80] In an age of intense consumption and of heightened competition over place distinctiveness, landscape has become more important over the years in defining cultural difference—a trend New Glarus exemplifies par excellence.

Conversely, with its rebirth and rediscovery of ethnic identity through tourism and landscape, the community has become, ironically but inevitably, less ethnic. It is more open to outsiders, and a running joke is that everyone can become Swiss for a day. Social stratification, once limited by intragroup norms, is now heightened by increasing capital control by a few local businesses—a trend set in motion by Robbie Schneider, and which persists even after his untimely departure. While most might favor tourism and would like to see efforts to attract visitors increased, an important minority feels that, as one resident put it, "tourism only benefits a few businessmen and is a pain in the neck for the residents." The result is an unavoidable ambivalence toward the tourist industry and concerns about the other-directedness of its landscape.[81]

Authenticity, Lionel Trilling noted years ago, emerges to consciousness when a doubt arises—a fact recognized by some in New Glarus today. One resident, a non-Swiss and self-described "resident alien," sees the conspicuously constructed landscapes in her adopted town as "Swiss-in-your-face" culture. "Building chalets is the best way to seem Swiss," she says, "but how genuine can it be when you have to try so hard?"[82] Such local reflexivity, while a minority view, is nonetheless important because it threatens to puncture the community's efforts to present an authentic, seamless front to the tourist. New Glarus's success thus far in attracting visitors relies on keeping dissonant voices backstage. Ethnic landscape, to a significant extent, *is* that seamless front. Ever separate from the everyday landscapes of the here and now, New Glarus's Swisscapes form the crucial attraction today in the village that has watched festival attendance level off in the years since Schneider's departure.

In the complex mixture of tourism, landscape, and ethnic memory, New Glarus's Swisscapes point to the negotiability and contingency of place-based identity. Where Rieder's self-directed chalets were accepted and copied by artisans like Don Ott as suitable icons of Swiss identity, Schneider's other-directed "Swiss Village"—considerably more "authentic" in design and materials—was rejected. At bottom, the difficulty stems not from the apparent inauthentic nature of the landscape, nor from merely its market orientation. "Keeping a heritage and making it a salable commodity are not necessarily contradictory goals," Calvin Trillin noted more then twenty years ago while on assignment in New Glarus. Rather, contrasting views of authenticity and heritage hinge on who has the authority and power to say what is really real. The *New Yorker* journalist got it exactly right: "a heritage, after all, is whatever those who treasure it say it is."[83]

Epilogue

Ethnic Place in an Age of High Modernity

After another scorching August day during the heat wave of 1995, a cool evening breeze refreshed the growing crowd near New Glarus's floral clock. The gathering of nearly two thousand on the temporary bleachers made the entrance to the village seem a little like the setting for a high school football game. With the full moon rising above the hills to the east, the first sounds of the ceremony were heard in the distance. The bells from the Swiss United Church of Christ behind the crowd brought immediate recognition, but the murmur that followed seemed otherworldly. Off in the distance, from the same direction of the church, came an increasing crackling and rumbling that resembled firecrackers set against thunder.

Only when the procession reached the makeshift ceremonial center did it become apparent that the strange sound belonged to the Glarus, Switzerland-based bell ringing group. Supported by a wooden yoke that carried two oversize cowbells, a dozen bell ringers trailed five more young men cracking whips (fig. E.1). The rich combination reverberated between the hills of the Little Sugar River Valley, creating an impressive atmosphere for the people who followed: the *Wilhelm Tell* usherettes, each carrying the flag of a different canton; aged veterans bearing the American Legion colors; the two "Ladies of Liberty," one representing Switzerland and one the United States (fig. E.2); and dignitaries from the local community and from Canton Glarus.

As part of New Glarus' week-long 150th anniversary celebration, the occasion marked the unveiling and dedication of a monument created and paid for by half the members of the audience—Swiss nationals from Canton Glarus. The canton's Governor, Christoph Stüssi, felt that the setting reminded him of the Landsgemeinde that he leads back

221

Figure E.1. Swiss bell ringers at the 150th. The Swiss bell ringers from Glarus, Switzerland, created an impressive atmosphere at the unveiling ceremony and, shown here, during the Sunday parade. (Photograph by author)

home every year, while other Swiss dignitaries spoke of the continuing relationship between "Old and New Glarus." Two local women, serving as masters of ceremonies, translated the speeches into Swiss German. Acting as a bridge between the two cultures they claim as their own, Marion Streiff and Esther Zgraggen later told of their surprise at being asked to lead such an important ceremony. "In Switzerland," Streiff, a third-generation Swiss American whose Schwyzerdütsch still recalls the origins of her family, says, "it would have been unthinkable for a woman to preside over this kind of event." Zgraggen, a recent Swiss immigrant, agrees: "Look, women only got the right to vote in recent years. No way would they let us do this back home."[1]

Such a critique enjoyed no place on the balmy August evening, for the message of the ceremony was not one of conflict and exclusion, but of commonality and mutual aid. Though the event was designed to celebrate the ties between the Switzerland and the Wisconsin community, Village President Richard Schmied was careful to acknowledge the other "area representatives of the many nationalities, Norway, Ireland, Scotland, Germany, Austria, France and those who came before us, the Native Americans" and their contribution to "our colorful local heritage and history." Echoing this sentiment, Swiss Foreign Minister

J. C. Joseph reminded the crowd two days later that multiculturalism is a hallmark of both countries and confronting diversity is nothing new. These perfunctory comments aside, the evening belonged to the Swiss on both sides of the Atlantic and their enduring relationship.

In order to solidify that relationship, the visiting Swiss presented the New Glarus Swiss with a monument aimed at perpetuating the memory of the emigration and, in the words of its sculptress, Tina Hauser, to foster a "link between Old Glarus and New Glarus." Like the pioneer monument similarly dedicated eighty years earlier, the 1995 monument served to anchor collective remembering—a process that, by its very nature, is dispersed, always changing, and ultimately intangible—in a fixed, condensed, and palpable site. But unlike the earlier site of memory, the new monument was not designed explicitly to connect with the receding pioneer past. Instead, the Swiss monument was to be a concrete marker of the enduring relationship between citizens of both countries. Such a monument would been unthinkable before the ease of trans-Atlantic crossings beginning in the 1960s, but by the 1990s much of the impetus for the New Glarus memory work came not from the village itself, but from Switzerland.

It was for this reason that the prospects of a new monument gave some people in the American community so much pause for concern.

Figure E.2. Miss Liberty and Helvetia. Set against the downtown Swisscape, two gendered representations of the United States and Switzerland sanctify the commemoration. (Photograph by author)

For one, there was concern that the statue, sent as it was from Switzerland and not created from within the community, would overshadow their own monument to the pioneer. Especially when it was discovered that the statue would be of Saint Fridolin, anxieties were heightened. As one member of the planning committee put it, "we were real worried that 'a second man' would overshadow the first." Another committee member put it more bluntly: "we just don't know what to do with a statue of Fridolin. Nobody knows who he is or was and it just seems kind of odd." Implicit in these concerns was a second issue, not felt by everyone, but by some of the more influential older residents who were involved in the Swiss United Church of Christ. Fridolin, the first planning committee member noted, "was Catholic, and the Swiss of New Glarus were followers of the Protestant reformer, Zwingli." A statue of St. Fridolin, in essence, might not speak to the committed Swiss of New Glarus, even though he appears on the cantonal flag that is flown throughout the village.[2]

Those responsible for the monument's creation in Switzerland must have been sensitive to these potential concerns, for the issue of a "second man" was resolved through the abstraction of Fridolin. Rather than depicting the bearded monk wandering through the rugged Alps, Tina Hauser created a monument of Fridolin's hiking staff, or *Wanderstab* (fig. E.3). Replete with cartooned etchings depicting Fridolin's own emigration from Ireland, the Wanderstab recalls the other modernist monument from Switzerland, the Hall of History. Both mediated tensions between conflicting views of the way to commemorate history in place. The sculpture, as the President of the Glarus, Switzerland, Art Society put it, is "young art, not dusty. The sculpture," Kaspar Marti told the audience during its unveiling at dusk, "will be a stimulating one. It will stimulate many interpretations." Many interpretations, indeed. One long time (non-Swiss) participant in *Wilhelm Tell* thinks "the statue is absolutely great. It seems Norwegian to me, don't you think? Just imagine it: a Norwegian sculpture in New Glarus! Now that is some real news!"[3]

The unveiling of Fridolin's Wanderstab was only one portion of a summer of ethnic commemoration in America's Little Switzerland. The week leading up to the unveiling brought roughly one thousand visiting Swiss, mostly from Canton Glarus, and they and their American hosts listened to more than a dozen different folk groups from Switzerland and witnessed a special performance of Schiller's *Wilhelm Tell*. And before that, in May, Wisconsin Governor Tommy Thompson, Swiss Ambassador to the U.S. Carlo Jagmetti, and Swiss Consul General Friedrich Vogel each lent an air of respectability through solemn speeches. Planning for such an ambitious event had begun three years

Figure E.3. Fridolin's Wanderstab. A gift to its daughter colony by the residents of Glarus, Switzerland, in commemoration of the emigration 150 years ago, Fridolin's Wanderstab is a concrete marker of the persisting trans-Atlantic relationship. The modern sculpture, though, is a world apart from the chalet-style entrance sign behind it. (Photograph by author)

earlier and brought together individuals across the small community's increasingly complex social structure: recent Swiss immigrants; fifth generation Swiss Americans whose ancestors arrived with the immigration of 1845; and people with no connection to Switzerland or Swiss heritage at all—bankers, lawyers, hotel keepers, shop owners, farmers, and commuting Madison professionals.

In many ways, the differences among planners of the upcoming event were greater than their commonalities, but no greater difference existed than between Glarus and New Glarus, a fact demonstrated by the differing views surrounding the Fridolin monument. The Swiss Friends of New Glarus, the organization that helped create the Hall of History and fund its permanent exhibits, in many ways pushed the sesquicentennial forward and acted as an initial impetus for local planning. In addition to presenting the Wanderstab, the Glarus-based Swiss Friends contributed ten thousand dollars to finance a new permanent exhibit at the Hall of History, published a book of essays on the community in both English and German, and funded, installed, and published an accompanying catalogue of a series of ambitious art exhibits located throughout the community. In total, more than 120 modern works of art were shown in seven different localities.[4] The art exhibit, representative of modern and avant-garde trends in Glarus, was designed explicitly to rattle the folkloric temper of the traditional Glarus-New Glarus relationship. "The contact between Glarus and New Glarus . . . has up to now taken place almost exclusively on the level of yodelers, men's choirs and Swiss wrestlers," Kaspar Marti, the exhibit's organizer, noted. He aimed to inform not only the American Swiss, but also the near "1,000 Glarners from Old Glarus . . . that Swiss culture is more than yodeling."[5]

Marti's plea for a more modern, dynamic Swiss culture did not exactly fall on deaf ears, but it was only one of several competing messages at the sesquicentennial. For the official culture of the Swiss national government, the event represented a space in which to escape contemporary dilemmas of modern, urban Swiss society. Kaspar Villiger, Switzerland's federal president, found that the New World celebration could provide "a major contribution for the recovery of our country from the fear and anxiety" present in the homeland. "In a time when Switzerland has some identity problems and many citizens almost seem ashamed about their history and heritage," the president continued, "it feels good to know that the picture of the homeland sparkles in all its glory from a distance" (fig. E.4).[6]

The President of the Swiss Friends of New Glarus, himself a cantonal judge in Basel and central organizer of the event, seconded this view. In trying to answer why so many of his countrywomen and men

Figure E.4. "The picture of the homeland sparkles from a distance." The New Glarus Dance Group seeks to maintain ethnic traditions largely vanished in Switzerland. Here they join the 150th anniversary parade. (Photograph by author)

traveled to New Glarus that summer, Hans Rhyner responded that "at a time when many of our children want nothing to do with yodeling and that sort of thing, I believe that many find it refreshing and reassuring to come here. It is as though you feel even more Swiss yourself here than at home. New Glarus," the judge continued, "is a way to get away from the self-centered and egotistical world. Here, there is a condensation of all the traditions that really no longer exist in Switzerland."[7] No longer just a regional "Swiss headquarters" serving the immigrants' need for reassurance during tremendous changes in the modernizing countryside as it was at the turn of the last century, the village had now taken on the role of international repository for the traditional Swiss culture that has all but vanished from the Old World (fig. E.5).

For many local residents, the sesquicentennial meant something entirely different, however. For Monument Committee Chair Kellie Engelke, the event provided an opportunity to participate in local civic culture. The Swiss culture that seemed so important for the European guests contained little relevance for this woman who, when asked by other New Glarners why she was "doing all this work? Are you Swiss?," replied, "'No. Do you have to be of Swiss heritage to partici-

227

Figure E.5. Memory tourism from Switzerland. Many tourists come from Switzerland to track down familial connections, and to find a repository of traditions long since vanished in modern Switzerland. This tourist is one of the hundreds who traveled from Glarus, Switzerland, to this international site of memory and tradition. (Photograph by author)

pate in your local community events? Where does it say you have to be of this village's founding heritage before you can get involved?" Such a matter of choice is contemporary ethnicity that you can become a part of the celebration, "no matter what your heritage."[8] This perspective would have been unimaginable sixty years ago during the 1935 historical pageant, much less during the turn of the century memorial occasions. During these earlier commemorations, legitimating a specific ethnic identity in the wake of widespread anti-immigration hysteria, coming to terms with a rapidly modernizing countryside, or reestablishing a connection with a fading public memory gave impetus for celebrating Swissness. No better example can be found of the historically variable nature of ethnic place than the conspicuous construction of heritage in this Old World community. From a solidifier of group allegiance to a voluntary and individualistic choice, ethnic identity persists today in Wisconsin, but in a radically symbolic form.[9]

But the ethnic commemorations, landscapes, and museums have always meant different things to different people, and the intense multivocality of such cultural displays was as evident in 1995 as in years earlier. For other people, the event remained stubbornly, and chauvinistically, ethnic. Several younger people saw the commemoration as "pretty heavy-handed" and "going overboard" with the unrestrained accent on Swiss culture. One Swiss American in his mid-twenties believed that he spoke for many residents of his age when he said that "a lot of people in my generation are sick of what they call all the 'Swiss shit.' And this weekend was the peak of that. I think it's kind of cool, but even I am a little tired of it and I'm Swiss." Another person of this generation pretty much agreed; a college graduate in her early twenties, she said that the community is more ethnically diverse than it used to be and overtures to the village's non-Swiss at the unveiling ceremony and parade were "pretty lame." She continues that "I'm Irish, not Swiss, and I felt that there was nothing for me the entire weekend. I can understand fifty years ago having an event like this, but the place just isn't that Swiss anymore. A lot has happened since the Swiss first came here and none of that is part of the 150th."[10]

A lot has happened, indeed, as more than one hundred years of conspicuously constructing identity has left behind a community, and a state, transformed. At once revelatory and obfuscatory, commemorations such as New Glarus's sesquicentennial are principal ways for local, vernacular cultures to rewrite history, to cash in on heritage, to trigger fading public memory, to solidify group consensus—in short, to invent ethnic place. Such ethnic cultural displays are intensely multivocal, giving reign to competing views even as they strive for unanimity. Dependent on mutually reinforcing legitimation, New Glarus and Wisconsin together welcome visitors to tour their Old World museums, participate in annual festive occasions, and rekindle their roots. Heritage has become big business at both the local and state levels and in the process has transformed the ethnic identity of both. Wisconsin— once vilified as the "58 percent American state"—now proudly proclaims its Old World heritage, and New Glarus—long a regional hub of ethnic commemoration with a drab, if neat landscape—opens its doors to visitors from around the world seeking tradition, stability, and a place that is "more Swiss than Switzerland."

The public memory emerging from this intersection of vernacular and official culture, I have argued throughout this book, matters. George Lipsitz notes that "what we choose to remember about the past, where we begin and end our respective accounts, and who we include and exclude from them—these do a lot to determine how we live and what decisions we make in the present."[11] Cultural leaders in New Glarus, and Wisconsin, have recognized this central fact and have chosen to tell certain stories about these places, stories that mean a great deal to a great many people and, inevitably, serve to alienate others. These stories—the root and substance of publicly constructed memory—have frequently taken on a performative character. Whether in a mock Landsgemeinde, a historical pageant depicting its history, or in annual performances of *Wilhelm Tell*, the fundamental issues facing the community have found their expression repeatedly on stage. Those issues changed with time as concerns over the pace of a modernizing countryside gave way to questions of loyalty to the nation-state which, in turn, shifted to worries about deindustrialization and authenticity. In every instance their resolution necessitated negotiation and effort by their principal players. With *Tell* in particular—an annual festival performed repeatedly for tourists—a larger political agenda melded with an individual need for ritual and shared communitas.

In other cases, Lipsitz's accounts about the past took concrete form, where local culture has come to mean ethnic culture in a most visible medium: landscape. The conspicuously Swiss landscape reflects a palimpsest of competing ideological visions of ethnic place. Where, for some contemporary Swiss nationals, the modernist Hall of History may be considered the "most authentically Swiss building in town," for local builders, a self-made chalet built from materials at hand and derived from stereotypical notions of Swiss building traditions makes for a deeply rooted vernacular material culture.

This new, global site of memory hardly exists in a vacuum, however, and just as Fred Holmes's *Old World Wisconsin* and George Hill's ethnic map legitimated the community's status during the 1940s as a place of *acceptable difference*, so too does contemporary official culture of the state reinforce the view of Wisconsin and its communities like New Glarus as Old World. Visitors to the State Historical Society of Wisconsin's flagship museum, Old World Wisconsin, are "whisked back, as by a time machine, to an enclave of the past" where immigrants from "France, Cornwall, Germany, Russia, Poland, Switzerland, Norway and Italy" all brought "their distinctive languages and religions, cultures and ways of life." Wisconsin, as seen at the open-air museum, is "a classic case study at the state level in the peopling of the nation."[12] Likewise, one of the National Trust for Historic Preservation's most successful recent Heritage Tourism Programs is Wisconsin's Ethnic Settlement Trail (WEST). Consisting of a series of nineteen separate motor and walking tours along Lake Michigan's western shore and covering more than two hundred miles, the multicultural WEST focuses on a melange of ethnic groups including Germans, Poles, Luxembourgers, Dutch, Irish, French Canadians, Belgians, and Swedes.[13] Finally, the Wisconsin sesquicentennial of 1998 featured a major collaborative effort to stage a folklife festival premised on the view of Wisconsin pioneered by Fred Holmes: the cultural geography of the state as "a mosaic of evolving but persisting ethnic cultures."[14]

Thus, where Ian McKay, in the strikingly similar context of twentieth-century Nova Scotia, found "no interplay" between the official culture of that province's elites and the vernacular culture of local inhabitants and an "outright appropriation of culture [with] a one-sided process of commodification," this study has demonstrated precisely the opposite.[15] I have argued that the politics of communication about history and place is rarely quite so straightforward. At critical junctures, the state appropriates and transforms vernacular memories into its official history. But so too does official culture acquire new readings and new meanings at the local level. Power is certainly unequal and those who possess it will be able to have their account of history accepted as the public version. In the case of New Glarus, a small ethnic community has fundamentally shaped the official version of the past and has had its provincial cosmopolitanism embraced as one of the dominant narratives of the state's public history. At a time of crisis in the wake of World War I, such a rewriting of history challenged a hegemonic view of nativist Americanism; more recently, it has accorded cultural and financial capital to the state's tourism coffers. The conspicuously constructed imagery of the past, in other words, both *reproduces* and *transforms* the political relationships at the local and state levels. Such imagery seldom imposes a single view of the past on the unsuspecting and passive "masses," but invites a dialogue between the many alternative visions accessible to audiences.[16]

Such a dialogical view of public memory and its role in shaping place is part and parcel of the age in which we live. Whether we see ours as a time of postmodernity (following David Harvey), high modernity (Anthony Giddens), or reflexive modernization (Ulrich Beck), invented ethnic places maintain a paradigmatic position in a context of globalization and the destruction of the local community.[17] Self-consciousness or reflexivity—a quality that defines places like New Glarus, Old World Wisconsin, or for that matter, Switzerland itself—pervades our thoroughly modern lives. As space shrinks to a "global village" and time speeds up so that the present is all there is, the local and the global conflate. Ethnic traditions do not entirely disappear in this "post-traditional" world and in some cases may find their standing enhanced. What Giddens notes for place in general is especially true for places poised as repositories of tradition. In the contemporary world "place becomes increasingly reshaped in terms of distant influences drawn upon in local terms. Thus local customs that continue to exist tend to develop altered meanings."[18] So it is with ethnic place.

From the mid-nineteenth century, when ethnic bonds were essential to frontier survival, to the late-twentieth century, when ethnic heritage has become the village's greatest capital asset (and therefore equally crucial for survival), New Glarus's Swissness has remained its one constant attribute. But the meaning of that ethnicity, its social and economic implications, and its symbolic representation—in short, the construction of Swissness—have changed radically over time.

It is possible to consider oneself "ethnic" outside ethnic place, but this is achieved only through considerable effort and with frequent visits to places like New Glarus.[19] Although ethnic identity has long been shaped by inventedness and constructedness—David Gerber and others have documented the American invention of ethnicity far into the nineteenth century—ethnic place stems increasingly from a heightened self-consciousness and a growing ability to commercialize on the recognizable symbols of that identity.[20] In the context of late-twentieth-century America, ethnic place has become tourist place. If New Glarus became a sort of regional Swiss commemorative headquarters in the late nineteenth century, channeling and focusing a receding ethnic identity, so has it become now an international and global guardian of Swiss heritage and values. Especially today, when Swissness is not easily found in Switzerland itself and when the country grapples with deeply rooted crises of its recent and highly contested history, people cross the Atlantic to find "genuine" Swiss culture in New Glarus.[21] The global meets the local in New Glarus, where the homeland happily sparkles from a distance.

That sparkle's luster is questioned, but not tarnished, by its very sophistication and reflexivity. Indeed, there is a tendency to regard as inauthentic anything where conscious thought intervenes. But such a premise neglects a fundamental characteristic of modernity whereby group identity is continually restructured by the information that a community receives about itself. Giddens, again, is helpful here when he writes that "the reflexivity of modern social life consists in the fact that social practices are constantly examined and reformulated in the light of incoming information about those very practices, thus constitutively altering their character."[22] In this way, a tradition such as the *Wilhelm Tell* drama is invented by a community outsider that, with time, comes to speak to the deepest levels of individual—and place—identity. The village and its inhabitants *become* ethnic precisely through the ritualized performance of a deliberately created tradition. The group, Victor Turner argues, "creates its identity by telling a story about itself." Accordingly, all socially dramatic performances contain some means of "public reflexivity" as they attempt to make social sense of schism, ambiguity, and division.[23] Recognizing such reflexively charged identities returns agency and a voice to the ordinary people who have formed the core of this book.

At the same time that a thoroughly reflexive and modern weltanschauung invented the *Wilhelm Tell* tradition, and virtually all the performative and landscape traditions that I have traced, these cultural productions were called upon to revolt against that very condition. One might argue, in fact, that ethnicity has long served as a surrogate for what has been called "antimodernism"—the "recoil from an 'overcivilized' modern existence to more intense forms of physical or spiritual experience."[24] The present, simply put, is rejected in favor of an idealized past. Arising in the late nineteenth century as a dissent from modern industrial society, antimodernism expresses skepticism about "progress" and fear that unprecedented social and economic change are destroying rootedness and tradition. In the turn of the century commemorations, however, antimodernist images often appeared as stages in the inevitable evolution of society. These pageants flourished, as did many such pageants throughout the country, "at the intersection of progressivism and antimodernism [by placing] nostalgic imagery in a dynamic, future-oriented reform context."[25] The Swiss, so the commemorations' rhetoric suggested, adhered to "timeless" traditions, while simultaneously becoming the most "progressive" and "forward looking" farmers in the state. But by the 1930s, the progressive contexts for these Gemeinschaft-, or community-laden images, faded and the cultural displays became more deeply embedded with antimodernism as a bulwark against modernity. Fred Holmes's *Old World Wisconsin* and George Hill's ethnic mapping, no less steeped in antimodernism than *Wilhelm Tell* or the Swiss Historical Village, conferred validity and authenticity to those projects.

The question remains: precisely how available and pervasive are the processes of ethnic place invention and conspicuous construction? On one level, while the forces directing the memory work among different groups may be distinct, the catalysts of the processes and their final goals arise from similar concerns. Thus, European American groups and African Americans, though experiencing vastly different places within the American socioeconomic hierarchy, share strikingly analogous approaches to commemorating the past. This is to contend that all ethnic inventions endeavor to confer a group of individuals having certain objective traits, like language and region of origin, with a "sense of common peoplehood through the inventive symbolization of elements of the past they come to see as a shared possession."[26] African Americans, no less than Finnish Americans or Native Americans, create, rework, employ festive culture to bolster esteem and unify diverse elements within the umbrella of a shared heritage.

Seen in this way, annual performances of *Wilhelm Tell* and yearly celebrations of Kwanzaa originate from remarkably similar concerns of identity and heritage. The prodigious genealogy movement that consumes members of all ethnic and racial groups had its beginning not with Polish Americans or Swedish Americans, after all, but with African Americans. Similarly, though Swiss (and any other Euro-American group) culture may be commodified, new ethnicities and older oppressed ones, like African Americans, are no less prone to place marketing. European tourists tour Harlem's jazz sites and listen to gospel choirs after visiting New Glarus and learning how to yodel.[27] The exploding interest in, and strategic use of, ethnic heritage is a defining feature of a increasingly complex and interdependent world, from which no group—no matter how priviledged or oppressed—is immune.

On a second and deeper level, social, political, and economic inequality persistently steer the use of heritage through quite different terrain. The ultimate goal of a pluralist society should be a situation that encourages for all groups the high degree of agency and choice implied in ethnic place invention. Unfortunately, and as is all too well known by blacks, Asian Americans, and other contemporary ethnic groups, such is not at all the case. While African Americans conspicuously construct identity through celebrations like Kwanzaa and by visiting such sites of memory as the Civil Rights Memorial in Birmingham, Alabama, and Muddy Waters's birth place in the Mississippi Delta, theirs is a heritage premised on an ongoing prejudice and injustice unimaginable to most whites. Indeed, an unintended consequence of the celebration of roots and ethnic place among Euro-Americans (including those in New Glarus) is an often implicit comparison between "good" ethnic places (the small town and inner city ethnic enclave) and "bad" racial places (the inner city ghetto). Through its antimodernist rhetoric, New Glarus proposes to transport visitors away from the messy, dirty, and dangerous city that, by implication, is inhabited by nonwhites. To put it differently, the social, economic, and political forces shaping memory, though they may lead to similar ends for a good many groups, carry considerable weight in the invention of ethnic place.

Even for the Swiss of New Glarus, surely a paradigmatic case of what Lawrence Fuchs calls "voluntary pluralism," larger structural forces weigh heavily on the inventiveness of ethnic identity: the anti-German hysteria and Americanization crusades of World War I eliminated the possibility of an uncontested dual identity; political rhetoric surrounding the Second World War provided a window of opportunity for a limited vision of ethnic identity that the Swiss, in turn, exploited to their advantage; and the shift from an industrial economy to one increasingly reliant on tourism further problematized the nature of ethnic identity, leading to conflict and seeming unending negotiation. Or, to paraphrase Gary Gerstle, and as this book has argued, coercion *and* liberty—social structure and individual agency—have been intrinsic to the ongoing process of shaping ethnic place.[28]

If it is the weight of class and racial oppression that tempers the memory work of marginalized groups, so is the possibility of inventing ethnic place and tradition among America's diverse cultures reason for optimism. New Glarus' 1995 sesquicentennial was premised on a vision of ethnic place in which affiliation of shared descent was more voluntary than prescribed. As such, it represented a performative display—however flawed—of what a "postethnic" America would look like, of a place balanced by an appreciation for a community of descent with the possibilities of other communities, of a place rooted but open to people with different ethnic and racial backgrounds.[29] For some, the cultural display offered the opportunity to experience communitas, while others found it another example of "Swiss shit." Either way, the choice of participation was theirs as it was for the non-Swiss. That some found it broad enough to encompass their own non-Swiss background, while for others it remained stubbornly and exclusively ethnic, attests to the event's imperfection and the incessant reality of boundaries. But it also points to a postethnic place where the community that gathers is not necessarily illusory, but a voluntary, personally constructed American creation.

Notes
Bibliography
Index

Notes

PROLOGUE: A PLACE MORE SWISS THAN SWITZERLAND

1. "And so, in an endlessly narrowing circle, I am moving ever so slowly toward the place most narrow and final, where all life stands still. I am only a shadow of my former self, and soon only my name will remain." Translation by author.

CHAPTER 1. INVENTED ETHNIC PLACES

1. In Switzerland, *Schwingfeste* are immensely popular, though that popularity varies by region and group. Paul Hugger, *Handbuch der Schweizerischen Volkskultur*, 3 vols. (Zürich: Offizin, 1992). For an American view of this "most authentic" Swiss sport, see Clare Nullis, "Schwinger Festival Shows 'Real' Face of Switzerland," *Wisconsin State Journal*, 21 August 1995. Though written a half century ago, Wayland D. Hand's excellent ethnography of a California Schwingfest provides a nice comparison to New Glarus' event: "Schweizer Schwingen: Swiss Wrestling in California," *California Folklore Quarterly* 2 (1943): 77–84.

2. Jennifer Roethe, "From Glarus to Berne, Switzerland," *New Glarus Post*, 5 August 1992; Karl Luod, "Schweizerischer als die Schweiz," in *Schweizer in Amerika: Karrieren und Misserfolge in der Neuen Welt* (Olten: Walter Verlag AG, 1979), 79–88; Marshall Cook, "Coming Home: From 'Beyond the Horizon,' Herbert Kubly has returned to New Glarus," *Wisconsin Trails*, July 1985, 26–29; and Raymond Stoiber, "Welcome from the Mayor," *New Glarus Post*, 2 September 1987.

3. Angus Gillespie, "Folk Festival and Festival Folk in Twentieth-Century America," in *Time out of Time: Essays on the Festival*, ed. Alessandro Falassi (Albuquerque: University of New Mexico Press, 1987), 152–61.

4. *Official Calendar of Events* (Madison: Wisconsin State Division of Tourism, 1993). This number represents the *least* number of such celebrations. Many more small, local events that range from an annual lutefisk dinner at a Norwegian Lutheran church to a smaller Italian neighborhood festival would count as ethnic heritage celebrations, but, due to their lack of promotion, were not included in this official list. There are, of course, considerable analytical and

political difficulties with lumping African Americans, Native Americans, and Swedish Americans into one category. In effect, by conflating race with ethnicity, we run the risk of ignoring the sharpest inequalities of treatment within America. Nevertheless, the tidal wave of heritage cuts across groups today such that no group is immune from the impulse to commemorate its distinct past. See Michael Omni and Howard Winant, *Racial Formation in the United States from the 1960s to the 1990s*, 2nd ed. (New York: Routledge, 1994), 14–24, 70.

5. Steven Hoelscher and Robert C. Ostergren, "Old European Homelands in the Middle West," *Journal of Cultural Geography* 13 (1993): 87–106. I am indebted to Michael Conzen for the metaphor of the ethnic archipelago; Michael P. Conzen, "Culture Regions, Homelands, and Ethnic Archipelagos in the United States: Methodological Considerations," *Journal of Cultural Geography* 13 (1993): 13–29.

6. Frederick Jackson Turner, "The Significance of the Frontier in American Life," in *The Turner Thesis: Concerning the Role of the Frontier in American History*, ed. George Rogers Taylor (Lexington, MA.: D. C. Heath, 1972 [1893]), 17. The literature on questions of ethnic assimilation in American culture is, of course, voluminous. One excellent starting point is the many articles by Philip Gleason, recently collected in *Speaking of Diversity: Language and Ethnicity in Twentieth-Century America* (Baltimore: Johns Hopkins University Press, 1992). Two recent and useful evaluations of the assimilation model, to which Turner was only one contributor, include: Russell A. Kazal, "Revisiting Assimilation: The Rise, Fall, and Reappraisal of a Concept in American Ethnic History," *American Historical Review* 100 (1995): 437–71; and Ewa Morawaka, "In Defense of the Assimilation Model," *Journal of American Ethnic History* 14 (1994): 76–87. For an especially relevant survey of Turner's impact on the study of rural ethnicity, see Kathleen Neils Conzen, "Historical Approaches to the Study of Rural Ethnic Communities," in *Ethnicity on the Great Plains*, ed. Frederick C. Luebke (Lincoln: University of Nebraska Press, 1980), 1–18.

7. Richard D. Alba, "The Twilight of Ethnicity among Americans of European Ancestry: The Case of Italians," *Ethnic and Racial Studies* 8 (1985): 134–58.

8. Stephen Frenkel, "Alluring Landscapes: The Symbolic Economy of 'Bavarian' Leavenworth," Paper presented at the annual meetings of the Association of American Geographers, Ft. Worth, Texas, 4 April 1997; Mira Engler, "Drive-Thru History: Theme Towns in Iowa," *Landscape* 32 (1993): 8–18; and Majorie R. Esman, "Tourism as Ethnic Preservation: The Cajuns of Louisiana," *Annals of Tourism Research* 11 (1984): 451–67. I use the term "traditionalization" explicitly to call attention to tradition as a process that involves continual work. See Dell Hymes, "Folklore's Nature and the Sun's Myth," *Journal of American Folklore* 88 (1975): 353–55.

9. Michael P. Conzen, "Ethnicity on the Land," in *The Making of the American Landscape*, ed. Michael P. Conzen (Boston: Unwin Hyman, 1990), 245. A similar observation is made by Wilbur Zelinsky: "America in Flux," in *The Cultural Geography of the United States*, rev. ed. (Englewood Cliffs, NJ: Prentice Hall, 1992), 143–85.

10. David Harvey, *The Condition of Postmodernity* (Oxford: Basil Blackwell, 1989), 292–93.

11. Herbert J. Gans, "Symbolic Ethnicity: The Future of Ethnic Groups and Cultures in America," *Ethnic and Racial Studies* 2 (1979): 1–20.

12. Mary C. Waters, *Ethnic Options: Choosing Identities in America* (Berkeley: University of California Press, 1990), 147. See also Richard D. Alba, *Ethnic Identity: The Transformation of White America* (New Haven: Yale University Press, 1990).

13. The work of David Lowenthal is most germane to the discussion of heritage's global reach: "Identity, Heritage, and History," in *Commemorations: The Politics of National Identity,* ed. John Gillis (Princeton: Princeton University Press, 1994), 41–60; *Possessed by the Past: The Heritage Crusade and the Spoils of History* (New York: The Free Press, 1996); and *The Past is a Foreign Country* (Cambridge: Cambridge University Press, 1985). Also insightful is Robert Hewison, *The Heritage Industry: Britain in an Age of Decline* (London: Methuen, 1987).

14. Michael Hout and Joshua Goldstein, "How 4.5 Million Irish Immigrants Became 40 Million Irish Americans: Demographic and Subjective Aspects of the Ethnic Composition of White Americans," *American Sociological Review* 59 (1994): 64–82. The ancestry question that originally appeared in the 1980 U.S. Census for the first time allows researchers to ask about ethnic identification. See also: Reynolds Farley, "The New Census Question about Ancestry: What Did It Tell Us?" *Demography* 28 (1991): 411–29; Ira Rosenwaike, "Ancestry in the United States Census, 1980–1990," *Social Science Research* 22 (1993): 383–90; Stanley Lieberson and Mary C. Waters, "The Ethnic Responses of Whites: What Causes Their Instability, Simplification, and Inconsistency?" *Social Forces* 72 (1993): 421–50; and Stanley Lieberson and Mary C. Waters, *From Many Strands: Ethnic and Racial Groups in Contemporary America* (New York: Russell Sage Foundation, 1988).

15. John Ibson, "Virgin Land or Virgin Mary? Studying the Ethnicity of White Americans," *American Quarterly* 33 (1981): 284–308; and Kathleen Neils Conzen, "Mainstreams and Side Channels: The Localization of Immigrant Cultures," *Journal of American Ethnic History* 11 (1991): 5–20.

16. On the rising assertion of local autonomy and identity, see Yi-Fu Tuan, *Cosmos and Hearth: A Cosmopolite's Viewpoint* (Minneapolis: University of Minnesota Press, 1996). Paradoxically, both geographic trends—an emphasis on nationality and globalization or on nostalgia and the local—are tied dialectically to the history of modernity. See Anthony Giddens, *The Consequences of Modernity* (Stanford: Stanford University Press, 1990); Robert David Sack, *Place, Modernity, and the Consumer's World: A Relational Framework for Geographical Analysis* (Baltimore: Johns Hopkins University Press, 1992), 5–11; and Benjamin Barber, *Jihad vs. McWorld* (New York: Ballantine Books, 1996).

17. David Hollinger, *Postethnic America: Beyond Multiculturalism* (New York: Basic Books, 1995).

18. Timothy J. Meagher, "Why Should We Care for a Little Tro or Walk in the Mud? St. Patrick, Columbus Day Parades in Worcester, Massachusetts, 1845–1915," *New England Quarterly* 58 (1985): 5–10. Mount Rushmore, of course, is itself a profoundly contested piece of real estate that dates only to the 1930s and reflects the changing role of government in commemorating

the past. See Michael Kammen, *Mystic Chords of Memory: The Transformation of Tradition in American Culture* (New York: Knopf, 1991), 448–55.

19. For an excellent critique of this tradition within cultural geography, see Deryck Holdsworth, "Revaluing the House," in *Place/Culture/Representation*, ed. James Duncan and David Ley (London: Routledge, 1993), 95–109. The "fakelore/folklore" dichotomy is deconstructed in Barbara Kirshenblatt-Gimblett, "Mistaken Dichotomies," *Journal of American Folklore* 101 (1988): 140–55. Also useful is Richard Bauman, "Folklore," in *Folklore, Cultural Performances, and Popular Entertainments*, ed. Richard Bauman (New York: Oxford University Press, 1992), 29–40; and Regina Bendix, *In Search of Authenticity: The Formation of Folklore Studies* (Madison: University of Wisconsin Press, 1997).

20. Umberto Eco, *Travels in Hyperreality* (New York: Harcourt, Brace, Jovanich, 1983), 8. Indeed, Hildegard Johnson, for one, speaks of New Glarus's "attempt to emulate a Swiss village," and Michael Conzen adds that the community "groans" under the weight of its sheer weightlessness. Johnson, *Order Upon the Land: The U.S. Rectangular Land Survey and the Upper Mississippi Country* (New York: Oxford University Press, 1976), 185; and Conzen, "Ethnicity on the Land," 245–47. See also Frank McGinn, "Ersatz Place," *Canadian Heritage*, February–March 1986, 25–29. A number of commentators recently have questioned this rather stark interpretation of the "heritage industry." In her ethnographic description of heritage tourism in Ireland, Nuala Johnson finds that "critiques of the heritage industry's representations of the past as little more than bogus history are too often overdrawn, not to mention overwrought." While such widespread criticisms of heritage and public history serve to highlight the importance of heritage sites, she correctly notes that they also contribute little in terms of sustained analysis. Nuala C. Johnson, "Where Geography and History Meet: Heritage Tourism and the Big House in Ireland," *Annals of the Association of American Geographers* 86 (1996): 546.

21. "The deliberate 'Swissification' of historic Swiss associated properties, and the distorted recreation of so-called 'Swiss' forms in an attempt to cater to tourism, threatens to destroy the integrity of a number of Swiss resources and obliterate authentic Swiss ethnic building traditions." Randall Wallar, *Settlement*, ed. Barbara Wyatt, 3 vols., vol. 1, *Cultural Resources Management in Wisconsin* (Madison: State Historical Society of Wisconsin, 1986), sections 1–4.

22. Werner Sollors, ed., *The Invention of Ethnicity* (New York: Oxford University Press, 1989). For an earlier treatment, see Werner Sollors, *Beyond Ethnicity: Consent and Descent in American Culture* (New York: Oxford University Press, 1986) as well as the following: Kathleen Neils Conzen, "German-Americans and the Invention of Ethnicity," in *America and the Germans: An Assessment of a Three-Hundred Year History*, ed. Frank Tommler and Joseph McVeigh, vol. 1 (Philadelphia: University of Pennsylvania Press, 1985), 131–47; and William L. Yancey, Eugene P. Erickson, and Richard N. Juliani, "Emergent Ethnicity: A Review and Reformulation," *American Sociological Review* 41 (1976): 391–402. In her recent treatment of the 1925 Norwegian American Immigration Centennial, April Schultz makes use of a similar model: *Ethnicity on Parade: Inventing the Norwegian American through Celebration* (Amherst: Univer-

sity of Massachusetts Press, 1994). A useful synthesis with a series of supporting case studies is Kathleen Neils Conzen et al., "The Invention of Ethnicity: A Perspective from the U.S.A.," *Journal of American Ethnic History* 12 (1992): 3–41.

23. Conzen, et al., "The Invention of Ethnicity," 4. For two relevant treatments of the reflexivity of modern life as it impinges on identity, see George Schöpflin, "Nationalism and Ethnic Minorities in Post-Communist Europe," in *Europe's New Nationalism: States and Minorities in Conflict,* ed. Richard Caplan and John Feffer (New York: Oxford University Press, 1996), 157; and Douglas Kellner, "Popular Culture and the Construction of Postmodern Identities," in *Modernity and Identity,* ed. Scott Lash and Jonathan Friedman (Oxford: Basil Blackwell, 1992), 153–54.

24. Michael Fischer, "Ethnicity and the Post-Modern Arts of Memory," in *Writing Culture: The Poetics and Politics of Ethnography,* ed. James Clifford and George Marcus (Berkeley: University of California Press, 1986), 20.

25. David Roediger, *The Wages of Whiteness: Race and the Making of the American Working Class* (London: Verso, 1991). Mary C. Waters is uncompromising in her view that choice is limited, a view seconded with somewhat less vigor by David Hollinger. My own view is closer to Waters, but hold out for Hollinger's vision of a "postethnic America." Waters, *Ethnic Options*; Hollinger, *Postethnic America.*

26. Kammen, *Mystic Chords of Memory,* 7. Eric Hobsbawm and Terence Ranger, eds., *The Invention of Tradition* (Cambridge: Cambridge University Press, 1983); cf. Richard Handler and Jocelyn Linnekin, "Tradition, Genuine or Spurious," *Journal of American Folklore* 97 (1984): 273–90.

27. As a critique of Hobsbawm's influential work, Giddens writes that "'invented tradition,' which at first sight seems almost a contradiction in terms, and is intended to be provocative, turns out on scrutiny to be something of a tautology. For *all* traditions, one could say, are invented traditions." Anthony Giddens, "Living in a Post-Traditional Society," in *Reflexive Modernization: Politics, Tradition and Aesthetics in the Modern Social Order,* ed. Ulrich Beck, Anthony Giddens, and Scott Lash (Stanford: Stanford University Press, 1994), 93. Emphasis in original.

28. Raymond Williams, *Marxism and Literature* (New York: Oxford University Press, 1977), 115–16. Emphasis in original. For a substantial elaboration on this perspective, see David Lowenthal, *The Past is a Foreign County.*

29. Giddens, "Living in a Post-Traditional Society," 100–104.

30. John Bodnar, *Remaking America: Public Memory, Commemoration, and Patriotism in the Twentieth Century* (Princeton: Princeton University Press, 1992). See also Maurice Halbwachs, *The Social Frameworks of Memory* (Chicago: University of Chicago Press, 1992); and Kammen, *Mystic Chords of Memory,* 3–14.

31. Bodnar, *Remaking America,* 13–14.

32. Sack, *Place, Modernity, and the Consumer's World,* 3, 166–67. Commodification has come to be seen as a chief characteristic of the (post)modern world in which we live. Accordingly, the literature surrounding such a significant process is almost as overwhelming as the phenomenon itself. The following provide useful overviews: John Urry, "The Consumption of Tourism" in *Con-*

suming Places (London: Routledge, 1990), 129–40; Jon May, "In Search of Authenticity off and *on* the Beaten Track," *Environment and Planning D: Society and Space* 14 (1996): 709–36; Steven Best, "The Commodification of Reality and the Reality of Commodification: Jean Baudrillard and Post-Modernism," *Current Perspectives in Social Theory* 9 (1989): 23–51; Erik Cohen, "Authenticity and Commoditization in Tourism," *Annals of Tourism Research* 15 (1988): 371–86; John B. Stephenson, "Escape to the Periphery: Commodifying Place in Rural Appalachia," *Appalachian Journal* 11 (1984): 187–200. Two complementary works that illuminate the experience of being in place are J. Nicholas Entrikin, *The Betweenness of Place: Towards a Geography of Modernity* (Baltimore: Johns Hopkins University Press, 1991); and Yi-Fu Tuan, *Space and Place: The Perspective of Experience* (Minneapolis: University of Minnesota Press, 1977).

33. Kammen, *Mystic Chords of Memory,* 13, 402–3; Zelinsky, "America in Flux," 176. The two other factors that Zelinsky delineates fit well with this book's central thesis: a self-consciousness; and a sharing and border crossing of identities.

34. Dean MacCannell, *The Tourist: A New Theory of the Leisure Class,* 2nd ed. (New York: Schocken Books, 1989).

35. By ethnic tourism I mean the sort of tourism where the principal allure is the "cultural exoticism of the local population and its artifacts," including its landscape, clothing, theater, music, and dance. Pierre van den Berghe and Charles Keyes, "Tourism and Re-Created Ethnicity," *Annals of Tourism Research* 11 (1984): 343–52. Also pertinent to this discussion are the following: Pierre van den Berghe, *The Quest for the Other: Ethnic Tourism in San Cristobal, Mexico* (Seattle: University of Washington Press, 1994); Pierre van den Berghe, "Tourism and the Ethnic Division of Labor," *Annals of Tourism Research* 19 (1992): 234–49; John Urry, *The Tourist Gaze: Leisure and Travel in Contemporary Societies* (London: Sage, 1990); and Dean MacCannell, "Reconstructed Ethnicity: Tourism and Cultural Identity in Third World Communities," in *Empty Meeting Grounds: The Tourist Papers* (London: Routledge, 1992), 158–71. David Glassberg, "Public History and the Study of Memory," *Public Historian* 18 (1996): 19.

36. Jonathan Culler, "Semiotics of Tourism," *American Journal of Semiotics* 1 (1981): 127–40.

37. Regina Bendix, "Tourism and Cultural Displays: Inventing Traditions for Whom?" *Journal of American Folklore* 102 (1989): 131–46.

38. Roger D. Abrahams, "An American Vocabulary of Celebrations," in Falassi, ed., *Time Out of Time,* 173–83. Emphasis in original. See also "Shouting Match at the Border: The Folklore of Display Events," in *"And Other Neighborly Names": Social Process and Culture Image in Texas Folklore,* ed. Richard Bauman and Roger D. Abrahams (Austin: University of Texas Press, 1981), 303–22; Alessandro Falassi, "Festival: Definition and Morphology," in Falassi, ed., *Time out of Time,* 1–10; and Frank Manning, ed., *The Celebration of Society: Perspectives on Contemporary Cultural Performance* (Bowling Green, OH: Bowling Green University Popular Press, 1983).

39. Timothy Oakes, "Place and the Paradox of Modernity," *Annals of the*

Association of American Geographers 87 (1997): 510. The classic work remains Yi-Fu Tuan, *Space and Place.*

40. Don Handelman, *Models and Mirrors: Towards an Anthropology of Public Events* (New York: Cambridge University Press, 1990), 9.

41. My use of theater as spatial and experiential metaphor is deliberate. The staging of heritage points not only to the many cultural displays that have played such a critical role in the invention of ethnicity, but also to their highly visible and self-conscious construction. Herein, landscape joins with bodily performance as a shaper of identity. See Denis Cosgrove, "Spectacle and Society: Landscape as Theater in Premodern and Postmodern Cities," in *Understanding Ordinary Landscapes,* ed. Paul Groth and Todd W. Bressi (New Haven: Yale University Press, 1997), 99–110. Paul Connerton, *How Societies Remember* (Cambridge: Cambridge University Press, 1989); and Deborah A. Kapchan, "Performance," *Journal of American Folklore* 108 (1995): 479.

42. Kathleen Neils Conzen, "Ethnicity as Festive Culture: Nineteenth-Century German America on Parade," in Sollors, ed., *The Invention of Ethnicity,* 44–76.

43. Victor Turner, *The Ritual Process: Structure and Anti-Structure* (Ithaca: Cornell University Press, 1969), 42–43, 93–97. Similar conclusions were reached by April Schultz in her study of Norwegians in Minneapolis. She sees the Norse-American celebrations of 1925 not as the swan song of the inevitable progression from a static Norwegian culture to a full acceptance of Americanization, but rather as part of an ongoing negotiation between host and ethnic society; Schultz, *Ethnicity on Parade.* See also Rudolph Vecoli, "Primo Maggio in the United States: An Invented Tradition of the Italian Anarchists," in *May Day Celebrations,* ed. Andrea Pannacione (Venice: Marsilio Editori, 1988), 55–88.

44. Bodnar, *Remaking America,* 13–20, 41–77. The most poignant example that Bodnar offers is the multivocal reading of the Vietnam Veterans Memorial in Washington D.C. The construction of this monument pitted "official culture" in the form of patriotism and legitimation of an unpopular war against "vernacular culture" in the form of tangible memories of ordinary people and their personal suffering.

45. J. D. B., "An Unaltered Swiss Colony," *Nation,* 2 August 1883, 93–94.

46. Frederick Jackson Turner, "Dominant Forces in American Life," in *The Frontier in American History,* 2nd ed. (Tucson: University of Arizona Press, 1992 [1897]), 235.

47. J. D. B., "An Unaltered Swiss Colony," 93. Useful period sources on these earliest years may be found in the following: Conrad Zimmerman, "Town of New Glarus," in *History of Green County, Wisconsin* (Springfield, IL: Union Publishing, 1884), 1023–45; J. J. Tschudy, "Additional Notes," *Wisconsin Historical Collections* 8 (1879): 30–35; Wilhelm Streissguth, "New Glarus in 1850," *Wisconsin Magazine of History* 16 (1935 [1851]): 328–44; John Luchsinger, "The Planting of the Swiss Colony at New Glarus, Wis.," *Wisconsin Historical Collections* 12 (1892): 335–82; John Luchsinger, "The Swiss Colony of New Glarus," *Wisconsin Historical Collections* 8 (1879): 1–29; D. Dürst, *Gründung und Entwicklung der Kolonie Neu-Glarus, Wisconsin, Nord-America, umfassend den Zei-*

traum von 1844–1892, nebst einer Reisebeschreibung (Zürich: Institut Drell Fuessli, 1894); and Helen Brigham, "New Glarus," in *History of Green County, Wisconsin* (Milwaukee: Burdick and Armitage, 1877), 247–57. For an outstanding secondary source see Dieter Brunnschweiler, *New Glarus: Gründung, Entwicklung und heutiger Zustand einer Schweizerkolonie im Amerikanischen Mittlewesten* (Zürich: Buchdruckerei Flutern, 1954); see also Leo Schelbert, ed., *New Glarus 1845–1970: The Making of a Swiss American Town* (Glarus: Kommissionsverlag Tschudi, 1970); Ernest Menolfi and Leo Schelbert, "The Wisconsin Swiss: A Portrait," *Clarion* 16 (1991): 57–63; and Frederick Hale, *The Swiss in Wisconsin* (Madison: State Historical Society of Wisconsin, 1984). Local historians have also played a significant role. See the following: Elda Schiesser and Linda Schiesser, *The Swiss Endure Year by Year: A Chronological History* (New Glarus, WI: n.p., 1994); Millard Tschudy, *New Glarus, Wisconsin: Mirror of Switzerland,* 2nd ed. (New Glarus, WI: n.p., 1995).

48. For useful comparisons of other contemporary nineteenth-century European transplantations that relied heavily on chain migration, see Walter D. Kamphoefner, *The Westfalians: From Germany to Missouri* (Princeton: Princeton University Press, 1987); Jon Gjerde, *From Peasants to Farmers: The Migration from Balestrand, Norway to the Upper Middle West* (Cambridge: Cambridge University Press, 1985); Robert C. Ostergren, *A Community Transplanted: The Trans-Atlantic Experience of a Swedish Immigrant Settlement in the Upper Middle West, 1835–1915* (Madison: University of Wisconsin Press, 1988); and Anne Kelly Knowles, *Calvinists Incorporated: Welsh Immigrants on Ohio's Industrial Frontier* (Chicago: University of Chicago Press, 1997).

49. Guy Serge Metraux, "Social and Cultural Aspects of Swiss Immigration into the United States in the Nineteenth Century," (Ph.D. diss., Yale University, 1949).

50. Jakob Winteler, *Geschichte des Landes Glarus: von 1638 bis zur Gegenwart,* band 2, (Glarus, Switzerland: Kommissionsverlag E. Baeschlin, 1954), 461–62; and Dominik Wunderlin, *Colorful Clothes, Natural Ice, Sap Sago, and Blue Gold: Exhibit for the 1995 New Glarus Sesquicentennial,* trans. Catherine De Capitani-Dolf (Basel: Swiss Museum of Folk Arts and Traditions, 1995).

51. Such an encounter with capitalism on the European side of the Atlantic is the leitmotiv of John Bodnar's influential synthesis of these migrations, *The Transplanted: A History of Immigrants in Urban America* (Bloomington: Indiana University Press, 1985). For a detailed background on the emigration itself and the crucial role played by the textile industry, see Otto Bartel and Adolf Jenny, *Glarner Geschichte in Daten* (Glarus, Switzerland: Buchdruckerei Neue Glarner Zeitung, 1936), 361–489; Jürg Davatz, "Die Glarner Textilindustrie," in *Glarus und die Schweiz: Streiflichter auf wechselseitige Beziehungen,* ed. Jürg Davatz (Glarus, Switzerland: Verlag Baeschlin, 1991), 128–36; Heinrich Stüssi, "Auswanderung," in Davatz, ed., *Glarus und die Schweiz,* 146–54; and Winteler, *Geschichte des Landes* Glarus, 461–64. Also useful is Brunnschweiler, *New Glarus,* 10–23. The significant source on the migration itself is from the diary of one of those immigrants, Matthias Dürst. See Dürst, "Matthias Dürst's Travel Diary," reprinted and translated into the English in Leo Schelbert, ed., *New Glarus*

1845–1970: The Making of a Swiss American Town (Glarus, Switzerland: Kommissionverlag Tschudi, 1970), 20–150.

52. Tschudy, *Mirror of Switzerland*, 26; and Brunnschweiler, *New Glarus*, 43–70. The best treatment of the Green County cheese industry remains Glenn T. Trewartha, "The Green County, Wisconsin, Foreign Cheese Industry," *Economic Geography* 2 (1926): 292–308.

CHAPTER 2. "THERE WAS A CONFUSION OF THE FOREIGN AND AMERICAN": PUBLIC MEMORY AT THE TURN OF THE TWENTIETH CENTURY

1. "Lands and Peoples: Swiss at New-Glarus," *Our Times: A Weekly Journal of Current Events*, 21 October 1905, 123–24.

2. Don Handelman, *Models and Mirrors: Towards an Anthropology of Public Events* (New York: Cambridge University Press, 1990), 15–16.

3. Victor Turner, "Introduction," in *Celebrations: Studies in Festivity and Ritual*, ed. Victor Turner (Washington, D.C.: Smithsonian Institution Press, 1982), 18.

4. J. J. Tschudy, "Additional Notes," *Wisconsin Historical Collections* 8 (1879): 30–35.

5. Quoted in Kathleen Neils Conzen, "Ethnicity as Festive Culture: Nineteenth-Century German America on Parade," in *The Invention of Ethnicity*, ed. Werner Sollors (New York: Oxford University Press, 1989), 51–52. Emphasis added.

6. Eric Hobsbawm, "Mass-Producing Tradition: Europe, 1870–1914," in *The Invention of Tradition*, ed. Eric Hobsbawm and Terence Ranger (New York: Cambridge University Press, 1983), 263–308. On the general indifference toward, and lack of consensus of, the Fourth of July throughout much of the nineteenth century, see Wilbur Zelinsky, *Nation into State: The Shifting Symbolic Foundations of American Nationalism* (Chapel Hill: University of North Carolina Press, 1988), 72–73; Michael Kammen, *Mystic Chords of Memory: The Transformation of Tradition in American Culture* (New York: Knopf, 1991), 49; and Susan G. Davis, *Parades and Power: Street Theatre in Nineteenth-Century Philadelphia* (Berkeley: University of California Press, 1988), 166–73.

7. Elizabeth Moore Wallace and Lillian Wallace Maynard, *This Side of the Gully*, 1926, typeset manuscript, subject files, New Glarus Historical Society Archives (hereafter cited as NGHSA).

8. Tschudy, "Additional Notes," 34.

9. Davis, *Parades and Power*, 40–45; and Roy Rosenzweig, *Eight Hours for What We Will: Workers and Leisure in an Industrial City, 1870–1920* (New York: Cambridge University Press, 1983), 65–92.

10. Phyl Anderson, "Confirmation Classes Roll Call to be Taken," *New Glarus Post*, 22 September 1965; and Millard Tschudy interview with author, New Glarus, WI, September 1994.

11. For general, secondary accounts of Kilbi in New Glarus, see Millard Tschudy, *New Glarus, Wisconsin: Mirror of Switzerland* (Monroe, WI: Monroe

Evening Times, 1965), 27; Miriam Theiler, *New Glarus' First 100 Years* (Madison: Campus Publishing Co, 1946), 106–7; and *The First One Hundred Years of the Swiss Evangelical and Reformed Church* (New Glarus, WI: n.p., 1950). The date of the 1920s as marking the demise of Kilbi as a major community festival is from Dieter Brunnschweiler, *New Glarus, Wisconsin: Gründung, Entwicklung und heutiger Zustand einer Schweizerkolonie im Amerikanishen Mittlewesten* (Zürich: Fluntern, 1954), 99. On the secular/sacred blend so common in festivals, see Alessandro Falassi, "Festival: Definition and Morphology," in *Time out of Time: Essays on the Festival*, ed. Alessandro Falassi (Albuquerque: University of New Mexico Press, 1987), 1–10.

12. *Swiss Church Constitution*, 31 January 1859, laws and church ordinances of the congregation in New Glarus, 1859, subject files, NGHSA. Translated by the author. Hereafter, unless otherwise noted, all translations from the German are my own.

13. Roger D. Abrahams, "An American Vocabulary of Celebrations," in Falassi, ed., *Time out of Time*, 173–83.

14. In his classic and comprehensive early twentieth-century study of Switzerland, Robert Clarkson Brooks suggests that these three cultural productions developed such an intense following in the nineteenth century to be properly called "cults." See Brooks, *Civic Training in Switzerland: A Study of Democratic Life* (Chicago: University of Chicago Press, 1930), 364–77. An excellent and more recent statement of the importance of these activities to Swiss nation building is Regina Bendix, "National Sentiment in the Enactment and Discourse of Swiss Political Ritual," *American Ethnologist* 14 (1992): 768–90.

15. Luchsinger notes that "Kilbi is a blending of all holidays into one." John Luchsinger, "The Swiss Colony of New Glarus," *Wisconsin Historical Collections* 8 (1879): 21. See also Brunnschweiler, *New Glarus*, 99.

16. Falassi, "Festival: Definition and Morphology," 5–6.

17. Luchsinger, "The Swiss Colony," 21.

18. Beverly J. Stoeltje, "Festival in America," in *Handbook of American Folklore*, ed. Richard M. Dorson (Bloomington: Indiana University Press, 1983), 239–46.

19. Daniel Dürst noted in 1894 that, because of the cost of wine, beer was the beverage of choice for celebrating New Glarners; D. Dürst, *Gründung und Entwicklung der Kolonie Neu-Glarus, umfassend den Zeitraum von 1844–1892, nebst einer Reisebeschreibung* (Zürich: Institut Drell Fuessli, 1894), 28–29.

20. Wardon Allan Curtis, "'The Light Fantastic in the Central West': Country Dances of Many Nationalities in Wisconsin," *Century Magazine*, 1907, 570–79.

21. Dürst, *Gründung und Entwicklung*, 29.

22. Curtis, "'The Light Fantastic,'" 577–79.

23. *Memoirs of Green County: Town and Village of New Glarus*, 315, n.p., n.d. Documents Boxes, NGHSA.

24. For one of the many sets of competing Kilbi advertisements, see *Amerikanischer Schweizer Nachrichten und Green County Herold*, 15 September 1892.

25. Curtis, "'The Light Fantastic,'" 577.

26. Ibid. Luchsinger, writing thirty years earlier, confirms this basic description; Luchsinger, "The Swiss Colony," 22.

27. Curtis, "'The Light Fantastic,'" 572. John Luchsinger notes that, al-

though crime was rare in New Glarus, conflict most often occurred with "neighbors . . . or the social rivalry among young men" [i.e. non-Swiss] and "too free indulgence in liquor"; Luchsinger, "The Planting of the Swiss Colony at New Glarus, Wis.," *Wisconsin Historical Collections* 12 (1892): 379.

28. Curtis, "'The Light Fantastic,'" 572, 577.

29. Werner Enninger, "Clothing," in *Folklore, Cultural Performances, and Popular Entertainments: A Communications-Centered Handbook,* ed. Richard Bauman (New York: Oxford University Press, 1992), 217–24.

30. Turner, "Introduction," 29. See also Victor Turner, *The Ritual Process: Structure and Anti-Structure* (Ithaca: Cornell University Press, 1969).

31. Beverly J. Stoeltje, "Festival," in Bauman, ed. *Folklore,* 261–71. For a good overview of commemorative activity across a wide spectrum of cultures and periods, see the essays in John Gillis, ed., *Commemorations: The Politics of National Identity* (Princeton: Princeton University Press, 1994).

32. "New Glarus' Great Day," *Monroe Sun,* 5 September 1891. For the relative concentrations of Swiss in different regions of Green County, see Brunnschweiler, *New Glarus,* 89–93.

33. *Amerikanischer Schweizer Nachrichten und Green County Herold,* 3 September 1891; and *Monroe Sun,* 5 September 1891.

34. "The Swiss Celebration," *Monroe Sentinel,* 9 September 1891.

35. Brunnschweiler, *New Glarus,* 87.

36. "Grand Celebration Commemorating the 600th Anniversary of the Republic and Independence of Switzerland," *Monroe Sentinel,* 26 August 1891.

37. *Monroe Sentinel,* 9 September 1891; and Dürst, *Gründung und Entwicklung,* 21: "Diese Straßen darf man aber nicht mit denjenigen unseres lieben Schweizerlandes vergleichen." Dürst makes the important point that only the church stands out as singularly unique in the community to which I will return later in the chapter.

38. "New Glarus' Great Day," *Monroe Sun,* 5 September 1891.

39. *Monroe Sentinel,* 9 September 1891; *Monroe Sun,* 5 September 1891.

40. *Programm für die 600 Jähre Gedenkfeier den Gründung der Eidgenossenschaft abgehalten in New Glarus, Wis., am 3 September, 1891* (Monroe: Green County Herold, 1891), documents boxes, NGHSA: "Heimath, Heimath, über Alles/ Über Alles in der Welt/Du allein bist uns're Welt/Einigkeit und Recht und Freiheit Für das Schweizer Vaterland." Benedict Anderson, *Imagined Communities: Reflections on the Origin and Spread of Nationalism* (London: Verso, 1983).

41. *Programm für die 600 Jähre Gedenkfeier.* This second song, in particular, was quite popular at the time in Switzerland as a nationalist call to arms for the protection of its borders in the wake of the Franco-Prussian War of 1870–1871. Georg Kreis, *Helvetia: im Wandel der Zeiten* (Zürich: Verlag der Neue Zürcher Zeitung, 1991), 36, 99. For a useful treatment of romantic nationalism and its use of folklore, see William A. Wilson, "Herder, Folklore, and Romantic Nationalism," in *Folk Groups and Folklore Genres: A Reader,* ed. Elliot Oring (Logan, UT: Utah State University Press, 1989), 23–44.

42. *Programm für die 600 Jähre Gedenkfeier; Monroe Sentinel,* 9 September 1891; and *Monroe Sun,* 5 September 1891.

43. For a discussion of the tableaux's popularity in the United States, see

David Glassberg's outstanding work, *American Historical Pageantry: The Uses of Tradition in the Early Twentieth Century* (Chapel Hill: University of North Carolina Press, 1990), 16–20.

44. Josef Troxler, et al. *Wilhelm Tell* (Chapelle-sur-Moudon: Verlag Ketty and Alexandre, 1985), 39. In chapter 4, I will elaborate the foundations of the Tell myth and history in more detail.

45. This section is based on the official program for the event, *Programm für die 600 Jähre Gedeukfeier,* and on the newspaper accounts cited above.

46. On the importance of Columbia in American pageantry and symbolism, see Michael Kammen, *Spheres of Liberty: Changing Perceptions of Liberty in American Culture* (Madison: University of Wisconsin Press, 1986); and Zelinsky, *Nation into State,* 22–24. For other allegorical representations that focused on women, see Kreis, *Helvetia;* Maurice Agulhon, *Marianne into Battle: Republican Imagery and Symbolism in France, 1789–1880* (New York: Cambridge University Press, 1981); and Lothar Gall, *Germania: Eine deutsche Marianne?* (Bonn: Bouvier Verlag, 1993). Significantly, these European representations carried power across the Atlantic. For the importance, and contestation, of the figure Germania for Germans in Milwaukee, see Steven Hoelscher et al., "Milwaukee's German Renaissance Twice-Told: Inventing and Recycling Landscape in America's German Athens," in *Wisconsin Land and Life,* ed. Robert C. Ostergren and Thomas Vale (Madison: University of Wisconsin Press, 1997), 376–409.

47. For a similar reading of folklore productions in a slightly different context, see Ian McKay, *The Quest of the Folk: Antimodernism and Cultural Selection in Twentieth-Century Nova Scotia* (Montreal and Kingston: McGill-Queens, 1994).

48. Quotes from *Programm für die 600 Jähre Gedenkfeier* and *Monroe Sentinel,* 9 September 1891. On the use of Tell tableaux in nineteenth-century Swiss commemorations, see Regina Bendix, *Backstage Domains: Playing "William Tell" in Two Swiss Communities* (Bern: Peter Lang, 1989), 36–37.

49. Textbook quoted in Carol L. Schmid, *Conflict and Consensus in Switzerland* (Berkeley: University of California Press, 1981), 78.

50. Quotes from *Monroe Sentinel,* 9 September 1891.

51. Rudolf Braun, *Sozialer und Kultureller Wandel in einem Ländlichen Industriegebiet im 19. und 20. Jahrhundert* (Erlenbach-Zürich: Eugen Rentsch Verlag, 1965), 297–361. Bendix, "National Sentiment," 768–90.

52. Luchsinger, "The Planting," 381–82.

53. John Bodnar found a similar dynamic at work in his study of nineteenth-century commemorations. See Bodnar, "Public Memory in Nineteenth-Century America: Background and Context," in his *Remaking America: Public Memory, Commemoration, and Patriotism in the Twentieth Century* (Princeton: Princeton University Press, 1992), 20–38. As Lewis Atherton points out in his classic work on community life in the Midwest, such memorial celebrations were held throughout the region at this time; Atherton, *Main Street on the Middle Border* (Bloomington: Indiana University Press, 1984 [1954]), 206–9.

54. "Came Fifty Years Ago," *Milwaukee Sentinel,* 17 August 1895; "Red Letter Day," *Monroe Sun,* 17 August 1895; John Theo. Etter, "Kurze Beschreibung des 'Golden Jubiläums Festes' am 16. Aug. 1895, etc., etc.," in *Bescheidenes Ged-*

ächtnißkränzchen gewunden der Goldenen Jubelbraut New Glarus, Wis., einer der älteste Schweizer Kolonien der Ver. Staaten (Milwaukee: Druck der Germania, 1897), 30, subject files, NGHSA; and *Amerikanischer Schweizer Nachrichten,* 21 August 1895: "das Städtchen von Hunderten allen Nationalitäten angehörend angefüllt."

55. Etter, "Kurze Beschreibung," 31: "die Pilgerschaar zuerst mit schmerzlichen Heimwehgefühl Auflucht genommen hatten."

56. *Amerikanischer Schweizer Nachrichten,* 21 August 1895; *Milwaukee Sentinel,* 17 August 1895.

57. Etter's influence in the community was both extremely strong and, for the Midwest, rather unique. In a region and time when American congregations expected preachers to remain only a few years, Etter's longevity (which was in turn followed to a lesser extent by his successors) reflected and, indeed, contributed to a deep conservatism and, as one local historian put it, "a distrust and dislike [for] any religious innovation." Theiler, *New Glarus' First 100 Years,* 26–27. New Glarus's unusual church culture in the nineteenth century and, in particular, its striking influence on matters of the community were noted by Lewis Atherton, *Main Street,* 170.

58. John Theo. Etter, "Memorial Gottesdienstbetrachtung," in *Bescheidenes Gedächtnißkränzchen,* 14–18: "Einfactheit der Sitte, Lust zur Häuslichkeit, Sparsamkeit, Genügsamkeit, Dankbarkeit, Peität und vor allem warme, brüderliche Liebe und Eintracht."

59. *Milwaukee Sentinel,* 17 August 1895.

60. *Amerikanischer Schweizer Nachrichten,* 21 August 1895; and *Milwaukee Sentinel,* 17 August 1895. Emphasis added. "Big Day for New Glarus," *Monroe Sentinel,* 21 August 1895.

61. Conrad Zimmerman, "Town of New Glarus," in *History of Green County, Wisconsin* (Springfield, IL: Union Publishing, 1884), 1023–45.

62. Ibid., emphasis mine. *Laegele* is roughly translated as a small wooden cask, used for carrying drinking water into the fields and *Schotte* and *Chrut* are whey and spinach.

63. "Das 60 Jährige Jubiläumsfest ein grosser Erfolg!," *Der Deutsch Schweizerische Courier,* 22 August 1905; "Swiss Hold Odd Celebration," *Milwaukee Sentinel,* 17 August 1905; "The Swiss Festival at New Glarus," *Milwaukee Free Press,* 20 August 1905; "New Glarus Celebrates," *Madison Democrat,* 19 August 1905; "Immense Crowd at New Glarus: Biggest Event in the History of Green County's Swiss Metropolis," *Monroe Daily Journal,* 17 August 1905; and "Big Crowd at New Glarus," *Monroe Evening Times,* 16 August 1905. The format of the day's events is recounted in a broadside from *Der Deutsch Schweizerische Courier,* 25 July 1905; documents boxes, NGHSA.

64. Benjamin R. Barber, *The Death of Communal Liberty: A History of Freedom in a Swiss Mountain Canton* (Princeton: Princeton University Press, 1974), 10–11. It should be noted, however, that Landsgemeinden increasingly play a diminished role in Swiss political culture. Only the cantons of Glarus, Obwalden, and potentially one of the Appenzells (Appenzell Ausserrhoden will soon vote whether to abolish theirs) make use of Landsgemeinden in Switzerland today. I would like to thank Regina Bendix for this observation.

65. *Der Deutsch Schweizerische Courier,* 22 August, 1905; For an account of the Schwanden Landsgemeinde that sent nearly two hundred emigrants to Wisconsin, see Matthias Dürst, "Matthias Dürst's Travel Diary," in *New Glarus 1845–1970: The Making of a Swiss American Town,* ed. Leo Schelbert (Glarus, Switzerland: Kommissionsverlag Tschudi, 1970), 20–150; and Luchsinger, "The Planting," 342.

66. *Milwaukee Free Press,* 20 August 1905; and *Madison Democrat,* 19 August 1905.

67. *Milwaukee Free Press,* 20 August 1905; and *Der Deutsch Schweizerische Courier,* 22 August 1905.

68. *Milwaukee Free Press,* 20 August 1905; *Der Deutsch Schweizerische Courier,* 22 August 1905; and *Monroe Evening Times,* 16 August 1905. On the importance of limburger cheese to the region, see Glenn T. Trewartha, "The Green County, Wisconsin, Foreign Cheese Industry," *Economic Geography* 2 (1926): 298–308; J. M. Emery, "The Swiss Cheese Industry in Wisconsin," *Wisconsin Magazine of History* 10 (1926–1927): 42–52; Emery A. Odell, *Swiss Cheese Industry* (Monroe: Monroe Evening Times, 1936); Emery A. Odell, *Eighty Years of Swiss Cheese in Green County* (Monroe: Monroe Evening Times, 1949); and Gordon R. Lewthwaite, "Midwestern Swiss Migrants and Foreign Cheese," *Yearbook of the Association of Pacific Coast Geographers* 34 (1972): 41–60.

69. Broadside, 25 July 1905, documents box, NGHSA; *Der Deutsch Schweizerische Courier,* 22 August 1905; and *Madison Democrat,* 19 August 1905.

70. Georg Meyer, *The German-American: Program Dedicated to the Celebration of German-American Day* (Milwaukee: Hake and Stern, 1890).

71. *Milwaukee Free Press,* 20 August 1905; and *Monroe Daily Journal,* 17 August 1905.

72. Dr. Geo. Seiler, "Nachklänge zum Koloniefest, ein Gruss den alten Grauen!," *Der Deutsch Schweizerische Courier,* 29 August 1905. Reporter from the *Madison Democrat,* 19 August 1905.

73. Luchsinger, quoted in *Madison Democrat,* 19 August 1905. Emphasis added.

74. For three relevant treatments of antimodernism as it impinged on events such as the New Glarus sixtieth anniversary, see T. J. Jackson Lears, *No Place of Grace: Antimodernism and the Transformation of American Culture* (New York: Pantheon, 1981); McKay, *The Quest of the Folk;* and Glassberg, *American Historical Pageantry,* 35–44.

75. As if to maintain further the authenticity of the Swiss Landsgemeinde, there is no evidence that women were permitted to take part in the commemorative ceremony. Women in Switzerland, of course, were only granted the right to vote on the federal level in 1971, with even more resistance at the local level that, in one canton, persists today. Kenneth D. McRae, *Conflict and Compromise in Multilingual Societies: Switzerland* (Waterloo, Ont.: Wilfred Laurier University Press, 1983), 99–100.

76. Paul Connerton, *How Societies Remember* (Cambridge: Cambridge University Press, 1989). For a general statement on the importance of performance, see Richard Bauman, "Performance," in Bauman, ed., *Folklore,* 41–49.

77. *Der Deutsch Schweizerische Courier,* 22 August 1905.

78. "Celebration of the 70th Anniversary of the Settlement of New Glarus," *New Glarus Post,* 9 July 1915; and "Coloniefest mit Erfolg durchgeführt," *Der Deutsch Schweizerische Courier,* 17 August 1915.

79. "Big Celebration Successful," *New Glarus Post,* 20 August 1915; and *Der Deutsch Schweizerische Courier,* 17 August 1915.

80. *New Glarus Post,* 20 August 1915.

81. *New Glarus Post,* 20 August 1915; and *Der Deutsch Schweizerische Courier,* 17 August 1915. Emphasis added.

82. *New Glarus Post,* 20 August 1915. Emphasis added.

83. On the use of a variety of sculptural forms for memorials, see the excellent work by James E. Young on the commemoration of the Holocaust, *The Texture of Memory: Holocaust Memorials and Meaning* (New Haven: Yale University Press, 1993).

84. David Lowenthal, *The Past is a Foreign Country* (New York: Cambridge University Press, 1985), 321–24; David Lowenthal, "Pioneer Museums," in *History Museums in the United States: A Critical Assessment,* ed. Warren Leon and Roy Rosenzweig (Urbana: University of Illinois Press, 1989), 115–27; Barry Schwartz, "The Social Context of Commemoration: A Study in Collective Memory," *Social Forces* 61 (1982): 374–402; and Bodnar, *Remaking America,* 28–35.

85. *Der Deutsch Schweizerische Courier,* 17 August 1915: "Trotzdem fast alle Nationen vertreten, und Getränke im Ueberfluß vorhalnden ware, verlief das Fest sehr ruhig und ohne Störungen." Two months later the community would not be so lucky: in late October of that year a group of fourteen "Russian Pollocks," working on a paving project, got into a skirmish with local Swiss men, resulting in the death of one of the "Pollock" street workers. The non-Swiss work gang was removed from town the next day. "Shooting Fray Ends in Fatality," *New Glarus Post,* 22 October 1915.

86. *Madison Democrat,* 19 August 1905; and *Milwaukee Free Press,* 20 August 1905. Walter Stuckey, *The Hundredth Anniversary of the Swiss Evangelical and Reformed Church, New Glarus, Wisconsin* (New Glarus, WI: n.p., 1950).

87. *Der Deutsch Schweizerische Courier,* 17 August 1915.

88. Lowenthal, *The Past,* 323.

89. For an excellent case study of Civil War monuments, many of which were built at roughly this same period and similar in both iconography and their role in public memory, see Kirk Savage, "The Politics of Memory: Black Emancipation and the Civil War Monument," in *Commemorations: The Politics of Public Memory,* ed. John Gillis (Princeton: Princeton University Press, 1994), 127–49. Also of importance is Pierre Nora, "Between Memory and History: Les Lieux de Memoire," *Representations* 26 (1989): 7–25.

90. "Town and Village of New Glarus," in *Memoirs of Green County,* n.d. [ca. 1900], documents box, NGHSA.

91. Joseph Schafer, *Wisconsin Domesday Book: Town Studies,* vol. 1 (Madison: State Historical Society of Wisconsin, 1924), 15.

92. Eric E. Lampard, *The Rise of the Dairy Industry in Wisconsin: A Study in Agricultural Change, 1820–1920* (Madison: State Historical Society of Wisconsin,

1963), 91–120; Frederick Merk, *Economic History of Wisconsin During the Civil War Decade* (Madison: State Historical Society of Wisconsin, 1916), 28; and Trewartha, "Foreign Cheese Industry." For a series of provocative essays on rural America's capitalist transformation, see Steven Hahn and Jonathan Prude, eds., *The Countryside in the Age of Capitalist Transformation: Essays on the Social History of Rural America* (Chapel Hill: University of North Carolina Press, 1985). Anne Kelly Knowles explores these changes in the context of ethnicity in her *Calvinists Incorporated: Welsh Immigrants on Ohio's Industrial Frontier* (Chicago: University of Chicago Press, 1997).

93. John Luchsinger, "Report," in *Transactions of the Wisconsin State Agricultural Society* (Madison: 1882–1883), 274. John Luchsinger, "The History of a Great Industry," *Proceedings of the State Historical Society of Wisconsin* (Madison: 1899), 226–30. In his memoirs, John Luchsinger recalled that many of the early cheesemakers, like himself, first learned the skill from their mothers. John Luchsinger, "What America Has Meant to Me," Typeset Manuscript, n.d., subject files, NGHSA.

94. Lampard, *The Rise*, 236–41, 286–88. For the plant manager's statement, see "Wrecking Crews Begin Pet Milk Demolition," *New Glarus Post*, n.d. (probably October 1973), clippings scrapbooks, NGHSA.

95. Confidential communication, New Glarus, WI, September 1994.

96. Luchsinger, "The Planting," 372. "Lands and Peoples," 123.

97. It is a common joke among New Glarners that, by eventually marrying into the community, "the Norwegians saved us from ourselves." Significantly, the date given for this interethnic rescue is the early 1950s. The generalization about the unusually low rates of Swiss intermarriage is based upon comparisons with other groups analyzed in the data-rich study by Richard M. Bernard, *The Melting Pot and the Altar: Marital Assimilation in Early Twentieth-Century Wisconsin* (Minneapolis: University of Minnesota Press, 1980); see especially chapter 2.

98. Luchsinger, "The Swiss Colony," 2.

99. Brunnschweiler, *New Glarus*, 87.

100. Glassberg, *American Historical Pageantry*, especially chapters one and two.

CHAPTER 3. ETHNICITY DISCOVERED AND RE-PRESENTED:
THE LOCALIZATION OF MEMORY BETWEEN THE WARS

1. "Wisconsin's Changing Population," *Bulletin of the University of Wisconsin*, publication 9, serial no. 2642, general series no. 2426. (Madison: University of Wisconsin, 1942), 3–4, 13. Emphasis in original.

2. Ibid., 81–82.

3. Samuel Hopkins Adams, "Wisconsin Joins the War," *Everybody's Magazine* 38 (January 1918): 28–33.

4. John D. Stevens, "Suppression of Expression in Wisconsin During World War I" (Ph.D. diss., University of Wisconsin, 1967); John D. Stevens, "When

Sedition Laws were Enforced: Wisconsin in World War I," *Transactions of the Wisconsin Academy of Sciences, Arts, and Letters* 58 (1970): 39–60; and LaVern J. Rippley, *The Immigrant Experience in Wisconsin* (Boston: Twayne, 1985), 113–15.

5. LaVern J. Rippley, "Ameliorized Americanization: The Effect of World War I on German Americans in the 1920s," in *America and the Germans: An Assessment of a 300-Year History,* ed. Frank Trommler and Joseph McVeigh (Philadelphia: University of Pennsylvania Press, 1985), 217–31. Joseph Roucek, Caroline F. Ware, and M. W. Royse, "Summary to the Discussion of Cultural Groups," in *The Cultural Approach to History,* ed. Caroline F. Ware (New York: Columbia University Press, 1940), 87. Emphasis in original.

6. The notion of ethnic localization is derived from Kathleen Neils Conzen, "Mainstreams and Side Channels: The Localization of Immigrant Cultures," *Journal of American Ethnic History* 11 (1991): 5–20. Within a slightly different, but nevertheless relevant, context, see Jane Nadel-Klein, "Reweaving the Fringe: Localism, Tradition, and Representation in British Ethnography," *American Ethnologist* 18 (1991): 500–517.

7. John Higham, *Strangers in the Land: Patterns of American Nativism, 1860–1925,* 2nd ed. (New York: Athenaeum, 1978), 195. Higham's book, a classic, remains the standard work on nativism during this period. See especially chapter eight, 194–233. For the German American experience, in which the Swiss were inevitably entangled, see Frederick C. Luebke, *Bonds of Loyalty: German Americans and World War I* (De Kalb, IL: Northern Illinois University Press, 1974); and Rippley, "Ameliorized Americanization."

8. Philip Gleason, "The Melting Pot: Symbol of Fusion or Confusion?" in his *Speaking of Diversity: Language and Ethnicity in Twentieth-Century America* (Baltimore: Johns Hopkins University Press, 1992), 17; Thomas J. Archdeacon, "The Movement Toward Restriction, 1865–1924," in his *Becoming American: An Ethnic History* (New York: Free Press, 1983), 143–72. Woodrow Wilson, quoted in Richard Weiss, "Ethnicity and Reform: Minorities and the Ambience of the Depression Years," *Journal of American History* 66 (1979): 568.

9. Karen Falk, "Public Opinion in Wisconsin During World War I," *Wisconsin Magazine of History* 25 (1942): 390; David L. Brye, *Wisconsin Voting Patterns in the Twentieth Century* (New York: Garland, 1979), 250–57; and Rippley, *The Immigrant Experience,* 104–6. For a Wisconsin-based response to this "national scrutiny," see Charles D. Stewart, "Prussianizing Wisconsin," *Atlantic Monthly,* January 1919, 99–105.

10. Ellis B. Usher, "Wisconsin's Ill Repute for Loyalty," unnamed newspaper (probably *Madison Democrat*), 13 October 1917, Papers of the War History Commission: Wisconsin in World War I, clipping file, 1917–1918, series 1699, box 1, State of Wisconsin Historical Society, hereafter cited as SHSW; Ellis B. Usher, "Why Loyalty of Wisconsin is Questioned," unnamed newspaper (probably *Madison Democrat*), 20 April 1918, papers of the War History Commission: Wisconsin in World War I, clipping file, 1917–1918, series 1699, box 1, SHSW.

11. According to the 1920 census, slightly more than 41 percent of Wisconsin's population was either born in a foreign country or had both parents born

abroad. *U.S. Census, 1920. Volume 3: Population,* 1118, 1120. On Wisconsin as a "58 percent American state," see untitled pamphlet, Wisconsin Loyalty Legion Correspondence and Miscellaneous Papers, 1917–1919, box 1, SHSW; Ellis B. Usher, "Why Loyalty of Wisconsin is Questioned"; and Falk, "Public Opinion," 390–94. On Wisconsin as an "ethnological laboratory," see Wardon Allan Curtis, "The Light Fantastic in the Central West," *Century Magazine,* 1907, 570–79. This metaphor was picked up by Richard Nelson Current who wrote that "for many years Wisconsin was something of a living ethnological museum." Current, *Wisconsin: A Bicentennial History* (New York: Norton, 1977), 61.

12. Falk, "Public Opinion," 390; and Lorin Lee Carey, "The Wisconsin Loyalty Legion, 1917–1918," *Wisconsin Magazine of History* 53 (1969): 33–50.

13. 7 December 1917. War History Commission: Wisconsin in World War I, clipping file, 1917–1918, series 1699, box 1, SHSW.

14. These incidents are reported, along with hundreds others, in Papers of the War History Commission: Wisconsin in World War I, clipping file, 1917–1918, series 1699, box 1, SHSW.

15. *Monroe Evening Times,* 21 March, 2 April, and 4 April 1917. "Monroe People to Vote on War," *Wisconsin State Journal,* 22 March 1917; Official Referendum, in John M. Becker Papers, manuscript 109, box 1, folder 1, SHSW.

16. *United States v. John M. Becker,* general proceedings, August 1918, Becker Papers, box 1, folder 5, SHSW.

17. Clipping from *New York Sun,* 21 March 1918, from War History Commission: Wisconsin in World War I, clipping file, 1917–1918, series 1699, box 1, SHSW.

18. George Kull to John Stover, 4 May 1918, Papers of the Wisconsin Loyalty Legion, Correspondence and Miscellaneous Papers, 1917–1919, box 1, SHSW; "Resolutions Adopted by Wisconsin Loyalty Legion," *Evening Wisconsin,* 23 March 1918 in War History Commission: Wisconsin in World War I, clipping file, 1917–1918, series 1699, box 1, SHSW.

19. On the power of maps to delineate, monitor, and control, see Denis Wood, *The Power of Maps* (New York: Guilford Press, 1992). More generally on the use of space for social control, or territoriality, see Robert David Sack, *Human Territoriality: Its Theory and History* (Cambridge: Cambridge University Press, 1986).

20. This section is based on random issues of the *New Glarus Post* and *Der Deutsch Schweizerische Courier* between 1914 and 1918.

21. "Are Melchoir Schmid and Paul Jackson German Spies?" *New Glarus Post,* 23 March 1917.

22. *Der Deutsch Schweizerische Courier,* 12 April 1917: "Was immer unsere individuelle Idee, order Gefühl in Betreff des Krieges sein mag, ist die Zeit für Argumente jetzt vorbei, und als Lokale amerikanische Bürger sollten wir davon Abstand nehmen über die Situation in privaten oder öffentlichen Plätzen zu disputerien."

23. April R. Schultz, *Ethnicity on Parade: Inventing the Norwegian American through Celebration* (Amherst: University of Massachusetts Press, 1994). See also John Bodnar, "The Construction of Ethnic Memory," in his *Remarking America: Public Memory, Commemoration, and Patriotism in the Twentieth Century*

(Princeton: Princeton University Press, 1992), 57–61. Constitution of the Milwaukee Turners (Milwaukee, n.p., ca. 1925), collections of the Max Kade Institute for German-American Studies, Madison, WI.

24. "80th Anniversary to be Observed," *New Glarus Post*, 12 August 1925. See John Higham, "Closing the Gates," in his *Strangers in the Land*, 300–330 for a useful description of the passage of the Immigration Restriction Act.

25. Michael Kammen, *Mystic Chords of Memory: The Transformation of Tradition in American Culture* (New York: Knopf, 1991), 436–40; and Chris Wilson, *The Myth of Santa Fe: Creating a Modern Regional Tradition* (Albuquerque: University of New Mexico Press, 1997).

26. Higham, *Strangers in the Land*, 195–295; Weiss, "Ethnicity and Reform."

27. Philip Gleason, "Americans All: World War II and the Shaping of American Identity," *Review of Politics* 43 (1981): 483–85.

28. For a similar argument, see Schultz, *Ethnicity on Parade*.

29. "Anniversary of the Founding of New Glarus," *New Glarus Post*, 19 August 1925.

30. Charles Linton Curtis, "How Our Ancestors Settled in Wisconsin," *The Wisconsin Magazine*, October 1924, 19–21, 29. For a follow-up story, see Mary L. Bauchle, "Last of the *Swiss* Pioneers," *The Wisconsin Magazine*, April 1927, 17.

31. Rudy Prusok to Elda Schiesser, Marquette, Michigan, 7 January 1987, subject files, NGHSA; Prusok to Schiesser, 4 January 1986, subject files, NGHSA; "Regional Results, Wilhelm Tell Schuetzenverein, New Glarus," *Shooting and Fishing*, 30 August 1906, 412; Results from the Central Sharpshooter's Union at the New Glarus Schuetzen Park, July 24–26, 1917, and July 19–23, 1923, subject files, NGHSA; and Chris T. Westergaard, "My Days in the Schützen Game," *American Single Shot Rifle News* 41 (1987 [originally written in 1951]): 1–5, 19.

32. Warren James Belasco, *Americans on the Road: From Autocamp to Motel, 1910–1945* (Cambridge, MA: MIT Press, 1979), 71–103; John Jakle, *The American Small Town: Twentieth-Century Place Images* (Hamden, CT: Archon Books, 1982), 155–63; and John Jakle, *The Tourist: Travel in Twentieth-Century North America* (Lincoln: University of Nebraska Press, 1985), 200–212.

33. Dorothy May, "Editorial," *Monroe Evening Times*, 3 September 1937; Perry C. Hill, "Swiss Village in New Wisconsin: New Glarus is 93 Years Old," *Milwaukee Sentinel*, 10 October 1938. Emphasis added.

34. *New Glarus: The 1933 Convention City, 46th Annual Convention of the Wisconsin State Fireman's Association* (Milwaukee: Mueller, Jefferson, Butler, 1933), 9–14, subject files, NGHSA.

35. *New Glarus Post*, 6 March 1935.

36. From 1851 until just before World War I, all services were conducted in German, with a minimal number of English-language services added in 1913 and again in 1924. The emphasis was changed in 1935 to three services per month conducted in English, and only one in German. Dieter Brunnschweiler, *New Glarus: Gründung, Entwicklung und heutiger Zustand einer Schweizerkolonie im Amerikanischen Mittlewesten* (Zürich: Buchdruckerei Flutern, 1954), 97.

37. Larry Danielson, "The Ethnic Festival and Cultural Revivalism in a Small Midwestern Town" (Ph.D. diss., Indiana University, 1972); and Larry Danielson, "St. Lucia in Lindsborg, Kansas," in *Creative Ethnicity: Symbols and Strategies of Contemporary Ethnic Life,* ed. Stephen Stern and John Allan Cicala (Logan, UT: Utah State University Press, 1991), 187–203.

38. *New Glarus Post,* 20 February and 13 March 1935.

39. Much of the biographical background on John Schindler is derived from a telephone interview with his daughter, Carol Brand, of Monroe, Wisconsin, October 1993.

40. John Schindler, "The Old Lead Trail," (Monroe, WI: Monroe Evening Times, 1934). Copy held at SHSW archives, SC 2540. For a useful treatment of the antimodernism of writers such as Schindler and its relation to the nation generally, see Robert L. Dorman, *Revolt of the Provinces: The Regionalist Movement in America, 1920–1945* (Chapel Hill: University of North Carolina Press, 1993).

41. Thomas H. Dickinson, *The Case of American Drama* (New York: Houghton-Mifflin, 1915), 111.

42. Ethel Theodora Rockwell, "Historical Pageantry: A Treatise and a Bibliography," *State Historical Society of Wisconsin Bulletin of Information* 84 (1916): 5–19. For a general overview of Rockwell's work and on the historical pageantry moment in general, see Glassberg, *American Historical Pageantry,* 116, 243, 263. Henry M. Schmid to Ethel Rockwell, 10 October 1939, Wilhelm Tell Guild Community Archive, in possession of Elda Schiesser (hereafter cited as WTCGA-ES).

43. John Schindler, "Historical Pageant Given at the Ninetieth Anniversary Celebration of the Settlement of New Glarus, by the Swiss Colonists," in *Official Program to the 90th Anniversary Pageant* (New Glarus, WI: n.p., 1935), 13, documents box, NGHSA.

44. Schindler, "Historical Pageant," 13. Such was also a prime impulse for the 1925 Norse–American Celebration in Minneapolis. Schultz, *Ethnicity on Parade.*

45. *New Glarus Post,* 1 May 1935. The irony here is that in forty more years' time, Swiss-made alpenhorns could eventually be purchased at the local drug store.

46. *New Glarus Post,* 24 July 1935.

47. Clayton Streiff; Marion Streiff; Millard Tschudy; Elda Schiesser; Hubert Elmer; Herbert Kubly interviews with the author, New Glarus, Wisconsin, 1991–1995.

48. *New Glarus Post,* 12 June, 31 July 1935. Manuscript notes for the ninetieth anniversary pageant, n.d., subject files, NGHSA. It is interesting to note that the pageant was written in English, then translated into Swiss German for the program.

49. Richard Schechner, *Between Theater and Anthropology* (Philadelphia: University of Pennsylvania Press, 1985), 74.

50. *New Glarus Post,* 31 July 1935. Emphasis added.

51. Ibid; Fred Ott, "Greetings," in *Official Program to the 90th Anniversary Pageant* (New Glarus, WI: n.p., 1935), 3. "Town Made by Cheese: Swiss Settle-

ment of Ninety Years Ago is Celebrated in Wisconsin," *New York Times*, 6 August 1935.

52. *New Glarus Post*, 7 August 1935.

53. Ibid. Emphasis in original.

54. The remainder of this section is based largely on the *Official Program to the 90th Anniversary Pageant*.

55. See Glassberg, *American Historical Pageantry*. For other pertinent analyses of ethnic commemorative festivals in the Midwest, see John Bodnar, *Remaking America*, 41–77; April Schultz, "'The Pride of the Race Had Been Touched': The 1925 Norse-American Immigration Centennial and Ethnic Identity," *Journal of American History* 77 (1991): 1265; Schultz, *Ethnicity on Parade;* and Danielson, "The Ethnic Festival."

56. W. Langdon, "The Pageant in America," *American Monthly*, March 1911, 102–3.

57. *Official Program to the 90th Anniversary Pageant*, 40. Emphasis added.

58. Ibid., 21.

59. Ibid., 25.

60. "New Glarus Blazes a Trail," *Monroe Times*, n.d., clipping scrapbooks, NGHSA. Thomas Bender, *Community and Social Change in America* (Baltimore: Johns Hopkins University Press, 1982).

61. For what has become the classic statement on carnival and its ability to invert and mock the dominant order, see Mikhail Bakhtin, *Rabelais and His World* (Bloomington: Indiana University Press, 1984).

62. *Official Program to the 90th Anniversary Pageant*, 13.

63. On the early schisms engendered by "Gross-Tal" and "Klein-Tal" settlements, see Mathias Dürst's illuminating travel diary, reprinted and translated in *New Glarus, 1845–1970: The Making of a Swiss American Town*, ed. Leo Schelbert (Glarus, Switzerland: Kommissionsverlag Tschudi, 1970), 20–156; for reports of the KKK meetings and cross burnings, see *New Glarus Post*, 28 June and 8 August 1924; and for the decline of small-scale cheesemaking with the rise of the industry's increasing rationalization, see Brunnschweiler, *New Glarus*, 65–85, and chapter 2 of this book.

64. "Green County Historical Society is Organized," *New Glarus Post*, 22 September 1937. Marcus Lee Hansen, "The Problem of the Third Generation Immigrant," in *American Immigrants and Their Generations: Studies and Commentaries on the Hansen Thesis after Fifty Years*, ed. Peter Kivisto and Dag Blanck (Urbana, IL: University of Illinois Press, 1990), 191–203. Essay first delivered to the Augustana Historical Society, Rock Island, Illinois, 15 May 1937. Pierre Nora, "Between Memory and History: Les Lieux de Memoire," *Representations* 26 (1989): 7–25.

65. *New Glarus Post*, 22 September 1937.

66. John A. Schindler to Esther Stauffacher, 30 September 1938, correspondence folder, NGHSA. "Local Historical Group is Organized," *New Glarus Post*, 26 September 1938, "Historical Society to Meet Next Monday Evening," *New Glarus Post*, 11 January 1939.

67. For more general statements on the ethnic historical society in this period, see John J. Appel, *Immigrant Historical Societies in the United States,*

1880–1950 (New York: Arno Press, 1980); and John Higham, "The Ethnic Historical Society in Changing Times," *Journal of American Ethnic History* 14 (1994): 30–44.

68. Heinz Meier, *The Swiss American Historical Society* (Norfolk, VA: Donning, 1977), 11.

69. The Swiss American Historical Society, *Prominent Americans of Swiss Origin* (New York: James T. White, 1932).

70. Meier, *Swiss American Historical Society,* 18.

71. Ibid. The Swiss American Historical Society's second publication did equally little to bridge the distance between the official culture of the umbrella organization and the local concerns of New Glarners. *The Swiss in the United States,* though attempting to be comprehensive, made only passing reference to the Green County settlement, devoting considerably more to California, "sacred training" in eastern Wisconsin, and the like. John Paul von Grueningen, *The Swiss in the United States* (Madison: Swiss-American Historical Society, 1940).

72. Minutes of the New Glarus Historical Society, 1938–1946, 22 September 1938, NGHSA; "Historical Society Adopts Constitution and By-Laws," *New Glarus Post,* 26 October 1938. Emphasis added.

73. Minutes of the New Glarus Historical Society, 1938–1946, 22 May, 22 June, 19 October 1939, NGHSA; *New Glarus Post,* 24 May, 5 July 1939; *Monroe Times,* n.d., NGHSA scrapbook.

74. Bodnar, *Remaking America,* 14–17.

75. Karal Ann Marling, *George Washington Slept Here: Colonial Revivals and American Culture, 1876–1986* (Cambridge: Harvard University Press, 1988), 44. For the role of women generally as "patrons of tradition," see Kammen, *Mystic Chords of Memory,* 266–69.

76. For a useful, if brief, biography of Esther Stauffacher, see Victoria Brown, "Esther Streiff Stauffacher," in *Uncommon Lives of Common Women: The Missing Half of Wisconsin History* (Madison: Wisconsin Feminists Project, 1975), 84. For useful obituaries, see Miriam B. Theiler, "My 'Aunt Esser' Stauffacher," *New Glarus Post,* 21 November 1945; "Mrs. W. W. Stauffacher Passes Away Sunday," *New Glarus Post,* 14 November 1945. I also owe a special debt to Elda Schiesser of New Glarus, a woman in many ways as remarkable as Esther Stauffacher.

77. "Swiss Museum and Village Proposed by Historical Society," *New Glarus Post,* 5 July 1939; "NYA to Aid Progress of Local Museum Project," *New Glarus Post,* 22 January 1941; "Historical Village NYA Labor is Approved," *New Glarus Post,* 2 April 1941; Minutes of the New Glarus Historical Society, 15 January 1941, subject files, NGHSA.

78. Esther Stauffacher, "Historical Society Sponsors Swiss Supper and Program," *New Glarus Post,* 3 April 1940. Emphasis in original. It would appear that Stauffacher's piercing editorial came more than a half year after the Historical Society first realized the extent of opposition to the open-air museum idea. "Unfavorable comments on the part of many citizens to such a [historic] village in our village park," led them to look for alternative sites. Minutes of the New Glarus Historical Society, 19 October 1939, subject files, NGHSA.

79. "Village Board Proceedings," *New Glarus Post,* 12 June 1940; "History of the Land the Museum Stands On," typeset manuscript, n.d.; and Minutes of the New Glarus Historical Society, June 1940, both from subject files, NGHSA.

80. Esther Stauffacher, quoted in "Historical Pilgrimage of Green County to Be Here on Sunday," *New Glarus Post,* 28 May 1941. Such memory retrieval concerns defined Artur Hazelius's creation of Europe's first open-air museum in Sweden, Skansen. See Barbara Kirshenblatt-Gimblett, "Objects of Ethnography," in *Exhibiting Cultures: The Poetics and Politics of Museum Display,* ed. Ivan Karp and Steven D. Lavine (Washington, D.C.: Smithsonian Institution Press, 1991), 401–2.

81. Ibid.; "Volunteer Help Needed For Historical Project," *New Glarus Post,* 12 March 1941.

82. Esther Stauffacher, "President of County Historical Society Asks Aid in Project," *New Glarus Post,* 23 July 1941. For a useful reading of *Main Street* and its relation to small town life as experienced in places like New Glarus, see Joel Fisher, "Sinclair Lewis and the Diagnostic Novel: *Main Street* and *Babbitt,*" *Journal of American Studies* 20 (1986): 421–33.

83. "Folk Fest Here Sunday Afternoon," *New Glarus Post,* 23 September 1943; "Pioneer Home Dedication Program Held in Gym," *New Glarus Post,* 30 September 1943; "Program to the Dedication of the Community Log Cabin, 1942," n.d., typeset manuscript, subject files, n.d., NGHSA.

84. "Work on Historical Village Now Progressing," *New Glarus Post,* 6 May 1942; "First Building for Historical Village Under Way," *New Glarus Post,* 15 July 1942; "Community House at New Glarus Park to be Memorial to Pioneers," *Monroe Times,* date unknown, 1942, clipping scrapbooks, NGHSA; "Swiss Village to Be Restored at New Glarus," *Wisconsin State Journal,* date unknown (1943), clipping scrapbooks, NGHSA; Carl Marty to Esther Stauffacher, Monroe, Wis., subject files, n.d., NGHSA.

85. Esther Stauffacher, "Editorial," *New Glarus Post,* 14 July 1943.

86. "First Building For Historical Village Underway," *New Glarus Post,* 15 July 1942; "Interest Shown in New Glarus by Many Persons," *New Glarus Post,* 5 November 1947.

87. "Swiss Museum Will be Opened Sunday P.M.s," *New Glarus Post,* 4 June 1947; "Many Visitors See Swiss Museum at New Glarus," *New Glarus Post,* 16 July 1947; "101 Years of Yodeling," *Time,* 26 August 1946, 21; and William H. Nicholas, "Deep in the Heart of 'Swissconsin'," *National Geographic,* June 1947, 781–800.

88. Esther Stauffacher, quoted in "First Building for Historical Village Under Way," *New Glarus Post,* 15 July 1942. Mike Wallace, *Mickey Mouse History and Other Essays on American Memory* (Philadelphia: Temple University Press, 1996). David Lowenthal's *Possessed by the Past: The Heritage Crusade and the Spoils of History* (New York: Free Press, 1996) provides a wealth of information about the inevitable tension endemic to heritage. See also Barbara Kirshenblatt-Gimblett, "Theorizing Heritage," *Ethnomusicology* 39 (1996): 367–80; Susan Porter Benson, Stephen Brier, and Roy Rosenzweig, eds., *Presenting the Past: Essays on History and the Public* (Philadelphia: Temple University Press, 1986); Don Mitchell, "Heritage, Landscape, and the Production of Community: Consensus

History and Its Alternatives in Johnstown, Pennsylvania," *Pennsylvania History* 59 (1992): 198–226; and Robert Hewison, *The Heritage Industry: Britain in a Climate of Decline* (London: Methuen, 1987).

89. Taking a stand against the all too easy assumption that heritage tourism is "invariably bogus history," Nuala C. Johnson's work in Ireland offers a similar reading of the intense multivocality and open-endedness of museum displays; Johnson, "Where Geography and History Meet: Heritage Tourism and the Big House in Ireland," *Annals of the Association of American Geographers* 86 (1996): 551–66.

90. The term "memory work" is derived from John Gillis, "Memory and History: The History of a Relationship," in *Commemorations: The Politics of National Identity*, ed. John Gillis (Princeton: Princeton University Press, 1994), 3.

CHAPTER 4. "REPLANTING THE SWISS SCENE": WILHELM TELL RETURNS TO AMERICA'S LITTLE SWITZERLAND

1. Anonymous letter to Edwin Barlow, *New Glarus Post*, date unknown (probably mid-August, 1938), Wilhelm Tell scrapbooks, New Glarus Historical Society (hereafter cited as WTS). Emphasis added.

2. Although there is considerable debate whether or not Wilhelm Tell ever lived, there is no question that the Tell story takes place in the canton of Uri, far removed from Interlaken and canton Bern.

3. "Wilhelm Tell Play" Official Program, 4 and 5 September 1938, WTS. Emphasis added. Of course Kilbi and, beginning in the late 1920s, Volksfest recurred annually, but neither attempted to communicate a self-conscious Swiss ethnic identity so ambitiously.

4. One of the best examples of writing from this perspective may be found in Davydd Greenwood's discussion of a Basque festival which, through the injection of "local color," devolved from a communitas-creating event to an example of "culture by the pound." By introducing money and exploiting locals, the cultural display is commodified and "the meaning is gone." His stark position is nuanced slightly by the postscript to the updated version of this essay. Davydd J. Greenwood, "Culture by the Pound: An Anthropological Perspective on Tourism as Cultural Commoditization," in *Hosts and Guests: The Anthropology of Tourism*, ed. Valene Smith, 2nd ed. (Philadelphia: University of Pennsylvania Press, 1989), 171–85.

5. Regina Bendix, *Backstage Domains: Playing "William Tell" in Two Swiss Communities* (Bern: Peter Lang, 1989), 25.

6. Alfred Berchtold, "Wilhelm Tell im 19. und 20. Jahrhundert," in *Tell: Werden und Wandern eines Mythos*, ed. Lilly Stunzi (Bern: Hallway Verlag, 1973), 167–311; and J. R. von Salis, "Ursprung, Gestalt und Wirkung des schweizerischen Mythos von Tell," in Stunzi, ed., *Tell*, 9–88.

7. Alan Dundes, "The Apple-Shot: Interpreting the Legend of William Tell," *Western Folklore* 50 (1991): 327–60; and Bendix, *Backstage Domains*, 25–28.

8. Max Frisch, *Wilhelm Tell für die Schule* (Frankfurt am Main: Suhrkamp, 1971); and Henry Kamm, "The Swiss Debunk William Tell and All That," *New York Times*, 30 March 1994.

9. Peter Utz, *Die ausgehöhlte Gasse: Stationen der Wirkungsgeschichte von Schillers "Wilhelm Tell"* (Königsten: Forum Academicum, 1984); Bendix, *Backstage Domains*, 28–36.

10. Goethe, quoted in Bendix, *Backstage Domains*, 34.

11. Charles Passage, "Introduction," in *Friedrich von Schiller's William Tell* (New York: Frederick Unger, 1962), vii; and Shiela Margaret Benn, *Pre-Romantic Attitudes to Landscape in the Writings of Friedrich Schiller* (Berlin: Walter de Gruyter, 1991), 186–87.

12. Leslie Sharpe, "Wilhelm Tell," in *Friedrich Schiller: Drama, Thought, and Politics* (Cambridge: Cambridge University Press, 1991), 293–309; Regina Bendix, "National Sentiment in the Enactment and Discourse of Swiss Political Ritual," *American Ethnologist* 14 (1992): 768–90; J. Christopher Herold, *The Swiss without Halos* (New York: Columbia University Press, 1948), 33–36; and Utz, *Die ausgehöhlte Gasse*, 81–100.

13. Eric Hobsbawm, "Mass-Producing Tradition: Europe, 1870–1914," in *The Invention of Tradition*, ed. Eric Hobsbawm and Terence Ranger (New York: Cambridge University Press, 1983), 263–308. Passage, "Introduction," iii; and Bendix, *Backstage Domains*, 33–40, 96–107.

14. Elda Schiesser, "History of New Glarus Wilhelm Tell Play," *New Glarus Post*, 30 August 1995.

15. Louis Wirth, "The Problems of Minority Groups," in *The Science of Man in the World Crisis*, ed. Ralph Linton (New York: Columbia University Press, 1944), 361. For a more general look at ethnic cultural brokers and leaders from anthropological and historical perspectives, respectively, see Anya Peterson Royce, *Ethnic Identity: Strategies of Diversity* (Bloomington: Indiana University Press, 1982), 135–36; and Victor Greene, *American Immigrant Leaders, 1800–1910: Marginality and Identity* (Baltimore: Johns Hopkins University Press, 1987).

16. Deborah Neff and Phillip B. Zarrilli, *Wilhelm Tell: In America's "Little Switzerland," New Glarus Wisconsin* (New Glarus, WI: Wilhelm Tell Guild, 1987), 9. Much of what is known about Barlow is nicely synthesized here; it therefore provides a useful point of departure for the next several paragraphs on Barlow's early life.

17. Memoirs of Edwin Barlow, handwritten manuscript, n.d. (most likely 1957), subject files, NGHSA; "Edwin Barlow, Tell Originator and Benefactor, Dies at Ripon," *New Glarus Post*, 25 September 1957. Apart from one photograph of an undated performance, Neff and Zarrilli find no evidence of Barlow's supposed theatrical experience in New York. *Wilhelm Tell*, 8–10, 40.

18. "Edwin Barlow, Benefactor, Dies," *Monroe Evening Times*, 24 September 1957; "Editorial, Tribute to a Friend: Edwin P. Barlow," *Monroe Evening Times*, 24 September 1957; and *New Glarus Post*, 25 September 1957.

19. We have an indication of Barlow's style in his handwritten memoirs, drafted shortly before his death in 1957. Writing characteristically in the third person, Barlow summarized his life in this most revealing passage: "Mr. Barlow was a traveler extraordinary [sic] of New York, London, Paris, Rome, Spain, and Switzerland. He traveled the Atlantic 63 times and circumnavigated the globe twice and was as much at home in the remote parts of the world as in the United States. His blue eyes had the look of far horizons, his nose

haughty and sensitive, was a magnificent one for the dramatic and drama of every sort. All his life he found his closest friends among the theater, which he loved. He wanted nothing for himself except to be free and travel light. He was note worthy for what he was, rather than anything he did, a brilliant, often irascible, high handed, high hearted gentleman in love with the world." Barlow, Memoirs.

20. "Former Green County Resident Makes Suggestion to Local Friends," *New Glarus Post,* 18 September 1935.

21. Edwin Barlow to Esther Stauffacher, New York, 13 May 1937, Wilhelm Tell Community Guild Archives, in possession of Elda Schiesser (hereafter cited as WTCGA-ES). For a more detailed look at the heritage activities of Esther Stauffacher, see chapter 3.

22. Henry M. Schmid to Ethel Rockwell, 10 October 1939, WTCGA-ES.

23. Paul Grossenbacher, a recent immigrant and one of the few who had actually seen the Interlaken *Tell,* felt that such a production could never be put on in the United States, and especially not in such an insignificant community of ordinary farmers and workers. "I told him it's too much for a small town," Grossenbacher relayed years later, "I don't think you realized how big the play is." *New Glarus Post,* 28 August 1991. See also Grossenbacher's autobiography "Looking Back: From Burgdorf, Canton Bern, to New Glarus, Wisconsin. An Autobiographical Sketch of a Twentieth Century Swiss Immigrant," *Swiss American Historical Review* 25 (1989): 5–49.

24. "Edwin Barlow Tells Origin of Pageant," *Monroe Evening Times,* 17 September 1941, WTS.

25. For similar findings of small-town festivals in Minnesota, see two chapters by Robert H. Lavenda: "Family and Corporation: Two Styles of Celebration in Central Minnesota," in *The Celebration of Society: Perspectives on Contemporary Cultural Performance,* ed. Frank H. Manning (Bowling Green, OH: Bowling Green University Popular Press, 1983), 55–61, and "Festivals and the Creation of Public Culture: Whose Voice(s)?," in *Museums and Communities: The Politics of Public Culture,* ed. Ivan Karp, Christine Mullen Kreamer, and Steven D. Lavine (Washington, D.C.: Smithsonian Institution Press, 1992), 76–104. For a well-argued analogy between daily life and festival behavior, see Roger Abrahams and Richard Bauman, "Ranges of Festival Behavior," in *The Reversible World: Symbolic Inversion in Art and Society,* ed. Barbara Babcock (Ithaca: Cornell University Press, 1978), 193–208.

26. Minutes of the Wilhelm Tell Community Guild, 28 July 1938, WTCGA-ES.

27. Schmid to Rockwell, 10 October 1939; Minutes of the Wilhelm Tell Community Guild, 28 July 1938, WTCGA-ES.

28. Minutes of the Wilhelm Tell Community Guild, 28 July, 20 August 1938, WTCGA-ES. Data on casting is from Neff and Zarrilli, *Wilhelm Tell,* 14, 43–48.

29. *Official Program of the Wilhelm Tell Play, Elmer's Grove, New Glarus* (New Glarus: n.p., 1938), WTS; Henry M. Schmid to Henry Platt, 17 May 1942, WTCGA-ES; and Neff and Zarrilli, *Wilhelm Tell,* 14.

30. Although Barlow and his most informed associates most likely knew of the Altdorf *Tell* production, there is no evidence that this indoor play served as a model in an way comparable to the Interlaken version.

31. Schmid to Rockwell, 10 October 1939; J. Ivan Elmer to Henry M. Schmid, 4 April 1942, WTCGA-ES; Hubie Elmer interview with author, New Glarus, Wisconsin, 29 August 1995.

32. "Plans for William Tell Play are Progressing," *New Glarus Post*, 24 August 1938, WTS; K. F. Mueller to Henry M. Schmid, 13 August 1940, WTCGA-ES; Fred B. Luchsinger to Henry M. Schmid, Monroe, 19 August 1939, WTCGA-ES; Schmid to Rockwell, 10 October 1939, WTCGA-ES; E. B. Dolfen to Henry M. Schmid, Madison, 19 August 1940, WTCGA-ES; Henry M. Schmid to Capital City Awning and Tent Co., 19 June 1942; K[atherine] T[heiler], "Wilhelm Tell Sidelights," *New Glarus Post*, 7 September 1938, WTS; Clayton Streiff, interview with author, New Glarus, Wisconsin, 27 February 1992.

33. Paul Grossenbacher, "Wilhelm Tell Recollections," handwritten manuscript, n.d., WTCGA-ES; Theiler, "Sidelights"; Mueller to Schmid, 13 August 1940.

34. Letters in *New Glarus Post*, 14 September 1938 by Jane Porter; 21 September 1938 by Muetti Ingold; and 21 September 1938 by Ida Mauer. Jeanne L. Fay to Wilhelm Tell Community Guild, Chicago, 5 August, 1940, WTCGA-ES.

35. Mueller to Schmid, 13 August 1940; and Theiler, "Sidelights." For merely one example, but a useful one, of ethnic leadership's opposition to the loss of language, see Einar Haugen, *Immigrant Idealist: A Literary Biography of Waldemar Ager, Norwegian American* (Northfield, MN: Norwegian-American Historical Association, 1989).

36. "Committees Are Busy on William Tell Play," *New Glarus Post*, 3 August 1938.

37. Introduction to *Official Program;* Schmid to Rockwell, October 10, 1939.

38. Minutes of the Wilhelm Tell Community Guild, 14 and 28 September and 7 October 1938. Emphasis added.

39. "The Tale of Tell: Picturesque Pageant at New Glarus," unidentified newspaper, n.d. [probably 1939], WTS.

40. Minutes of the Wilhelm Tell Community Guild, 14 and 28 September and 7 October 1938.

41. "William Tell Contest Judged Friday Evening," *New Glarus Post*, 17 May 1939.

42. Herbert Kubly, interview with the author, New Glarus, Wisconsin, 16 February 1992; "Eighth Annual Production of Wilhelm Tell," *New Glarus Post*, 29 August 1945.

43. "Editorial," *New Glarus Post*, 9 September 1944.

44. James P. Zollinger to Edwin Barlow, New York, 22 June 1940, WTCGA-ES; Grossenbacher, "Wilhelm Tell Recollections." "New Glarus Girl Shows Sketch for N.Y. 'Beaux Arts Ball,'" *New Glarus Post*, n.d.; "New Glarus Prepared Wilhelm Tell Pageant," unidentified newspaper, 15 August 1941. "Swiss Glamor Girls Wear Native Costumes at Wilhelm Tell Fete," unidentified newspaper, n.d. [probably 1940], all from WTS.

45. "Folk Dances Due for Revival Here: Authentic Old Steps by Costumed Participants Would Add Color," *New Glarus Post,* 7 September 1940. Kay Gmur interview with the author, New Glarus, Wisconsin, August 1995.

46. Figures calculated from data in Cedric Parker, "Swiss Fight for Freedom in New Glarus Play," *Capital Times,* 24 August 1941.

47. *New Glarus Post* articles, 1938–1941, WTS.

48. Peter Etter, interview with the author, New Glarus, Wisconsin, August 1994.

49. "Publicity Spread on Tell Play Thru Midwest," *New Glarus Post,* 28 August 1941.

50. *Official Program of the 1942 New Glarus Presentations of Schiller's Wilhelm Tell* (New Glarus, WI: Wilhelm Tell Community Guild, 1942). Emphasis added.

51. "New Glarus Presents Third Annual Costume Pageant of Tell, Gains Nation-Wide Notice," *Wisconsin State Journal,* 30 August 1941; "The Forbidden Story," *Capital Times,* 1 November 1941; "Fourth Annual Festival Will Start Today," *Chicago Tribune,* 31 August 1941; and "Swiss Defense of Freedom Symbolized in 'Wilhelm Tell,'" *Christian Science Monitor,* 3 May 1941. The extent of *Tell's* outward legitimation was revealed when festival organizers boasted that their pageant possessed a more germane lesson for the country than even the founders of New England. A Swiss American living in East Milton, Massachussetts, made the pilgrimage to Green County for the 1940 production of *Tell* and, after the "moving experience," began a congenial correspondence with the Tell Guild. G. A. Salzberger was struck by the similarity between the Swiss drama and a play that his community had staged for a number of years entitled "The Old Homestead." The drama was "a typical New England play about the people of a typical New England town," not too different for Americans from the way that *Wilhelm Tell* "is of great interest to all Swiss and Swiss loving people." While the Guild secretary thanked the visitor for the information and comparison between the two pageants, he made it clear that "ours is more noteworthy in the general run of conditions as they are now." G. A. Salzburger to Henry M. Schmid, East Milton, Massachusetts, 3 August 1940, and 9 August 1942; Henry M. Schmid to G. A. Salzburger, 19 August 1940, and 19 August 1942, WTCGA-ES.

52. *Articles of Incorporation,* 10 July 1939, WTCGA-ES.

53. Quoted in Cedric Parker, "Swiss Fight for Freedom."

54. *Articles of Incorporation.*

55. Thieler, "Wilhelm Tell Sidelights."

56. Millard Tschudy interview with the author, New Glarus, Wisconsin, September 1994; Lila Kubly Dibble quoted in Ann Schmidt, "History Brings Tradition to Tell Festival," *New Glarus Post,* 28 August 1991; and Clayton Streiff interview with the author, New Glarus, Wisconsin, 27 February 1992.

57. Robert L. Dorman, *Revolt of the Provinces: The Regionalist Movement in America, 1920–1945* (Chapel Hill: University of North Carolina Press, 1993) offers an excellent point of comparison for the nationwide phenomenon. For the way that nationalism shaped ethnic relations during the war, and especially how it differed from the previous world war, see Philip Gleason, "Americans

All: World War II and the Shaping of American Identity," *Review of Politics* 43 (1981): 483–518.

58. Werner Sollors, *Beyond Ethnicity: Consent and Descent in American Culture* (New York: Oxford University Press, 1986), 241–43.

59. *New Glarus Post,* 17 August 1938.

60. Regina Bendix, "Tourism and Cultural Displays: Inventing Traditions for Whom?," *Journal of American Folklore* 102 (1989): 131–46. See also her *Backstage Domains.*

61. "Publicity Spread on Tell Play Thru Midwest," *New Glarus Post,* 28 August 1941.

62. Lawrence Eldred to Henry M. Schmid, Elgin, Illinois, 1, 11, 18, and 21 August 1941, WTCGA-ES.

63. Sterling Sorensen, "4,000 Witness 'Wilhelm Tell' at New Glarus," *Capital Times,* 4 September 1946.

64. "Petitions Ask Special Stamp for Centennial," *Monroe Evening Times,* 7 February 1945; "An Honor Richly Deserved," *Monroe Evening Times,* 9 February 1945; "Now It's Our Turn," *Monroe Evening Times,* 12 February 1945.

65. Emma Becker to Frank Walker, Monroe, Wisconsin, 22 July 1945; Edwin Fred to Frank Walker, Madison, 15 February 1945; Frank Walker, Postmaster General, to Robert M. La Follette, Jr., Washington, D.C., 16 March 1945; Robert M. La Follette, Jr. to Edwin Barlow, Washington, D.C., 12 February 1945; State of Wisconsin Senate Resolution No. 60, S. (4 June 1943); and State of Wisconsin Senate Resolution no. 37, S. (21 February 1945). All from NGHSA.

66. Emma Becker to Frank Walker, Monroe, Wisconsin, n.d. [probably 1945]; Edwin Barlow to Robert M. La Follette, Jr., 8 and 19 February 1945, emphasis in original. Both from NGHSA.

67. Henry M. Schmid to *Life* magazine, 12 August 1939; Charlotte Case to Henry M. Schmid, New York, 24 August 1939; Henry M. Schmid to Charlotte Case, 13 July 1940; Charlotte Case to Henry M. Schmid, New York, 18 July 1940. WTCGA-ES.

68. "101 Years of Yodeling," *Time,* 26 August 1946, 21; and William H. Nicholas, "Deep in the Heart of 'Swissconsin'," *National Geographic,* June 1947, 781–800.

69. For a critical reading of the well-known magazine that highlights its portrayal of the exotic, see Catherine Lutz and Jane L. Collins, *Reading National Geographic* (Chicago: University of Chicago Press, 1993).

70. "Editorial," *Monroe Evening Times,* 26 August 1941; *Dixon Evening Telegraph,* 30 August 1941; and *Chicago Tribune,* 31 August 1941.

71. E. A. Rowles to Gilbert Ott, Mount Morris, Illinois, 1 August 1942; Sohei Kawakatsu to Wilhelm Tell Community Guild, Madison, 14 September 1954; and Fred Kupfer to Wilhelm Tell Community Guild, St. Joseph, Missouri, 15 September 1954, all from WTS.

72. *New Glarus Post,* 17 August 1938, WTCGA-ES; "New Glarus Achievement Properly Supplemented," *Monroe Evening Times,* 20 August 1940.

73. "Swiss in Wisconsin to Stage Pageant of Oldest Democracy," *Chicago Sun,* 2 September 1942; and "New Glarus Honors its Heritage," *The Richland Democrat,* 11 September 1941.

74. Benn, *Pre-Romantic Attitudes to Landscape,* 190–99.

75. *The Richland Democrat,* 11 September 1941.

76. A sampling of the sort of heartfelt connections to a vernacular memory may be found in the following: E. Krumholz to Henry M. Schmid, Fountain City, Wisconsin, 12 September 1941; Mrs. E. Riemenschmied to Mayor of New Glarus, Milwaukee, 22 August 1939; and Mrs. Wilson Hoppe to "Sirs," Spring Green, Wisconsin, 24 August 1939; E. R. Wohlgemuth to Chamber of Commerce, Mellen, Wisconsin, 17 July 1940; and Jeanne L. Fay to Manager of "William Tell" Play, Chicago, 5 August 1940. All from WTS.

77. For a useful account of the historic bifurcation of culture into these categories, with particular reference to staged drama, see Lawrence Levine, "William Shakespeare in America," in *Highbrow/Lowbrow: The Emergence of Cultural Hierarchy in America* (Cambridge, MA.: Harvard University Press, 1988), 11–82.

78. George H. Willet, "About It and About," *Grant County Herold,* September 1950.

79. "Large Crowds See 12th Pageant Both Days," *New Glarus Post,* 7 September 1949. A measure of *Tell's* high-brow appeal was found in the appearance this year by Frank Lloyd Wright.

80. *Monroe Evening Times,* n.d. [probably August 1950], WTS.

81. On "folklorism" in a strikingly similar context see Bendix, *Backstage Domains,* 243. For case studies that concentrate on the official construction of local culture, see David E. Whisnant, *All That is Native and Fine: The Politics of Culture in an American Region* (Chapel Hill: The University of North Carolina Press, 1983); and Ian McKay, *The Quest of the Folk: Antimodernism and Cultural Selection in Twentieth-Century Nova Scotia* (Montreal and Kingston: McGill-Queens, 1994).

82. David Lowenthal, *Possessed by the Past: The Heritage Crusade and the Spoils of History* (New York: The Free Press, 1996). Barbara Kirshenblatt-Gimblett, "Mistaken Dichotomies," *Journal of American Folklore* 101 (1988): 140–155. Streiff interview with author, New Glarus, Wisconsin, 27 February 1992.

83. Barbara Myerhoff, *Number Our Days* (New York: Simon and Schuster, 1978), 32.

84. J. P. von Grueningen to Edwin Barlow, Madison, 9 August 1939, WTCGA-ES.

85. Herbert Kubly, interview with the author, New Glarus, Wisconsin, February 1992.

86. Although the themes and scope of this chapter are largely historical and rely on period sources, it could not have been written without the benefit of extensive fieldwork. For four years, I took part in the German-language production of *Tell* and, as much as the archives, this direct experience brought the early years of the pageant to life. During the many hours of rehearsal, picnics, the several performances, and committee meetings, I learned of *Tell's* history in two ways: through stories and reminisces of older cast members I heard the oral history of the pageant; and from the experience of "taking on the role," I got a fleeting sense of its grip on audiences and cast members. For sharing their many stories, and for welcoming me so kindly into their commu-

nity, I want to thank the members of the Wilhelm Tell Guild and particularly the members of the German-language cast. I am especially grateful to Doris Arn, Peter Etter, and Marilyn and Howard Christensen.

CHAPTER 5. PROVINCIAL COSMOPOLITANISM:
MAPPING AND TRAVELING THROUGH OLD WORLD WISCONSIN

1. E. Giradela Wackler to Fred Holmes, Oakland, California, 25 July 1945, clipping scrapbook number 6, Fred L. Holmes Papers, State Historical Society of Wisconsin (hereafter cited as FHP).

2. M. M. Quaife, 15 September 1944, typeset manuscript review of *Old World Wisconsin*, in clipping scrapbook number 6, FHP. Quaife's review was eventually published in the *American Historical Review*.

3. George L. Mosse, *The Nationalization of the Masses: Political Symbolism and Mass Movements in Germany from the Napoleonic Wars Through the Third Reich* (New York: H. Fertig, 1975); George L. Mosse, *Fallen Soldiers: Reshaping the Memory of the World Wars* (New York: Oxford University Press, 1990); Michel Foucault, *Language, Countermemory, Practice: Selected Essays and Interviews* (Ithaca: Cornell University Press, 1977); Ian McKay, *The Quest of the Folk: Antimodernism and Cultural Selection in Twentieth-Century Nova Scotia* (Montreal and Kingston: McGill-Queens, 1994); and John Bodnar, *Remaking America: Public Memory, Commemoration, and Patriotism in the Twentieth Century* (Princeton: Princeton University Press, 1992).

4. The essays in John Gillis's important collection all point to this central theme; John Gillis, ed., *Commemorations: The Politics of National Identity* (Princeton: Princeton University Press, 1994).

5. Susan G. Davis, *Parades and Power: Street Theatre in Nineteenth-Century Philadelphia* (Berkeley: University of California Press, 1988), 17.

6. Material on the ethnic studies of the Chicago school and Jane Addams is voluminous. Useful secondary sources include: Stow Persons, *Ethnic Studies at Chicago, 1905–45* (Urbana: University of Illinois Press, 1987); Martin Bulmer, *The Chicago School of Sociology: Institutionalization, Diversity, and the Rise of Sociological Research* (Chicago: University of Chicago Press, 1984); David Ward, *Poverty, Ethnicity, and the American City, 1840–1925: Changing Conceptions of the Slum and the Ghetto* (Cambridge: Cambridge University Press, 1989); Mary Jo Deegan, *Jane Addams and the Men of the Chicago School, 1892–1918* (New Brunswick, NJ: Transaction Books, 1988). Several important monographs provide the best entry points into the Chicago school's writing on immigration and ethnicity: William I. Thomas and Florian Znaniecki, *The Polish Peasant in Europe and America: Monograph of an Immigrant Group*, 5 vols. (Chicago: University of Chicago Press, 1918; and Boston: Badger Press, 1919–1920); Robert E. Park and Herbert A. Miller [orig. author, William I. Thomas] *Old World Traits Transplanted* (New York: Harper and Row, 1921); Robert E. Park and Ernest W. Burgess, *Introduction to the Science of Sociology*, 2nd ed. (Chicago: University of Chicago Press, 1924); and Robert E. Park, Ernest W. Burgess, and Roderick D. McKenzie, *The City* (Chicago: University of Chicago Press, 1925). For Jane Addams, see *Hull-House Maps and Papers: A Presentation of Nationalities and Wages in a Congested District of Chicago* (New York: T. Y. Crowell, 1895).

7. Olaf Larson, "The Contribution of Wisconsin Rural Sociology: The Early Years," in *Rural Sociology: The Wisconsin Contribution, Current Status and Future Directions* (Madison: College of Agriculture and Life Sciences, University of Wisconsin, 1981), 1–34; and Arthur F. Wileden, "Early History of the Department of Rural Sociology," (Madison: Department of Rural Sociology, University of Wisconsin, 1979). One measure of Wisconsin's leadership role in rural studies at this time lies in the fact that John H. Kolb, together with Edmund de S. Brunner, was appointed by Herbert Hoover's Research Committee on Social Trends in 1929 to investigate the state of America's rural affairs as the nation was on the brink of its most significant rural crisis in a generation. J. H. Kolb and Edmund de S. Brunner, "Rural Life," in *Recent Social Trends in the United States* (New York: McGraw-Hill, 1933), 497–552.

8. Charles [Josiah] Galpin, "The Social Anatomy of an Agricultural Community," Research Bulletin No. 34 (Madison: Agricultural Experiment Station, College of Agriculture, University of Wisconsin, 1915). Lowry Nelson quoted in Larson, "The Contribution of Wisconsin Rural Sociology," 15. On Robert Park, see John H. Kolb, "Dr. Galpin at Wisconsin," *Rural Sociology* 13 (1948): 130–45.

9. John H. Kolb, "Rural Primary Groups," Research Bulletin No. 51 (Madison: Agriculture Experiment Station, College of Agriculture, University of Wisconsin, 1921), 5–6. See also by John H. Kolb, "Trends in Country Neighborhoods," Research Bulletin No. 120 (Madison: Agriculture Experiment Station, College of Agriculture, University of Wisconsin, 1933); "Neighborhood-Family Relations in Rural Society: A Review of Trends in Dane County, Wisconsin over a 35 Year Period," Research Bulletin No. 201 (Madison: Agriculture Experiment Station, College of Agriculture, University of Wisconsin, 1957); and his capstone work, *Emerging Rural Communities: Group Relations in Rural Society. A Review of Wisconsin Research in Action* (Madison: University of Wisconsin Press, 1959).

10. On the ethnic historians and literary critics, see respectively, Fred Matthews, "Paradigm Changes in Interpretations of Ethnicity, 1930–80: From Process to Structure," in *American Immigrants and Their Generations: Studies and Commentaries on the Hansen Thesis After Fifty Years*, ed. Peter Kivisto and Dag Blanck (Urbana: University of Illinois Press, 1990), 167–90; and David Hollinger, "Ethnic Diversity, Cosmopolitanism, and the Emergence of the American Liberal Intelligentsia," in his *In the American Province: Studies in the History and Historiography of Ideas* (Bloomington: Indiana University Press, 1985), 56–73.

11. Hjalmer J. Proope, U.S. Minester to Finland, to George W. Hill, Washington, D.C., 23 December 1941; and George W. Hill to Hjalmer J. Proope, 24 January 1942, box 1, folder 3, George Hill Records, Nationality Study, University of Wisconsin Archives (hereafter cited as GHR-NS). John I. Kolehmainen and George Hill, *Haven in the Woods: The Story of the Finns in Wisconsin* (Madison: State Historical Society of Wisconsin, 1951).

12. George W. Hill, "The People of Wisconsin: A Research Monograph," n.d., box 2, A-5, GHR-NS, emphasis added; and George W. Hill, "An Estimate of Differences in Farm Life Among Wisconsin Culture Types," 1939, Box 2, A-9, GHR-NS.

13. "Nation Watches U.W. Study of Political, Social, and Economic Folkway Institutions in Wisconsin," *University of Wisconsin Press Bulletin*, 4 September 1946.

14. George W. Hill, "Forward, Cultural-Ethnic Backgrounds in Wisconsin," 30 July 1940, I-iv, box 1, Wisconsin State Census Studies for 1905, State Historical Society of Wisconsin Archives; George Hill to J. H. Kolb, n.d., memo of Project 352, "Nationalities of Wisconsin," box 2, A-20, GHR-NS; and Hill, "The People of Wisconsin: A Research Monograph." The project was multifaceted and extended over a five-year period. At different times, and through different reports, notes, and memos, the project was variously known as "The People of Wisconsin," "Cultural-Ethnic Backgrounds in Wisconsin," "Study of Wisconsin Nationality Groups," "Nationalities of Wisconsin," and "Wisconsin Nationality Project." Hill and his colleagues apparently used these terms interchangeably, using the last most frequently. Therefore, and to eliminate confusion, I will refer to the various projects under the umbrella term, "Wisconsin Nationality Project" or, more simply, "Nationality Project."

15. Michel Foucault, *The Archaeology of Knowledge*, trans. A. M. Sheridan Smith (New York: Pantheon, 1972).

16. Project summary, "Study of Wisconsin Nationality Groups," WPA project 465–53–3-170, 21 May 1938, box 3, B-4, GHR-NS.

17. Project outline, "Study of Wisconsin Nationality Groups," WPA project 465–53–3-170, 21 May 1938, box 3, B-4, GHR-NS. See also Hill's methodological and theoretical statement of the project: George W. Hill, "The Use of the Culture-Area Concept in Social Research," *American Journal of Sociology* 47 (1941): 39–47.

18. Ibid.; Project summary, "Study of Wisconsin Nationality Groups."

19. Zetta Bankert, "Wisconsin Immigration Propaganda," carbon copy manuscript, 1938, box 2, A-12, GHR-NS; Walter Slocum, "Ethnic Stocks as Culture Types in Rural Wisconsin" (Ph.D. diss., University of Wisconsin, 1940); Rockwell Smith, "Church Affiliation as Social Differentiator in Rural Wisconsin" (Ph.D. diss., University of Wisconsin, 1942).

20. George W. Hill to Bankert, Slocum, Smith, and Smith, memo, 14 September 1938, box 2, A-11, GHR-NS. George W. Hill to A.J. Haas, memo, n.d., Budget for A Study of Wisconsin Nationality Groups, box 3, B-7, GHR-NS; George W. Hill, "Ethnic Backgrounds in Wisconsin: A Retabulation of Population Schedules from the Wisconsin State Census of 1905;" and project outline, "Study of Wisconsin Nationality Groups."

21. Hill, "The Culture-Area Concept," 43. Emphasis added.

22. Useful background on Hill's mapping may be found in Karl B. Raitz, "Ethnic Maps of North America," *Geographical Review* 68 (1978): 338–39; and Anne K. Knowles, "On the Trail of Ethnic Mapping," photocopied manuscript, 15 December 1989, vertical files, map collection, State Historical Society of Wisconsin Archives.

23. Hill, "The People of Wisconsin: A Research Monograph."

24. Douglas G. Marshall, "Cultural Background of Wisconsin People (Nationality Background): Source of Information, 1938," n.d., documentation file, State Historical Society of Wisconsin Archives.

25. George W. Hill to Register of Copyrights, 29 October 1940, box 7, methods folder, GHR-NS.

26. "Wisconsin's Changing Population," Science Inquiry Report no. 9, serial no. 2642, general series no. 2426 (Madison: University of Wisconsin, 1942), 4. George W. Hill to M. L. Freidlander (Commercial Director of Rand McNally and Company, Chicago), 24 September 1942; and Edward J. Mayland to George W. Hill, 14 July 1942, box 1, folder 3, GHR-NS.

27. George W. Hill to Dean (of the University of Wisconsin) Chris Christensen, 15 December 1941, box 2, A-10, GHR-NS. Emphasis added.

28. On the general ability of maps to be seen as "proof," see Denis Wood, *The Power of Maps* (New York: Guilford Press, 1992).

29. George W. Hill, "Works Progress Administration Project Proposal: A Study of Wisconsin Nationality Groups," 21 May 1938, box 3, B-4, GHR-NS. Emphasis added.

30. George W. Hill, "The Cultural Influence in the Agriculture of Price County, unpublished report," 15 January 1942, box 1, file 2, GHR-NS. George W. Hill, J. H. Kolb, and Ronald A. Smith, "A Study of Rural Relief in Wisconsin, Unpublished Report of WPA Project no. 344," n.d., box 2, A-14, GHR-NS.

31. George W. Hill, "Untitled Project Prospectus, A Study of Wisconsin Nationality Groups," 1937, box 3, B-4, GHR-NS.

32. George W. Hill, "An Estimate of Differences in Farm Life Among Wisconsin Culture Types, Unpublished Report," 3–6, 1939, box 2, A-9, GHR-NS; and Hill, Kolb, and Smith, "A Study of Rural Relief in Wisconsin." The results of this research were never fully elaborated, but the study that comes closest is George W. Hill and Ronald A. Smith, "Man in the 'Cut-Over': A Study of Family-Farm Resources," Research Bulletin No. 139 (Madison: Agricultural Experiment Station, University of Wisconsin, 1941). This report is a more localized case study of the larger, statewide phenomena that so interested Hill.

33. Clifford Butcher, "Little League of Nations is Wisconsin's Biggest Asset," *Milwaukee Journal*, 29 June 1941.

34. "Wisconsin's Changing Population," 4. George Hill served as the report's primary author.

35. Ibid., 13. Emphasis in original.

36. George W. Hill, "Fields and Faces in Wisconsin, Unpublished Lecture," 1941, box 2, A-10, GHR-NS.

37. George W. Hill, "A Review of 'Cultural Factors in Land Use Planning' by John Province," 29 December 1940, box 2, A-20, GHR-NS.

38. George Hill quoted in Butcher, "Little League of Nations," *Milwaukee Journal*, 29 June 1941.

39. Information on the life of Fred Holmes is gleaned from the impressive holdings at the State Historical Society of Wisconsin's manuscript archives. The Fred L. Holmes Papers (FHP) contain nine archival boxes and six scrapbooks detailing his extensive writing and public life. This short description of his life is based also on obituaries and appreciations in the *Capital Times*, 28 July 1946; Lawrence Whittet, "Frederick Lionel Holmes, 1883–1946," *Wisconsin Magazine of History* 30 (December 1946): 184–85; and "Frederick Lionel Holmes," in *Dic-*

tionary of Wisconsin Biography (Madison: State Historical Society of Wisconsin, 1960), 174–75.

40. Fred L. Holmes, *Side Roads: Excursions into Wisconsin's Past* (Madison: State Historical Society of Wisconsin, 1949). Fred L. Holmes, *Regulation of Railroads and Public Utilities in Wisconsin* (New York: D. Appleton and Co., 1915). Fred L. Holmes, *Old World Wisconsin: Around Europe in the Badger State* (Eau Claire, WI: E. M. Hale, 1944), 10.

41. Fred L. Holmes, *Alluring Wisconsin* (Milwaukee: E. M. Hale, 1937), 9–10. Fred L. Holmes, *Badger Saints and Sinners* (Milwaukee: E. M. Hale, 1939), 9–11.

42. Fred L. Holmes to E. M. Hale, 18 March 1943, box 2, folder 2, FHP.

43. Holmes, *Old World Wisconsin,* 9.

44. Ibid., 9–10. For travel writing as a genre, see Paul Fussell, "Travel Books as Literary Phenomena," in his *Abroad: British Literary Traveling Between the Wars* (Oxford: Oxford University Press, 1980), 202–15; and Sallie Tisdale, "Never Let the Locals See Your Map: Why Most Travel Writers Should Stay Home," *Harper's,* September 1995, 66–74.

45. Fred L. Holmes to C. H. Thordarson, 20 June 1941, box 7, folder 10, FHP.

46. The book's title needed to reflect this sense of dual purpose. Notes written during the early stages reveal the succession of names that alternated between travelogue and ethnography: "Around the World in Wisconsin"; "Trips About in Wisconsin"; "European Visibility in Wisconsin"; and "Old World Ways in Wisconsin." In the end, he settled on a blend that seemed to strike the right tone: *Old World Wisconsin: Around Europe in the Badger State.* The title was derived from a series of articles published weekly during the late 1930s in the *Milwaukee Journal,* which highlighted different groups throughout the state. Untitled, undated handwritten manuscript, box 5, folder 1, FHP.

47. Fred L. Holmes to Carl H. Daley, 8 July 1941; Fred L. Holmes to John Chapple, 8 July 1941; Ilmar Kauppinen to Fred L. Holmes, 2 October 1941; George Halonen to Fred L. Holmes, Ashland, Wisconsin, 15 September 1941; Fred L. Holmes to R. J. Holvenstot, 28 August 1941; Thomas A. Hippaka to Fred L. Holmes, Iowa City, 24 June 1941; Fred L. Holmes to George W. Hill, October 21, 1941; Fred L. Holmes to William Kinnuscese, 24 December 1941; Fred L. Holmes to David Wentela, 27 October 1941; Fred L. Holmes to Walter Frankie, 9 March 1942; and Fred L. Holmes to State Senator Phil Nelson, 26 June 1942, all from box 7, folder 12, FHP.

48. Fred L. Holmes to J. J. Figi, 27 May 1941; Fred L. Holmes to Esther Stauffacher, 6 and 13 June 1941; Fred L. Holmes to Henry M. Schmid, 25 August 1941; handwritten notes from interview with Emery Odell, Monroe, Wisconsin, 14 January 1941, all from box 6, folder 6, FHP.

49. Typeset questionnaire with handwritten responses, n.d., and handwritten question list, n.d., box 6, folder 6, FHP.

50. Ilmar Kauppinen to Fred L. Holmes, 2 October 1941; George Halonen to Fred L. Holmes, Ashland, Wisconsin, 15 September 1941; and Fred L. Holmes to S. W. Rahr, 14 September 1942 all from box 7, folder 12, FHP.

51. Corrected manuscript to chapter 12 of *Old World Wisconsin,* n.d., box

7, folder 12, FHP; and George W. Hill to Fred L. Holmes, Madison, 28 October 1941, box 7, folder 12, FHP.

52. Corrected manuscript to chapter 12 of *Old World Wisconsin,* n.d., box 7, folder 12, FHP.

53. Holmes, *Old World Wisconsin,* 131–50. Most of the background material from this paragraph is covered in chapters 1 and 2 of this book. The "resemblance myth" is worthy of a study all its own. One of the first discussions came from Glen Trewartha's work on the Green County cheese industry. Aware that such views were long on rhetoric and short on fact, he wrote: "Geographic, probably more than the sentimental reasons that are so frequently given, caused the Swiss colony to locate at New Glarus in Green County. Canton Glarus in Switzerland, from which the emigrants came, is one of the wildest and most mountainous states in that country, and any resemblance between the mountains of Old Glarus and the hills of Green County is small indeed." Glen Trewartha, "The Green County, Wisconsin, Foreign Cheese Industry," *Economic Geography* 2 (1926): 292–308. Likewise, the existence of Pet Milk and the nonexistence of operating cheese factories were there for anyone, including Holmes, to see on the landscape. For a more detailed discussion, see chapter 2 of this book and Dieter Brunnschweiler, *New Glarus: Gründung, Entwicklung und heutiger Zustand einer Schweizerkolonie im Amerikanischen Mittlewesten* (Zürich: Buchdruckerei Flutern, 1954).

54. Holmes opens himself to criticism by his own admission to "publish the truth only," and by the nearly one hundred scholarly footnotes that underline his text. The State Historical Society's executive committee called Holmes "a historian who . . . appreciated the opportunity to broaden and deepen that knowledge of the lives of our pioneers." Publisher William Evjue believed that the book brought together "between two covers a scholarly study of the many nationalities which migrated to the state and contributed to its development," and the *Chicago Tribune* noted that, in preparation for the book, Holmes "thoroly [sic] studied the many nationalities which have helped to people . . . his beloved state." Finally, the editor for the *Oshkosh Northwestern* found that "in the volume . . . he had placed in type a scholarly and thorough study of the many nationalities that migrated to this state." Whittet, "Frederick Lionel Holmes"; *Capital Times,* 3 May 1944; *Chicago Tribune,* 28 May 1944; *Oshkosh Northwestern,* 8 May 1944.

55. David Lowenthal, "Identity, Heritage, and History," in *Commemorations: The Politics of National Identity,* ed. John Gillis (Princeton: Princeton University Press, 1994), 41–60; and David Lowenthal, *Possessed by the Past: The Heritage Crusade and the Spoils of History* (New York: The Free Press, 1996).

56. Holmes, *Old World Wisconsin,* 68, 88, 49.

57. Ibid., 26, 68, 84, 124.

58. For an excellent treatment of the regionalist movement(s) of this period and into which *Old World Wisconsin* comfortably fits, see Robert L. Dorman, *Revolt of the Provinces: The Regionalist Movement in America, 1920–1945* (Chapel Hill: University of North Carolina Press, 1993), quote by Walter Lippmann, 24. See also Michael Steiner and Clarence Mondale, *Region and Regionalism in the United States: A Sourcebook for the Humanities and Social Sciences* (New York: Gar-

land, 1988). The connection between *Old World Wisconsin* and the regionalist movement was not lost on Chris Christensen, a Danish American businessman from Chicago. Upon reading the book, he wrote to author Holmes that it "is truly unique in its analysis of the customs and traditions of the different nationality groups that make up Wisconsin. In other words, you have made a regional approach in literature in the same way that Grant Wood and John Steuart Curry paint people and the environment in which they live." Christensen to Fred L. Holmes, Chicago, 10 May 1944, clipping scrapbook number 6, FHP.

59. Philip Gleason, "American Identity and Americanization," in *Harvard Encyclopedia of American Ethnic Groups,* eds. Stephan Thernstrom, Ann Orlov, and Oscar Handlin (Cambridge, MA: Harvard University Press, 1980), 31–58; Philip Gleason, "The Melting Pot: Symbol of Fusion or Confusion?," and "The Odd Couple: Pluralism and Assimilation," in his *Speaking of Diversity: Language and Ethnicity in Twentieth-Century America* (Baltimore: Johns Hopkins University Press, 1992), 3–31, 47–90; David Hollinger, "Pluralism, Cosmopolitanism, and the Diversification of Diversity," in his *Postethnic America: Beyond Multiculturalism* (New York: Basic Books, 1995), 79–104; David Hollinger, "Ethnic Diversity, Cosmopolitanism, and the Emergence of the American Liberal Intelligentsia" and "Democracy and the Melting Pot Reconsidered," in his *In the American Province: Studies in the History and Historiography of Ideas,* (Bloomington: Indiana University Press, 1985), 56–73, 92–102; John Higham, "Ethnic Pluralism and Modern American Thought," in his *Send These to Me: Jews and Other Immigrants in Urban America* (Baltimore: Johns Hopkins University Press, 1984), 196–230; and Yi-Fu Tuan, *Cosmos and Hearth: A Cosmopolite's Viewpoint* (Minneapolis: University of Minnesota Press, 1996).

60. See Hollinger, "Ethnic Diversity, Cosmopolitanism," 56–73, and *Postethnic America,* 79–104.

61. Philip Gleason, "Americans All: World War II and the Shaping of American Identity," *Review of Politics* 43 (1981): 493–94.

62. Barbara Kirshenblatt-Gimblett, "Objects of Ethnography," in *Exhibiting Cultures: The Poetics and Politics of Museum Display,* ed. Ivan Karp and Steven D. Lavine (Washington, D.C.: Smithsonian Institution Press, 1991), 433.

63. Walter S. Goodland to E. M. Hale, Madison, 8 June 1944, clipping scrapbook number 6, FHP. Emphasis added. *Milwaukee Journal,* 14 May 1944, clipping scrapbook number 6, FHP.

64. Charles Gillen to E. M. Hale, 3 May 1944, clipping scrapbook number 6, FHP. Emphasis added.

65. Grace Bloom to Fred L. Holmes, Osceola, Wisconsin, 20 August 1944; and Stolen to Fred L. Holmes, 19 June 1944, both from clipping scrapbook number 6, FHP.

66. Jonathan Culler, "Semiotics of Tourism," *American Journal of Semiotics* 1 (1981): 127–40. See also Barbara Kirshenblatt-Gimblett and Edward M. Bruner, "Tourism," in *Folklore, Cultural Performances, and Popular Entertainments,* ed. Richard Bauman (New York: Oxford University Press, 1992), 300–307; Kirshenblatt-Gimblett, "Objects of Ethnography"; and John Frow, "Tourism and the Semiotics of Nostalgia," *October* 57 (1991): 123–51. For an excellent case study

that argues for a roughly similar process at exactly this same time, see Ian McKay, "Tartanism Triumphant: The Construction of Scottishness in Nova Scotia, 1933–1954," *Acadiensis* 21 (1992): 5–47.

67. Louis Adamic, typeset manuscript review of *Old World Wisconsin*, 1944, clipping scrapbook number 6, FHP. Adamic's review was published subsequently in the *Wisconsin Magazine of History* 28 (1944): 87–89.

68. Nicholas V. Montalto, *A History of the Intercultural Educational Movement, 1924–1941* (New York: Garland, 1982), 68; and Stephan Thernstrom, Ann Orlov, and Oscar Handlin, eds., Introduction to *Harvard Encyclopedia of American Ethnic Groups* (Cambridge, MA: Harvard University Press, 1980). For other useful treatments of Adamic's work, see Robert F. Harney, "E Pluribus Unum: Louis Adamic and the Meaning of Ethnic History," *Journal of Ethnic Studies* 14 (1986): 29–46; Philip Gleason, "Minorities (Almost) All," in his *Speaking of Diversity: Language and Ethnicity in Twentieth-Century America* (Baltimore: Johns Hopkins University Press, 1992), 91–122; Richard Weiss, "Ethnicity and Reform: Minorities and the Ambience of the Depression Years," *Journal of American History* 66 (1979): 579–82; and Carey McWilliams, *Louis Adamic and Shadow-America* (Los Angeles: A. Whipple, 1935). Adamic's writing on ethnicity is approached best through three books, the first two of which contain significant collections of his popular writing and the third offers a more comprehensive appreciation of thirteen different groups: Louis Adamic, *My America, 1928–1938* (New York: Harper and Brothers, 1938); Louis Adamic, *From Many Lands* (New York: Harper and Brothers, 1940); and Louis Adamic, *A Nation of Nations* (New York: Harper and Brothers, 1945).

69. Louis Adamic, "Plymouth Rock and Ellis Island," in *From Many Lands*, 299.

70. Louis Adamic, "Thirty Million New Americans," in *My America*, 210–32.

71. Montalto, *Intercultural Educational Movement*, 71–73.

72. Louis Adamic, "Common Council for American Unity," in *From Many Lands*, 347.

73. Adamic, *A Nation of Nations*, 6.

74. Holmes's sanguine optimism allows him to put the most favorable spin imaginable on difficult situations. One of the most spectacular was his explanation for farm tenancy among Norwegians. Rather than lack of capital (the factor that Allan G. Bogue cites for tenancy among farmers on the corn belt), Holmes puts a racialist explanation of culture to work: "It was probably because of this fine physique, coupled with an innate sense of duty to give service for pay, that the Norwegian immigrants made such excellent farm hands. . . . Nearly every young Norwegian immigrant had to begin as a hired man." Although we read that many were "frightfully exploited," the larger message is that they were "strong enough to endure and regarded the conditions imposed upon them as necessary to their final winning of a place in America." *Old World Wisconsin*, 107, 108. Cf. Allan G. Bogue, *From Prairie to Cornbelt: Farming on the Illinois and Iowa Prairies in the Nineteenth Century* (Chicago: Quadrangle Books, 1968), 56–57.

75. Adamic, *A Nation of Nations*, 12.

76. Adamic, *From Many Lands*, 301, and *A Nation of Nations*, 5. David Hollinger's recent suggestion of a "rooted cosmopolitanism" and Yi-Fu Tuan's "cosmopolitan hearth" bear considerable resemblance to what I call provincial cosmopolitanism. Hollinger, *Postethnic America*, 5; and Tuan, *Cosmos and Hearth*, 182–88.

77. Herbert Kubly, "An American Finds America," *Common Ground* 3 (1943): 54. Herbert Kubly, "Everyone We Call 'Du'," *Common Ground* 3 (1943): 82–85. So indebted was Kubly, that he took the title and spirit for one his books directly from Adamic. Where Adamic described his 1933 trip back to Slovenia, Kubly wrote of his return visit to Switzerland. Louis Adamic, *The Native's Return: An American Immigrant Visits Yugoslavia and Discovers His Old Country* (New York: Harper and Brothers, 1934). Herbert Kubly, *Native's Return: An American of Swiss Descent Unmasks an Enigmatic Land and People* (New York: Stein and Day, 1981).

78. James Gray, "A Goodly Heritage: Souvenirs of Europe in Wisconsin," *St. Paul Dispatch*, 10 May 1944.

79. The most useful secondary source on the Old World Wisconsin Outdoor History Museum was written by its most recent director, Thomas A. Woods. See Woods, "Filiopietism at Historic Sites and Contemporary Alternatives" (paper presented at the annual meeting of the of the American Association of State and Local History, Miami, 22 September 1992). I would like to thank Director Wood for supplying me a copy of this paper. Primary sources include: *An Historic Structure Inventory for Wisconsin, Including the Old World Wisconsin Research Project* (Madison: State Historical Society of Wisconsin, 1969); *Old World Wisconsin: An Outdoor, Ethnic Museum* (Madison: Department of Landscape Architecture, University of Wisconsin, 1973); *Out of Many . . . One* (Madison: State Historical Society of Wisconsin, 1976); William H. Tishler, *Final Environmental Impact Statement for Old World Wisconsin, Bicentennial Park* (Madison: State Historical Society of Wisconsin, 1973); and William H. Tishler, *Old World Wisconsin: A Study Prepared for the State Historical Society of Wisconsin* (Madison: Department of Landscape Architecture, University of Wisconsin, 1972). It might be noted that George Hill's map was used as a key source in the planning of the Old World Wisconsin Outdoor History Museum. Douglas Marshall lent the State Historical Society of Wisconsin a copy of the manuscript atlas of Wisconsin counties in 1972 as part of that project. State Historical Society of Wisconsin Contact Card, "The People of Wisconsin According to Ethnic Stocks, 1940," 19 July 1972. I would like to thank archivist Harry Miller for bringing this connection to my attention.

80. *Old World Wisconsin: An Outdoor, Ethnic Museum*, 3.

81. Mary Tuohy Ryan (Assistant Supervisor of Wisconsin School Libraries) to E. M. Hale, Madison, 3 June 1943, box 2, folder 2, FHP. The master plan for the Old World Wisconsin Museum notes that: "In 1944 Fred Holmes published a book entitled *Old World Wisconsin*, a popular study of the ethnic folkways and customs of immigrants to the state. That book, issued four years before the centennial of statehood, stimulated a renewed interest in the ethnic history of Wisconsin and in the preservation of historic buildings." *Old World Wisconsin: An Outdoor, Ethnic Museum*, 17.

82. Maude Maxwell Munroe to Fred L. Holmes, n.d., scrapbook number 6, FHP. Rave reviews appeared in newspapers across the state and region: *New Richmond Leader, Fond du Lac Commonwealth Reporter, Wisconsin State Journal, Capital Times, Antigo Journal, Ashland Press, Milwaukee Journal,* and *St. Paul Dispatch* to name just a handful. Most saw the book, like the *Green Bay Press Gazette* (1 July 1944), as "a fascinating story of old customs in Wisconsin" and was pleased that "Green Bay is mentioned quite a bit as the center of a French community in Wisconsin." William Evjue (*Capital Times,* 3 May 1944) thought *Old World Wisconsin* to be "a book to be treasured in every Wisconsin library, a volume which will provide young and old with hour upon hour of entertaining and educational reading." August Derleth (*Capital Times,* 14 June, 1944) found it to be "a book to buy and keep on your shelves, informative, appreciative, and enthusiastic, a book to help you understand your native or adopted state." All reviews from clipping scrapbook number 6, FHP.

83. "Wisconsin's Luxemburgers Object," *Ozaukee Press,* reprinted in the *Milwaukee Journal,* 11 July 1944, clipping scrapbook number 6, FHP.

84. *Eau Claire Leader,* 18 June 1944, clipping scrapbook number 6, FHP.

85. Richard March, "Statewide Folk Arts Fieldwork in Preparation for the Wisconsin Folklife Festival," National Endowment for the Arts Grant Application WAB FY95–97 (Madison: Wisconsin Arts Board, 1994), 2. I would like to thank Rick March for sharing his proposal with me and for a fruitful conversation about the Old World in Wisconsin.

86. Hill, "Fields and Faces in Wisconsin," 4.

87. Ethel Theodora Rockwell, *A Century of Progress Cavalcade in Wisconsin: A Pageant Drama Based on Research in Wisconsin History Through the Century* (Madison: Wisconsin State Centennial Committee, 1948). John Bodnar puts the message of the centennial somewhat differently, as "unity over diversity." I would prefer to put see the former as a subset of the latter, hence unity *in* diversity. John Bodnar, "Memory in the Midwest After World War II," in his *Remaking America,* 138–68.

CHAPTER 6. "SWISSCAPES" ON MAIN STREET: LANDSCAPE, ETHNIC TOURISM, AND THE COMMODIFICATION OF PLACE

1. Herbert Kubly, "An American Finds America," *Common Ground* 3 (1943): 49–56, and "Everyone We Call 'Du'," *Common Ground* 3 (1943): 82–85; and "101 Years of Yodeling," *Time,* 26 August 1946, 21. This section is based largely on Kubly's memoir of this return visit: Herbert Kubly, "A Village of Wilhelm Tells," in *At Large* (Garden City, NY: Doubleday, 1964), 39–52, and from a taped interview with this author, 6 February 1992.

2. Kubly, "A Village of Wilhelm Tells," 41–42.

3. Ibid., 45.

4. Ibid., 50.

5. Mark Gottdiener, *The Theming of America: Dreams, Visions, and Commercial Spaces* (Boulder, CO: Westview Press, 1997).

6. J. B. Jackson, "Other-Directed Houses," in *Landscapes: Selected Writings of J. B. Jackson,* ed. Ervin H. Zube (Amherst: University of Massachusetts Press, 1970), 55–72. For a model case study into the intentionality of landscape pro-

duction, see Don Mitchell, *The Lie of the Land: Migrant Workers and the California Landscape* (Minneapolis: University of Minnesota Press, 1996).

7. Pierre van den Berghe and Charles Keyes, "Tourism and Re-Created Ethnicity," *Annals of Tourism Research* 11 (1984): 343–52. Useful case studies of this approach include Pierre van den Berghe, "Tourism and the Ethnic Division of Labor," *Annals of Tourism Research* 19 (1992): 234–49; Susan R. Pitchford, "Ethnic Tourism and Nationalism in Wales," *Annals of Tourism Research* 22 (1994): 35–52; Sylvia Rodriguez, "Ethnic Reconstruction in Contemporary Taos," *Journal of the Southwest* 32 (1990): 541–55; Majorie R. Esman, "Tourism as Ethnic Preservation: The Cajuns of Louisiana," *Annals of Tourism Research* 11 (1984): 451–67; Ian McKay, "Tartanism Triumphant: The Construction of Scottishness in Nova Scotia, 1933–1954," *Acadiensis* 21 (1992): 5–47; and Martha Norkunas, *The Politics of Public Memory: Tourism, History, and Ethnicity in Monterey, California* (Albany: State University of New York Press, 1993).

8. Dean MacCannell, "Reconstructed Ethnicity: Tourism and Cultural Identity in Third World Communities," in *Empty Meeting Grounds: The Tourist Papers* (London: Routledge, 1992), 158–171. Emphasis added. Mira Engler, "Drive-Thru History: Theme Towns in Iowa," *Landscape* 32 (1993): 8–18.

9. Edward M. Bruner, "Abraham Lincoln as Authentic Reproduction: A Critique of Postmodernism," *American Anthropologist* 96 (1994): 407; and Arjun Appadurai, ed., *The Social Life of Things: Commodities in Cultural Perspective* (Cambridge: Cambridge University Press, 1986), 44–45. See also Erik Cohen, "Authenticity and Commoditization in Tourism," *Annals of Tourism Research* 15 (1988): 371–86; Donald Redfoot, "Tourist Authenticity, Tourist Angst, and Modern Reality," *Qualitative Sociology* 7 (1984): 291–309; and Robert David Sack, *Place, Modernity, and the Consumer's World: A Relational Framework for Geographical Analysis* (Baltimore: Johns Hopkins University Press, 1992); 134–76. For a powerful deconstruction of authenticity as a discursive formation that bears considerable relevance to this study, see Regina Bendix, *In Search of Authenticity: The Formation of Folklore Studies* (Madison: University of Wisconsin Press, 1997).

10. A survey conducted for New Glarus's Comprehensive Plan revealed that 88 percent of the respondents agreed that "tourism is an important source of revenue for New Glarus," and 82 percent agreed that "the community's image as a 'Swiss Tourist Village' should be continually promoted." Rob Vetter, "New Glarus, Wisconsin Comprehensive Planning Program," Planning Report No. 105 (Platteville, WI: Southwest Wisconsin Regional Planning Commission, 1995), A-3.

11. This point is noted in William H. Tishler, "Built From Tradition: Wisconsin's Rural Ethnic Folk Architecture," *Wisconsin Academy Review* 30 (1984): 14–18.

12. D. Dürst, *Gründung und Entwicklung der Kolonie Neu-Glarus, Wisconsin, Nord-America, umfassend den Zeitraum von 1844–1892, nebst einer Reisebeschreibung* (Zürich: Institut Drell Fuessli, 1894), 21. This and all translations are my own.

13. Walter Bosshard, "New Glarus: Die Schweizer Mustersiedlung in U.S.A.," *Züricher Illustrierte,* 30 June 1933, 820–21. For a similar, but more detailed account, see Andreas Baumgartner, *Bei den Glarnern in Wisconsin* (Glarus, Switzerland: Glarner Nachrichten, 1934).

14. Dieter Brunnschweiler, *New Glarus: Gründung, Entwicklung und heutiger Zustand einer Schweizerkolonie im Amerikanischen Mittlewesten* (Zürich: Buchdruckerei Flutern, 1954), 85.

15. John Luchsinger, "The Swiss Colony of New Glarus," *Wisconsin Historical Collections* 8 (1879): 1–29. Basel traveler Emanuel Dettweiler, quoted in Ernest Menolfi and Leo Schelbert, "The Wisconsin Swiss: A Portrait," *Clarion* 16 (1991): 57–63.

16. *The First One Hundred Years of the Swiss Evangelical and Reformed Church* (New Glarus, WI: n. p., 1950). For more details on the "old stone church" and the modern church that replaced it, see chapter 2.

17. The metaphor "time out of time" is derived from the useful collection of essays in Alessandro Falassi, ed., *Time Out of Time: Essays on the Festival* (Albuquerque: University of New Mexico Press, 1987).

18. Most of the details of Jacob Rieder's life are derived from interviews with his acquaintances in New Glarus (Don Ott, Herbert Kubly, Elda Schiesser, and Millard Tschudy) and from his obituary, *New Glarus Post*, 4 March 1953.

19. Don Ott interview with author, New Glarus, Wisconsin, August 1994.

20. "Former Green County Resident Makes Suggestion to Local Friends," *New Glarus Post*, 10 September 1935.

21. Ibid.

22. The best resource of Swiss building traditions remains Richard Weiss's classic, *Häuser und Landscaften der Schweiz* (Zürich: Eugen Rentsch Verlag, 1959).

23. Canton Glarus, it should be noted, has a long history of vernacular building traditions, of which chalets are not included. Ibid.

24. Clayton Streiff, interview with the author, New Glarus, Wisconsin, August 1993.

25. William H. Nicholas, "Deep in the Heart of 'Swissconsin'," *National Geographic*, June 1947, 781–800. A postcard of the Emmentaler Chalet from the 1950s reveals that Thierstein was especially fond of his full-page photograph that appeared in the *National Geographic* article, as he adorned his favorite room with it.

26. Miriam T. Abplanalp, "Chalet on the Hill—A Bit of Old World in New Glarus," *Wisconsin State Journal*, 20 March 1949.

27. Penny Kubly interview with the author, New Glarus, Wisconsin, July 1994.

28. The figures on the plant closure come from the *New Glarus Post*, 17 January 1962. The calculation of employment data was based on 1970 data derived from Donald Rosenbrook, "New Glarus, Wisconsin Comprehensive Planning Program," Planning Report No. 23 (Platteville, WI: Southwestern Wisconsin Regional Planning Commission, 1977), 27.

29. "Editorial," *New Glarus Post*, 17 January 1962.

30. Ibid.; Marcia Crowley, "New Glarus Businesses Working on 'New Look,'" *Wisconsin State Journal*, 26 May 1963.

31. *New Glarus Post*, 24 January and 7 February 1962. Don Mitchell, "Heritage, Landscape, and the Production of Community: Consensus History and its Alternatives in Johnstown, Pennsylvania," *Pennsylvania History* 59 (1992):

198–226. See also Briaval Holcomb, "Revisioning Place: De- and Reconstructing the Image of the Industrial City," in *Selling Places: The City as Cultural Capital, Past and Present*, ed. Gerry Kearns and Chris Philo (Oxford: Pergamon Press, 1993), 133–44.

32. "Gottfried Wants New Glarus to be Even More Swiss-Like," *Wisconsin State Journal*, n.d. [probably 1956], NGHSA.

33. Millard Tschudy interview with the author, New Glarus, Wisconsin, August 1994. For reports to the community by experts at the Wisconsin Dells, the Wisconsin Chamber of Commerce and the University Extension, see William E. Millard, "The Sale of Culture" (Ph.D. diss., University of Minnesota, 1969); Crowley, "New Glarus Businesses"; "Lions See Film on Tourism Money," *New Glarus Post*, 3 April 1963; and Gerald Hannon, "Retail Trade and Services Provide Major Economic Base," *New Glarus Post*, 27 June 1962.

34. See the following *New Glarus Post* articles: "Report of the New Glarus Business Association Meeting," 2 May 1962; "Betterment Group Plans for Summer," 2 May 1962; and "Post Chips," 16 May 1962.

35. Schneider quoted in Crowley, "New Glarus Businesses."

36. Roland Moss, "Community Aesthetics in New Glarus," *New Glarus Post*, 9 August 1962. Emphasis added.

37. *Capital Times*, 23 July 1970.

38. In a 1977 survey, 39.6 percent of those questioned did not favor architectural controls as opposed to 14.2 percent who were in favor of control; 46.2 percent needed more information. Rosenbrook, "Planning Program," 231. The percentage opposing such guidelines has remained constant since 1977. A follow up survey in 1995 found that 38 percent did not favor architectural controls: Vetter, "Planning Program," A-3. On Iowa theme towns, see Engler, "Drive-Through History."

39. Don Ott interview with the author, New Glarus, Wisconsin, August 1994. The following section is based on this interview.

40. John Brinkerhoff Jackson, *Discovering the Vernacular Landscape* (New Haven: Yale University Press, 1984), xii, 86, 149. This theme is carried into Jackson's last book, *A Sense of Place, A Sense of Time* (New Haven: Yale University Press, 1994).

41. Similar findings characterized the work of Benetta Jules-Rosette, who argues, in a different context, that the longer an artisan works in producing "tourist art" the more she develops an aesthetic that satisfies her own cultural identity. Benetta Jules-Rosette, *The Messages of Tourist Art: An African Semiotic System in Comparative Perspective* (New York: Plenum Press, 1984).

42. Paul Grossenbacher, "Looking Back: From Burgdorf, Canton Bern, to New Glarus, Wisconsin. An Autobiographical Sketch of a Twentieth Century Swiss Immigrant," *Swiss American Historical Review* 25 (1989): 48.

43. Hans Lenzlinger interview with the author, New Glarus, Wisconsin, July 1994; Emmy Gallaher (widow of Stuart M. Gallaher) telephone interview, November 1995; "Olbrich Architect Gallaher Dies at 60," *Wisconsin State Journal*, 31 July 1991.

44. Lenzlinger interview with the author, New Glarus, Wisconsin, July 1994. Weiss, *Häuser und Landscaften der Schweiz*.

45. Lenzlinger interview with the author, New Glarus, Wisconsin, July 1994.

46. David Harvey, *The Condition of Postmodernity* (Oxford: Basil Blackwell, 1989), 83. For a detailed and useful discussion of postmodern architecture, see Charles Jencks, *The Language of Post-Modern Architecture*, 4th ed. (New York: Rizzoli, 1984).

47. Gottlieb Brändli interview with the author, Monroe, Wisconsin, September 1994.

48. Jencks, *The Language of Post-Modern Architecture*. Pauline Ott interview with the author, New Glarus, Wisconsin, August 1994.

49. This finding is based on an unpublished visitor survey administered during the summer of 1993 in which one thousand tourists were asked their opinion of their trip to New Glarus. For the results of this survey and its methodology, see Steven Hoelscher, "The Invention of Ethnic Place" (Ph.D. diss., University of Wisconsin–Madison, 1995), 443–62, Appendix 2.

50. "Plans Going Ahead for Historic Village," *New Glarus Post*, 24 May 1939.

51. For the creation of the Swiss Historical Village, see chapter 3.

52. "'Kühermutzen' mit Amerikanischen Akzent: 158 Neu-Glarner Besuchen die alten Heimat," *Zürich Illustrierte Rundschau*, 1 October 1960; "Tour Guests Receive Warm Swiss Welcome," *New Glarus Post*, 16 September 1960; Millard Tschudy, *New Glarus, Wisconsin: Mirror of Switzerland*, 7th ed. (New Glarus, WI: n.p., 1995), 35. Clayton Streiff and Marion Streiff interview with the author, New Glarus, Wisconsin, February 1992.

53. Millard Tschudy interview with the author, New Glarus, Wisconsin, 21 February 1992. Tschudy was the chair of the 120th anniversary celebration and largely responsible for the idea of creating a Hall of History. "Glarus will Contribute $1,000 for History Hall," *New Glarus Post*, 23 June 1965; and "Hall of History: Idee, Realisierung, Finanzierung," n.d. [probably 1970], report produced by the Swiss Friends of New Glarus, Glarus, Switzerland, copy in author's possession; I would like to thank Kaspar Marti, of Engi, Switzerland for this copy. "Hall of History," n.d. [probably 1970], NGHSA; Jakob Winteler, "Glarus Sends Greetings on 120th Anniversary," *New Glarus Post*, 4 August 1965; and "2,500 at Finale of Swiss Celebration," *Capital Times*, 2 August 1965.

54. David Lowenthal notes that such a presentation of the past is common throughout the region; Lowenthal, "Pioneer Museums," in *History Museums in the United States: A Critical Assessment*, ed. Warren Leon and Roy Rosenzweig (Urbana: University of Illinois Press, 1989), 115–27.

55. Elda Schiesser, "Hall of History: Funding of it as Views Herd [*sic*] from Paul Grossenbacher explained to me," 14, March 1976, typed manuscript in possession of Elda Schiesser. Grossenbacher, "Looking Back," 48–49.

56. "Hall of History," 3; Schiesser, "Hall of History."

57. Jakob Winteler to Paul Grossenbacher, Glarus, Switzerland, n.d., translated by Paul Grossenbacher, NGHSA correspondence notebook. Emphasis added.

58. Lukas F. Burkhardt (Cultural Counselor of the Embassy of Switzer-

land) to Wayne Duerst, Washington, D.C., 22 November 1968, NGHSA correspondence notebook.

59. Kaspar Marti interview with the author, New Glarus, Wisconsin, August 1994.

60. Jakob Zweifel, Willi Marti, and Heinrich Strickler, "Hall of History New Glarus: Vorschlag, Gestalterische Ueberlegungen," November, 1966, NGHSA, emphasis in original, my translation. For a recent account, see Jakob Zweifel, "Die Enstehung der 'Hall of History,'" in *Amerikas Little Switzerland Erinnert Sich*, ed. Hans Rhyner (Glarus, Switzerland: Schweizerischer Verein der Freunde von New Glarus, 1996), 281–88.

61. Useful introductions to the modernist architectural framework that Zweifel's Hall of History adopts are Henry-Russell Hitchcock, *Architecture: Nineteenth and Twentieth Centuries*, 4th ed. (Harmondsworth, England: Penguin, 1977), 419–30; and Edward Relph, *The Modern Urban Landscape* (Baltimore: Johns Hopkins University Press, 1987), 98–118, 190–210.

62. "Hall of History: Idee, Realisierung, Finanzierung," 9; and "Hall of History," 7. Funds raised by the Glarus-based Swiss Friends came from major private sector and governmental donations at both the cantonal and federal level.

63. Grossenbacher, "Looking Back," 48–49.

64. "Das einzige echte Schweizer Gebäude in New Glarus." Claudia Kock and Kaspar Marti, "Geschichte(n) und Mythen: Vertraut und dennoch Fremd: Swiss-American Architecture in New Glarus," *Fridolin*, 28, September 1995.

65. In addition to interviews with eight local residents who knew Schneider well and all of whom wish to remain anonymous, background information for this section was gleaned from the following sources: Louise C. Marston, "The Hills are Alive with the Sound . . . " *Wisconsin State Journal*, 14 December 1969; Walter R. Tetzlaff, Brian J. Thorsen, and Gordon Mickelson, "Demand and Economic Analysis, Swiss Village, New Glarus," Research Report 4-0028 (Madison: Recreation Resources Center, University of Wisconsin Extension, 1973), 9; and Calvin Trillin, "U.S. Journal: New Glarus, WI: Swissness," *New Yorker*, 20 January 1975, 51–57. For a general history of the Swiss musical heritage into which Schneider fit, see James P. Leary, *Yodeling in Dairyland: A History of Swiss Music in Wisconsin* (Mount Horeb, WI: Wisconsin Folk Museum, 1991).

66. "New Glarus Hotel Gets New Manager," *New Glarus Post*, 10 December 1960; "New Glarus 'Swiss Face,'" *Monroe Evening Times*, 13, March 1962; "Open House Planned," *Capital Times*, 30 June 1962; Richard Vesey, "Make a Delicious Beef Fondue: Serve a Swiss Fun Meal," *Wisconsin State Journal*, 31 March 1963; and "Robbie's Yodel Club Has Grand Opening Festivities," *New Glarus Post*, 6 March 1966.

67. Tetzlaff, Thorsen, and Mickelson, "Demand and Economic Analysis," 9.

68. "Schneider and Hoffmann Get New Appointments," *New Glarus Post*, 4 May 1966; Confidential communications with author, New Glarus, Wisconsin, May and November 1993, August 1994; Trillin, "Swissness," 55–56.

69. Kerwin Steffen interview with the author, Madison, May 1993. Official attendance data supplied by Millard Tschudy, New Glarus, Wisconsin.

70. Robbie Schneider and Phyl Anderson, press release, the Swiss Village, n.d. [probably 1974], NGHSA; Trillin, "Swissness," 57. The list of participants in the research included: Michigan State University; U.S. Department of Commerce; Wisconsin Bureau of Commercial Recreation; Wisconsin Department of Natural Resources; Wisconsin Department of Resource Development; Wisconsin Department of Transportation; U.S. Department of Agriculture; University of Wisconsin Extension; Wisconsin Southwest Regional Planning Commission; Swiss National Museum in Zürich; Green County Agricultural and Soil Conservation Service; U.S. Department of Interior; Wisconsin Department of Labor and Human Relations; Texas A & M University; the University of Illinois; Wisconsin Department of Agriculture; Swiss National Tourist Office; and the University of Wisconsin's School of Landscape Architecture, School of Engineering, and Department of Business.

71. Tetzlaff, Thorsen, and Mickelson, "Demand and Economic Analysis," 1. Other studies include: L. G. Monthey, "Evaluation of Robert Schneider Land, with Respect to Developing a Swiss-Theme Village near New Glarus, Wis.," (Madison: University of Wisconsin Extension, n.d.); and Donald Schink, "Economic Impact of the Proposed Swiss Village, New Glarus," (Madison: Recreation Resources Center, University of Wisconsin Extension, 1975).

72. Tetzlaff, Thorsen, and Mickelson, "Demand and Economic Analysis," 6–7; Schneider and Anderson, press release.

73. "Swiss Village Design Finalized, Authentic Swiss Architecture Planned," *New Glarus Post,* 26 June 1974; "Swiss Building Specialists to Work on Schneider's Resort," *New Glarus Post,* 8 May 1974; "Schneider Unveils Swiss Village, Says Construction to Begin Soon," *New Glarus Post,* 1 May 1974; and Schneider and Anderson, press release. The comparison to Colonial Williamsburg is instructive and, unwittingly, points to similar controversies over authenticity, power, and the problematic of representing the past. Richard Handler and Eric Gable, *The New History in an Old Museum: Creating the Past at Colonial Williamsburg* (Durham: Duke University Press, 1997).

74. Sack, *Place, Modernity, and the Consumer's World,* 160–62.

75. Millard Tschudy quoted in Trillin, "Swissness," 56.

76. Mike Wallace, "Disney's America," in *Mickey Mouse History and Other Essays on American Memory* (Philadelphia: Temple University Press, 1996), 159–74. Donald Schink telephone interview with the author, 11 August 1994.

77. "Schneider Leases Out New Glarus Hotel," *New Glarus Post,* 4 April 1975; "Swiss Village Deal Falls Through," *Monroe Evening Times,* 19 October 1978; "Foreclosure on Schneider Home Almost Complete," *New Glarus Post,* 24 December 1986.

78. Trillin, "Swissness," 57.

79. Rosenbrook, "Planning Program," 226.

80. David Harvey, "From Space to Place and Back Again: Reflections on the Condition of Postmodernity," in *Mapping the Futures: Local Cultures, Global Change,* ed. Jon Bird et al. (London: Routledge, 1993), 3–29.

81. Vetter, "Planning Program," A-9-11. Confidential communication with author, New Glarus, Wisconsin, August 1995.

82. Confidential communication with the author, New Glarus, Wisconsin,

August 1995. Lionel Trilling, *Sincerity and Authenticity* (Cambridge, MA: Harvard University Press, 1972).

83. Trillin, "Swissness," 48.

EPILOGUE. ETHNIC PLACE IN AN AGE OF HIGH MODERNITY

1. Marion S. Streiff interview with author, New Glarus, Wisconsin, August 1995; Esther Zgraggen interview with author, New Glarus, Wisconsin, August 1995. This section is based largely on field notes, interviews, and brochures from the 1995 New Glarus sesquicentennial, 4–14 August 1995.

2. Confidential communications with author, New Glarus, Wisconsin, August 1994 and August 1995.

3. Ann Schmidt, "Swiss Bring Much to New Glarus 150th; Receive Much in Turn," *New Glarus Post*, 16 August 1995. Confidential communication with author, New Glarus, Wisconsin, August 1995.

4. Ann Schmidt, "Old Glarus Contributes to Sesquicentennial," *New Glarus Post*, 26 April, 1995; Ann Schmidt, "Art from Switzerland: A View from Modern Glarus," *New Glarus Post*, n.d.,—special sesquicentennial issue, 1995.

5. Kaspar Marti, "Von Kultur and Kulturen, von Kunst and Künsten" in *Kunst Glarus-Schweiz*, ed. Kaspar Marti, Madeleine Schuppli, and Susann Wintsch (Glarus, Switzerland: Glarner Kunstverein, 1995), 28–31.

6. Kaspar Villiger to Tony Zgraggen, Bern, Switzerland, n.d., 1995. Letter reprinted in *New Glarus Post*, 6 September, 1995.

7. Hans Rhyner interview with the author, New Glarus, Wisconsin, 12 August 1995.

8. Kellie Engelke, letter to the editor, *New Glarus Post*, 6 September, 1995.

9. These conclusions are roughly in agreement with Mary C. Waters's excellent work on white ethnics in nonethnic communities: *Ethnic Options: Choosing Identities in America* (Berkeley: University of California Press, 1990) and "The Construction of a Symbolic Ethnicity: Suburban White Ethnics in the 1980s," in *Immigration and Ethnicity: American Society—"Melting Pot" or "Salad Bowl"?*, ed. Michael D'Innocenzo and Josef P. Sirefman (Westport, CT: Greenwood, 1992), 75–90.

10. Confidential communications with the author, New Glarus, Wisconsin, August 1995.

11. George Lipsitz, *Time Passages: Collective Memory and American Popular Culture* (Minneapolis: University of Minnesota Press, 1990), 34.

12. *Old World Wisconsin: An Outdoor, Ethnic Museum* (Madison: Department of Landscape Architecture, University of Wisconsin, 1973), 8; and *Out of Many . . . One: Old World Wisconsin* (Madison: State Historical Society of Wisconsin, 1976), n.p.

13. *Wisconsin's Ethnic Settlement Trail, A Travel Brochure* (Sheboygan Falls, WI: WEST, 1995).

14. Richard March, "Statewide Folk Arts Fieldwork in Preparation for the Wisconsin Folklife Festival," National Endowment for the Arts Grant Application WAB FY95-97 (Madison: Wisconsin Arts Board, 1994).

15. Ian McKay, *The Quest of the Folk: Antimodernism and Cultural Selection in*

Twentieth-Century Nova Scotia (Montreal and Kingston: McGill-Queens, 1994), 298. Martha Norkunas presents a similar argument in the case of Monterey, California: *The Politics of Public Memory: Tourism, History, and Ethnicity in Monterey, California* (Albany: State University of New York Press, 1993).

16. My understanding of the complex nature of public memory and history owes much to David Glassberg, "Public History and the Study of Memory," *The Public Historian* 18 (1996): 7–23. Such a conclusion is also found in Rodger Lyle Brown's account of contemporary festivals in the South. See his *Ghost Dancing on the Cracker Circuit: The Culture of Festivals in the American South* (Jackson: University Press of Mississippi, 1997).

17. David Harvey, *The Condition of Postmodernity* (Oxford: Basil Blackwell, 1989); Anthony Giddens, *The Consequences of Modernity* (Stanford: Stanford University Press, 1990); and Ulrich Beck, "The Reinvention of Politics: Towards a Theory of Reflexive Modernization," in *Reflexive Modernization: Politics, Tradition and Aesthetics in the Modern Social Order,* ed. Ulrich Beck, Anthony Giddens, and Scott Lash (Stanford: Stanford University Press, 1994), 1–55. Also useful is Marshall Berman, *All That Is Solid Melts into Air: The Experience of Modernity* (New York: Simon and Schuster, 1982).

18. Anthony Giddens, "Living in a Post-Traditional Society," in Beck, Giddens, and Lash, eds., *Reflexive Modernization,* 101.

19. For the difficulty of "being ethnic" outside ethnic place, see Waters, *Ethnic Options.*

20. David A. Gerber, *The Making of an American Pluralism: Buffalo, New York, 1825–1860* (Urbana: University of Illinois Press, 1989). See also: Kathleen Neils Conzen, "German-Americans and the Invention of Ethnicity," in *America and the Germans: An Assessment of a Three-Hundred Year History,* ed. Frank Tommler and Joseph McVeigh (Philadelphia: University of Pennsylvania Press, 1985), 131–47; and Kathleen Neils Conzen et al., "The Invention of Ethnicity: A Perspective from the U.S.A.," *Journal of American Ethnic History* 12 (1992): 3–41.

21. A Swiss identity crisis, though brewing for decades, boiled over in the late 1990s with the Holocaust banking scandal. For a sampling of the troubling questions that have called unwelcome attention to the country and its struggle with public memory and heritage, see Alan Cowell, "Swiss, Irked by Critics, Ask 'Why Single Us Out?,'" *New York Times,* 6 June 1997, A3; Thomas Sanction, "A Painful History: Swiss Are Rethinking Their Role in the Nazi Era," *Time,* 24 February 1997, 40–41; Michael Hirsh, "Switzerland's Reckoning," *Newsweek,* 9 June 1997, 58; and "More Questions, More Squirming," *Economist,* 1 May 1997, 49. One indication of the increasingly contested nature of Swiss identity is demonstrated by the resignation of Switzerland's Ambassador to the U.S., Carlo Jagmetti, less than eighteen months after he praised New Glarus for maintaining traditions lost in "the homeland." David Sanger, "Swiss Envoy to U.S. Resigns after His Report on Holocaust Dispute Is Disclosed," *New York Times,* 28 January 1997, A4.

22. Giddens, *The Consequences of Modernity,* 38.

23. Victor Turner, "Are There Universals of Performance in Myth, Ritual, and Drama?" in *By Means of Performance: Intercultural Studies of Theatre and Rit-*

ual, ed. Richard Schechner and Willa Appel (New York: Cambridge University Press, 1990), 9.

24. T. J. Jackson Lears, *No Place of Grace: Antimodernism and the Transformation of American Culture* (New York: Pantheon, 1981), xv.

25. David Glassberg, *American Historical Pageantry: The Uses of Tradition in the Early Twentieth Century* (Chapel Hill: University of North Carolina Press, 1990), 5.

26. David Gerber, Ewa Morawska, and George E. Pozzetta, "Response to Comments on 'The Invention of Ethnicity'," *Journal of American Ethnic History* 12 (1992): 60.

27. Tamara K. Hareven, "The Search for Generational Memory: Tribal Rites in Industrial Society," *Daedalus* 107 (1978): 137–49. Mariellen R. Sandford, "Tourism in Harlem: Between Negative Sighteeing and Gentrification," *Journal of American Culture* 10 (1987): 99–106. On Kwanzaa, see Gerald Lyn Early, "Dreaming of a Black Christmas," *Harper's*, January 1997, 55–61; Judith Waldrop, "Happy Kwanzaa: African American Celebration of Culture and Heritage," *American Demographics*, December 1994, 4; and Eric V. Copage, *Kwanzaa: An African American Celebration of Culture and Cooking* (New York: William Morrow, 1991). I would like to thank Joyce Jackson for sharing this reference and her own experiences of Kwanzaa with me.

28. Lawrence Fuchs, *The American Kaleidoscope: Race, Ethnicity, and the Civic Culture* (Hanover, NH: Wesleyan University Press, 1990). Gary Gerstle, "Liberty, Coercion, and the Making of Americans," *Journal of American History* 84 (1997): 524–58.

29. David Hollinger, *Postethnic America: Beyond Multiculturalism* (New York: Basic Books, 1995). For a similar view, see: Yi-Fu Tuan, *Cosmos and Hearth: A Cosmopolite's Viewpoint* (Minneapolis: University of Minnesota Press, 1996). The distinction between the descriptive and normative is central to my argument, as it is for David Hollinger's. He notes that *"not enough Americans are now, or ever have been, as free as they should be to decide how much or how little emphasis to place on their communities of descent"*; Hollinger, "National Solidarity at the End of the Twentieth Century: Reflections on the United States and Liberal Nationalism," *Journal of American History* 84 (1997): 560. Emphasis in original.

Bibliography

MANUSCRIPT COLLECTIONS

New Glarus Historical Society Archives, Swiss Historical Village
(New Glarus, Wisconsin)

Of all the manuscript collections used during the course of this research, that of the New Glarus Historical Society at the Swiss Historical Village proved most useful. The archives there are extensive, diverse, and valuable, but, until recently, in considerable disarray. In 1995, the Historical Society had launched a project to organize and catalog the collection, resulting in the series of holdings listed below. I have labeled the holding for each manuscript document cited in the endnotes. Additionally and importantly, I have benefited from the personal collections of two local historians, Elda Schiesser and Millard Tschudy, who graciously opened their own considerable archives for my perusal; these also are listed as such in the endnotes.
Clipping Boxes
Correspondence Notebooks
Documents Boxes
Subject Files
Wilhelm Tell Scrapbooks

State Historical Society of Wisconsin Manuscript Archives
(Madison, Wisconsin)

John M. Becker Papers
Ethnic History of Wisconsin Project Recordings, William J. Schereck
Frederick L. Holmes Papers, Correspondence, *Old World Wisconsin* Manuscripts and Notes, Clipping Scrapbooks
Herbert Kubly Papers
Loyalty Legion Correspondence and Miscellaneous Papers, 1917–1919
Maps of the Cultural Background of Wisconsin People, Douglas G. Marshall
War History Commission: Wisconsin in World War I, Papers and Clipping File, 1917–1918
Wisconsin State Census Studies for 1905

Bibliography

Mills Music Library Archives, University of Wisconsin
(Madison, Wisconsin)

Helen Stratman-Thomas Papers

Max Kade Institute for German-American Studies
(Madison, Wisconsin)

Milwaukee Turners Constitution, ca. 1925

University of Wisconsin Archives
(Madison, Wisconsin)

George W. Hill Nationality Study Papers (8 boxes)
George W. Hill General Correspondence (6 boxes)

Wilhelm Tell Community Guild Vertical File,
New Glarus Elementary School
(New Glarus, Wisconsin)

Account Books
Secretary Notes
Treasury Notes

INTERVIEW DATA

The research for this book relies heavily on the information derived from formal interviews, less formal conversations, and direct observation of, and participation in, many community events between 1991 and 1995. Mentioned here are the people with whom I have conducted formal interviews as well as those with whom I spoke during the many hours of practice and performances of *Wilhelm Tell*, at various board meetings, and at dozens of other festivals. In most cases, the date (month and year) of the interview accompanies the name of the informant in the endnotes for each chapter. In numerous instances, however, when an informant explicitly wished to remain anonymous and to protect his/her identity, the citation was entered as "confidential communication." In each case, I coded that interview into my field notes, indicating the month and year of the interview. All the informants reside in New Glarus, Wisconsin, except where indicated otherwise.

Joyce Alderman, tourism promoter
Hans Anderegg, construction worker
Barbara Anderson, elementary school librarian
Doris Arn, insurance saleswoman
Richard Arn, businessman
Verla Babler, retired teacher
Janeen Joy Babler, folk artist (Monroe, Wisconsin)
Carol Brand, retired (Monroe, Wisconsin)
Gottlieb Brändli, cabinetmaker (Monroe, Wisconsin)
Julie Buehler, student
Keith Buehler, student

Sepp Candinas, sports store owner (Madison, Wisconsin)
Marilyn Christensen, German teacher
Howard Christensen, university employee
Howard Cosgrove, businessman
Janice Danz, homemaker (Janesville, Wisconsin)
Victor Danz, businessman (Janesville, Wisconsin)
Mary Dibble, businesswoman
Margaret Duerst, musem curator
Jackie Elmer, government worker
Hubert Elmer, retired farmer
Peter Etter, school district administrator
Toni Frank, bank manager
Dan Gartzke, lawyer
Roland Hagmann, chef (Zürich, Switzerland)
Jim Haldiman, gas station manager
Hans Hauser, farmer
Peter Herdegg, businessman (Chicago, Illinois)
Linda Hiland, teacher
Bill Hoesly, dairy worker
Robert (Buzz) Holland, lumber dealer
Ernest Jaggi, retired farmer, hotel keeper, and yodeler
Teresa Jaggi, retired hotel keeper and yodeler
Jeff Klassy, filmmaker
Claudia Kock, anthropologist (Engi, Switzerland)
Penny Kubly, homemaker
Herbert Kubly, writer (now deceased)
Hans Lenzlinger, chef and restaurant owner
Karl Luescher, farmer
Ruth Luescher, farmer
Kaspar Marti, architect (Engi, Switzerland)
Rodney Marty, lawyer
Phillip Marty, music teacher (Madison, Wisconsin)
Heinz Mattmann, butcher
Shannon McGuire, student
Jerome Mooney, photographer and publisher (La Crosse, Wisconsin)
Tom Nielson, pastor
Karen Nodorft, bank manager
Chuck Oldenburg, bank manager
Pauline Ott, hotel worker
Don Ott, social worker
Chuck Phillipson, photographer
Otto Puempel, bar owner
Hans Rhyner, judge (Glarus, Switzerland)
Phyllis Richert, museum curator
Jennifer Roethe, reporter (Monroe, Wisconsin)
Willi Ruef, butcher
Hanni Ruef, waitress

Elda Schiesser, folk artist
Todd Schneider, student
Kerwin Steffen, businessman
Clayton Streiff, retired (now deceased)
Dennis Streiff, retired insurance salesman
Marion Streiff, homemaker
Marion Streiff-Frederickson, museum worker
Kim Tschudy, university worker
Millard Tschudy, retired store owner
Virginia Tshudy, hotel worker
Rene Weber, cheesemaker (Monroe, Wisconsin)
Edward Willi, lawyer (now deceased)
Esther Zgraggen, farmer
Tony Zgraggen, farmer

OTHER PRINT SOURCES

Abrahams, Roger D. "An American Vocabulary of Celebrations." In Falassi, ed., *Time Out of Time*, 173–83.

Abrahams, Roger D. "The Discovery of Marketplace Culture." *Intellectual History Newsletter* 10 (1988): 23–32.

Abrahams, Roger D. "Shouting Match at the Border: The Folklore of Display Events." In *"And Other Neighborly Names": Social Process and Cultural Image in Texas Folklore,* edited by Richard Bauman and Roger D. Abrahams, 303–22. Austin: University of Texas Press, 1981.

Abrahams, Roger D., and Richard Bauman. "Ranges of Festival Behavior." In *The Reversible World: Symbolic Inversion in Art and Society,* edited by Barbara Babcock, 193–208. Ithaca: Cornell University Press, 1978.

Abramson, Harold. "Assimilation and Pluralism Theories." In Thernstrom, Orlov, and Handlin, eds., *The Harvard Encyclopedia of American Ethnic Groups,* 150–60.

Adamic, Louis. *From Many Lands*. New York: Harper and Brothers, 1940.

Adamic, Louis. *My America, 1928–1938*. New York: Harper and Brothers, 1938.

Adamic, Louis. *A Nation of Nations*. New York: Harper and Brothers, 1945.

Adamic, Louis. *The Native's Return: An American Immigrant Visits Yugoslavia and Discovers His Old Country*. New York: Harper and Brothers, 1934.

Adamic, Louis. "Review of Fred Holmes' *Old World Wisconsin.*" *Wisconsin Magazine of History* 28 (1944): 87–89.

Adams, Kathleen M. "Come to Tana Toraja, 'Land of the Heavenly Kings': Travel Agents as Brokers in Ethnicity." *Annals of Tourism Research* 11 (1984): 469–85.

Agnew, Jean-Christophe. "Coming Up for Air: Consumer Culture in Historical Perspective." *Intellectual History Newsletter* 12 (1990): 3–21.

Agulhon, Maurice. *Marianne into Battle: Republican Imagery and Symbolism in France, 1789–1880*. New York: Cambridge University Press, 1981.

Alba, Richard D. *Ethnic Identity: The Transformation of White America*. New Haven: Yale University Press, 1990.

Bibliography

Alba, Richard D. "The Twilight of Ethnicity among Americans of European Ancestry: the Case of Italians." *Ethnic and Racial Studies* 8 (1985): 134–58.

Anderson, K. J. "Cultural Hegemony and the Race-Definition Process in Chinatown, Vancouver: 1880–1980." *Environment and Planning D: Society and Space* 6 (1988): 127–49.

Anderson, Kay, and Fay Gale. *Inventing Places: Studies in Cultural Geography.* Melbourne: Longman, 1992.

Anderson, Phyl, and Elda Schiesser. *The History of the New Glarus Historical Society, Inc.* New Glarus, WI: New Glarus Historical Society, 1976.

Appadurai, Arjun, ed. *The Social Life of Things: Commodities in Cultural Perspective.* Cambridge: Cambridge University Press, 1986.

Appel, John J. *Immigrant Historical Societies in the United States, 1880–1950.* New York: Arno Press, 1980.

Archdeacon, Thomas J. *Becoming American: An Ethnic History,* New York: Free Press, 1983.

Atherton, Lewis. *Main Street on the Middle Border* (1954). Bloomington: Indiana University Press, 1984.

Bakhtin, Mikhail. *Rabelais and His World.* Bloomington: Indiana University Press, 1984.

Barber, Benjamin R. *The Death of Communal Liberty: A History of Freedom in a Swiss Mountain Canton.* Princeton: Princeton University Press, 1974.

Barber, Benjamin. *Jihad vs. McWorld.* New York: Ballantine Books, 1996.

Barnes, Trevor J., and James Duncan. "Introduction: Writing Worlds." In Barnes and Duncan, eds., *Writing Worlds: Discourse, Text and Metaphor in the Representation of Landscape,* 1–17. London: Routledge, 1992.

Bartel, Otto, and Adolf Jenny. *Glarner Geschichte in Daten.* Glarus, Switzerland: Buchdruckerei Neue Glarner Zeitung, 1936.

Bauchle, Mary L. "Last of the *Swiss* Pioneers." *The Wisconsin Magazine,* April 1927, 17.

Bauman, Richard, ed. *Folklore, Cultural Performances, and Popular Entertainments.* New York: Oxford University Press, 1992.

Bauman, Richard. "Folklore" and "Performance." In Bauman, ed., *Folklore,* 29–40, 41–49.

Baumgartner, Andreas. *Bei den Glarnern in Wisconsin.* Glarus, Switzerland: Glarner Nachrichten, 1934.

Bausinger, Hermann. "Organisierte Volkskunde als Objekt volkskundlicher Forschung." *Rheinische Heimatpflege* 4 (1968): 311–20.

Beck, Ulrich. "The Reinvention of Politics: Towards a Theory of Reflexive Modernization." In Beck, Giddens, and Lash, eds., *Reflexive Modernization,* 1–55.

Beck, Ulrich, Anthony Giddens, and Scott Lash, eds. *Reflexive Modernization: Politics, Tradition and Aesthetics in the Modern Social Order.* Stanford: Stanford University Press, 1994.

Belasco, Warren James. *Americans on the Road: From Autocamp to Motel, 1910–1945.* Cambridge, MA: MIT Press, 1979.

Bender, Thomas. *Community and Social Change in America.* Baltimore: Johns Hopkins University Press, 1982.

Bendix, Regina. *Backstage Domains: Playing "William Tell" in Two Swiss Communities.* Bern: Peter Lang, 1989.

Bendix, Regina. "Tourism and Cultural Displays: Inventing Traditions for Whom?" *Journal of American Folklore* 102 (1989): 131–46.

Bendix, Regina. "National Sentiment in the Enactment and Discourse of Swiss Political Ritual." *American Ethnologist* 14 (1992): 768–90.

Bendix, Regina. *In Search of Authenticity: The Formation of Folklore Studies.* Madison: University of Wisconsin Press, 1997.

Benn, Shiela Margaret. *Pre-Romantic Attitudes to Landscape in the Writings of Friedrich Schiller.* Berlin: Walter de Gruyter, 1991.

Bennett, John. "The Exhibitionary Complex." *New Formations* 4 (1988): 73–102.

Benson, Susan Porter, Stephen Brier, and Roy Rosenzweig, eds. *Presenting the Past: Essays on History and the Public.* Philadelphia: Temple University Press, 1986.

Berchtold, Alfred. "Wilhelm Tell im 19. und 20. Jahrhundert." In *Tell: Werden und Wandern eines Mythos,* edited by Lilly Stunzi, 167–311. Bern: Hallway Verlag, 1973.

Berman, Marshall. *All That Is Solid Melts into Air: The Experience of Modernity.* New York: Simon and Schuster, 1982.

Bernard, Richard M. *The Melting Pot and the Altar: Marital Assimilation in Early Twentieth-Century Wisconsin.* Minneapolis: University of Minnesota Press, 1980.

Best, Steven. "The Commodification of Reality and the Reality of Commodification: Jean Baudrillard and Post-Modernism." *Current Perspectives in Social Theory* 9 (1989): 23–51.

Billigmeier, Robert Henry. *A Crisis in Swiss Pluralism: The Romansh and their Relations with the German- and Italian-Swiss in the Perspective of a Millennium.* The Hague: Mouton Publishers, 1979.

Bodnar, John. "Collective Memory and Ethnic Groups: The Case of Swedes, Mennonites, and Norwegians." *Swenson Swedish Immigration Research Center Occasional Papers* (1991): 7–39.

Bodnar, John. *Remaking America: Public Memory, Commemoration, and Patriotism in the Twentieth Century.* Princeton: Princeton University Press, 1992.

Bodnar, John. "Remembering the Immigrant Experience in American Culture." *Journal of American Ethnic History* 15 (1995): 3–27.

Bodnar, John. *The Transplanted: A History of Immigrants in Urban America.* Bloomington: Indiana University Press, 1985.

Bogue, Allan G. *From Prairie to Cornbelt: Farming on the Illinois and Iowa Prairies in the Nineteenth Century.* Chicago: Quadrangle Books, 1968.

Boniface, Brian G., and Chris Cooper. *The Geography of Travel and Tourism.* 2nd ed. Oxford: Butterworth-Heinemann, 1994.

Boorstin, Daniel. "From Traveler to Tourist: The Lost Art of Travel." In *The Image: A Guide to Pseudo-Events in America,* (1961), 77–117. New York: Atheneum, 1973.

Bosshard, Walter. "New Glarus: Die Schweizer Mustersiedlung in U.S.A." *Züricher Illustrierte,* 30 June 1933, 820–21.

Bourne, Randolph. "Trans-National America" (1916). In *War and the Intellectuals: Essays of Randolph S. Bourne, 1915–1919,* edited by Carl Resek, 107–23. New York: Harper Torchbooks, 1964.

Braun, Rudolf. *Industrialization and Everyday Life,* translated by Sarah Hanbury Tenison. Cambridge: Cambridge University Press, 1990.

Braun, Rudolf. *Sozialer und Kultureller Wandel in einem Ländlichen Industriegebiet im 19. und 20. Jahrhundert.* Zürich: Eugen Rentsch Verlag, 1965.

Brigham, Helen. "New Glarus." In *History of Green County, Wisconsin,* 247–57. Milwaukee: Burdick and Armitage, 1877.

Britton, S. "Tourism, Capital, and Place: Towards a Critical Geography of Tourism." *Environment and Planning D: Society and Space* 9 (1991): 451–78.

Brooks, Robert Clarkson. *Civic Training in Switzerland: A Study in Democratic Life.* Chicago: University of Chicago Press, 1930.

Brown, Francis J., and Joseph Slabey Roucek, eds. *One America: The History, Contributions, and Present Problems of our Racial and National Minorities.* New York: Prentice-Hall, 1937.

Brown, Rodger Lyle. *Ghost Dancing on the Cracker Circuit: The Culture of Festivals in the American South.* Jackson: University Press of Mississippi, 1997.

Brown, Victoria. "Esther Streiff Stauffacher." In *Uncommon Lives of Common Women: The Missing Half of Wisconsin History,* 84. Madison: Wisconsin Feminists Project, 1975.

Bruner, Edward M. "Abraham Lincoln as Authentic Reproduction: A Critique of Postmodernism." *American Anthropologist* 96 (1994): 397–415.

Brunnschweiler, Dieter. *New Glarus: Gründung, Entwicklung und heutiger Zustand einer Schweizerkolonie im Amerikanischen Mittlewesten.* Zürich: Buchdruckerei Flutern, 1954.

Brye, David L. *Wisconsin Voting Patterns in the Twentieth Century.* New York: Garland, 1979.

Bulmer, Martin. *The Chicago School of Sociology: Institutionalization, Diversity, and the Rise of Sociology Research.* Chicago: University of Chicago Press, 1984.

Burckhardt, Lukas. "Swissconsin." In *America Unser Spiegel Bild?: Amerikanische Reiseberichte,* 45–52. Basel: Volksdruckerei, 1954.

Cameron, Catherine M., and John B. Gatewood. "The Authentic Interior: Questing Gemeinschaft in Post-Industrial Society." *Human Organization* 53 (1994): 21–32.

Carey, Lorin Lee. "The Wisconsin Loyalty Legion, 1917–1918." *Wisconsin Magazine of History* 53 (1969): 33–50.

Coates, Helen R. *The American Festival Guide.* New York: Exposition Press, 1956.

Cohen, Abner. "A Polyethnic London Carnival as a Contested Cultural Performance." *Ethnic and Racial Studies* 5 (1982): 23–41.

Cohen, Erik. "Authenticity and Commoditization in Tourism." *Annals of Tourism Research* 15 (1988): 371–86.

Cohen, Erik. "A Phenomenology of Tourist Experiences." *Sociology* 13 (1979): 179–201.

Connerton, Paul. *How Societies Remember.* Cambridge: Cambridge University Press, 1989.

Conzen, Kathleen Neils. "Ethnicity as Festive Culture: Nineteenth-Century German America on Parade." In Sollors, ed., *The Invention of Ethnicity,* 44–76.

Bibliography

Conzen, Kathleen Neils. "German-Americans and the Invention of Ethnicity." In *America and the Germans: An Assessment of a Three-Hundred Year History,* edited by Frank Tommler and Joseph McVeigh, 131–47. Philadelphia: University of Pennsylvania Press, 1985.

Conzen, Kathleen Neils. "Historical Approaches to the Study of Rural Ethnic Communities." In *Ethnicity on the Great Plains,* edited by Frederick C. Luebke, 1–18. Lincoln: University of Nebraska Press, 1980.

Conzen, Kathleen Neils. "Mainstreams and Side Channels: The Localization of Immigrant Cultures." *Journal of American Ethnic History* 11 (1991): 5–20.

Conzen, Kathleen Neils. *Making Their Own America: Assimilation Theory and the German Peasant Pioneer.* German Historical Institute Annual Lecture Series, vol. 3. New York: Berg, 1990.

Conzen, Kathleen Neils, et al. "The Invention of Ethnicity: A Perspective from the U.S.A." *Journal of American Ethnic History* 12 (1992): 3–41.

Conzen, Michael P. "Culture Regions, Homelands, and Ethnic Archipelagos in the United States: Methodological Considerations." *Journal of Cultural Geography* 13 (1993): 13–29.

Conzen, Michael P. "Ethnicity on the Land." In *The Making of the American Landscape,* edited by Michael P. Conzen, 221–48. Boston: Unwin Hyman, 1990.

Cook, Marshall. "Coming Home: From 'Beyond the Horizon,' Herbert Kubly has returned to New Glarus." *Wisconsin Trails,* July 1985, 26–29.

Copage, Eric V. *Kwanzaa: An African American Celebration of Culture and Cooking.* New York: William Morrow, 1991.

Cosgrove, Denis. "Spectacle and Society: Landscape as Theater in Premodern and Postmodern Cities." In *Understanding Ordinary Landscapes,* edited by Paul Groth and Todd W. Bressi, 99–110. New Haven: Yale University Press, 1997.

Cosgrove, Denis. "Culture Area." In *The Dictionary of Human Geography,* edited by R. J. Johnston, Derek Gregory, and David M. Smith, 92. Oxford: Basil Blackwell, 1986.

Cosgrove, Denis, and Stephen Daniels, eds. *The Iconography of Landscape: Essays on the Symbolic Representation, Design, and Use of Past Environments.* Cambridge: Cambridge University Press, 1988.

Cosgrove, Denis, and Peter Jackson. "New Directions in Cultural Geography." *Area* 19 (1987): 95–101.

Craig, Gordon A. "Zurich." In *Geneva, Zurich, Basel,* edited by Nicholas Bouvier, Gordan A. Craig, and Lionel Gossman, 41–62. Princeton: Princeton University Press, 1994.

Crispino, James A. *The Assimilation of Ethnic Groups: The Italian Case.* Staten Island, NY: Center for Migration Studies, 1980.

Culler, Jonathan. "Semiotics of Tourism." *American Journal of Semiotics* 1 (1981): 127–40.

Current, Richard Nelson. *Wisconsin: A Bicentennial History.* New York: Norton, 1977.

Curtis, Charles Linton. "How Our Ancestors Settled in Wisconsin." *The Wisconsin Magazine,* October 1924, 19–21, 29.

Curtis, Wardon Allan. "'The Light Fantastic in the Central West': Country

Dances of Many Nationalities in Wisconsin." *Century Illustrated Monthly Magazine*, 1907, 570–79.

Daniels, Stephen. "Marxism, Culture, and the Duplicity of Landscape." In *New Models in Geography*, edited by Richard Peet and Nigel Thrift, 196–220. London: Unwin Hyman, 1989.

Danielson, Larry. "The Ethnic Festival and Cultural Revivalism in a Small Midwestern Town." Ph.D. diss., Indiana University, 1972.

Danielson, Larry. "St. Lucia in Lindsborg, Kansas." In *Creative Ethnicity: Symbols and Strategies of Contemporary Ethnic Life*, edited by Stephen Stern and John Allan Cicala, 187–203. Logan, UT: Utah State University Press, 1991.

Davatz, Jürg. "Die Glarner Textilindustrie." In *Glarus und die Schweiz: Streiflichter auf wechselseitige Beziehungen*, edited by Jürg Davatz, 128–36. Glarus, Switzerland: Verlag Baeschlin, 1991.

Davis, Susan G. *Parades and Power: Street Theatre in Nineteenth-Century Philadelphia*. Berkeley: University of California Press, 1988.

Deegan, Mary Jo. *Jane Addams and the Men of the Chicago School, 1892–1918*. New Brunswick, NJ: Transaction Books, 1988.

DeSantis, Grace, and Richard Benkin. "Ethnicity without Community." *Ethnicity* 7 (1980): 137–43.

Dickinson, Thomas H. *The Case of American Drama*. New York: Houghton-Mifflin, 1915.

Dorman, Robert L. *Revolt of the Provinces: The Regionalist Movement in America, 1920–1945*. Chapel Hill: University of North Carolina Press, 1993.

Duncan, J., and N. Duncan. "(Re)reading the Landscape." *Environment and Planning D: Society and Space* 6 (1988): 117–26.

Duncan, James, and David Ley, eds. *Place/Culture/Representation*. London: Routledge, 1993.

Dundes, Alan. "The Apple-Shot: Interpreting the Legend of William Tell." *Western Folklore* 50 (1991): 327–60.

Dürst, Matthias. "Matthias Dürst's Travel Diary." In Schelbert, ed., *New Glarus*, 20–150.

Eagan, Richard. *Green County, Wisconsin: A History of the Agricultural Development*. Monroe, WI: Monroe Evening Times, 1929.

Early, Gerald Lyn. "Dreaming of a Black Christmas." *Harper's*, January 1997, 55–61.

Eco, Umberto. *Travels in Hyperreality*. New York: Harcourt, Brace, Jovanich, 1983.

Engler, Mira. "Drive-Thru History: Theme Towns in Iowa." *Landscape* 32 (1993): 8–18.

Enninger, Werner. "Clothing." In Bauman, ed., *Folklore*, 217–24.

Esman, Majorie R. "Tourism as Ethnic Preservation: The Cajuns of Louisiana." *Annals of Tourism Research* 11 (1984): 451–67.

Falassi, Alessandro, ed. *Time Out of Time: Essays on the Festival*. Albuquerque: University of New Mexico Press, 1987.

Falk, Karen. "Public Opinion in Wisconsin During World War I." *Wisconsin Magazine of History* 25 (1942): 389–407.

Farley, Reynolds. "The New Census Question about Ancestry: What Did It Tell Us?" *Demography* 28 (1991): 411–29.

Fischer, Michael. "Ethnicity and the Post-Modern Arts of Memory." In *Writing Culture: The Poetics and Politics of Ethnography*, edited by James Clifford and George Marcus, 194–233. Berkeley: University of California Press, 1986.

Fisher, Joel. "Sinclair Lewis and the Diagnostic Novel: *Main Street* and *Babbitt.*" *Journal of American Studies* 20 (1986): 421–33.

Fitzgerald, Thomas. "Media and Changing Metaphors of Ethnicity and Identity." *Media, Culture, and Society* 13 (1991): 193–214.

Foucault, Michel. *The Archaeology of Knowledge.* New York: Pantheon, 1972.

Foucault, Michel. *Language, Countermemory, Practice: Selected Essays and Interviews.* Ithaca: Cornell University Press, 1977.

"Frederick Lionel Holmes." In *Dictionary of Wisconsin Biography*, 174–75. Madison: State Historical Society of Wisconsin, 1960.

Frenkel, Stephen. "Alluring Landscapes: The Symbolic Economy of 'Bavarian' Leavenworth," Paper presented at the annual meeting of the Association of American Geographers, Ft. Worth, Texas, 4 April 1997.

Fricker, Yves. "Switzerland's Image Abroad." In *Switzerland in Perspective*, edited by Janet Eve Hilowitz, 207–22. New York: Greenwood 1990.

Frisch, Max. *Wilhelm Tell für die Schule.* Frankfurt am Main: Suhrkamp, 1971.

Frow, John. "Tourism and the Semiotics of Nostalgia." *October* 57 (1991): 123–51.

Fuchs, Lawrence H. *The American Kaleidoscope: Race, Ethnicity, and the Civic Culture.* Hanover, NH: Wesleyan University Press, 1990.

Fussell, Paul. "Travel Books as Literary Phenomena." In *Abroad: British Literary Travelling Between the Wars*, 202–15. Oxford: Oxford University Press, 1980.

Gable, Eric, and Richard Handler. "After Authenticity at an American Heritage Site." *American Anthropologist* 98 (1996): 568–78.

Gall, Lothar. *Germania: Eine Deutsche Marianne?* Bonn: Bouvier Verlag, 1993.

Galpin, Charles. "The Social Anatomy of an Agricultural Community." Madison: Agricultural Experiment Station, College of Agriculture, University of Wisconsin, 1915.

Gamper, Josef. "Reconstructed Ethnicity: Comments on MacCannell." *Annals of Tourism Research* 12 (1985): 250–52.

Gans, Herbert J. "Symbolic Ethnicity: the Future of Ethnic Groups and Cultures in America." *Ethnic and Racial Studies* 2 (1979): 1–20.

Geertz, Clifford. *The Interpretation of Cultures: Selected Essays.* New York: Basic Books, 1973.

Gerber, David A. *The Making of an American Pluralism: Buffalo, New York, 1825–1860.* Urbana: University of Illinois Press, 1989.

Gerber, David, Ewa Morawska, and George E. Pozzetta. "Response to Comments on 'The Invention of Ethnicity.'" *Journal of American Ethnic History* 12 (1992): 60.

Gerstle, Gary. "Liberty, Coercion, and the Making of Americans." *Journal of American History* 84 (1997): 524–558.

Giddens, Anthony. "Living in a Post-Traditional Society." In Beck, Giddens, and Lash, eds., *Reflexive Modernization*, 56–109.

Giddens, Anthony. *The Consequences of Modernity.* Stanford: Stanford University Press, 1990.

Gillespie, Angus. "Folk Festival and Festival Folk in Twentieth-Century America." In Falassi, ed., *Time Out of Time*, 152–61.

Gillis, John. "Memory and History: The History of a Relationship." In Gillis, ed., *Commemorations: The Politics of National Identity,* 3–26. Princeton: Princeton University Press, 1994.

Gjerde, Jon. *From Peasants to Farmers: The Migration from Balestrand, Norway to the Upper Middle West.* Cambridge: Cambridge University Press, 1985.

Glassberg, David. *American Historical Pageantry: The Uses of Tradition in the Early Twentieth Century.* Chapel Hill: University of North Carolina Press, 1990.

Glassberg, David. "Public History and the Study of Memory." *The Public Historian* 18 (1996): 7–23.

Glassie, Henry. "Tradition." *Journal of American Folklore* 108 (1995): 395–412.

Glazer, Nathan, and Daniel P. Moynihan. *Beyond the Melting Pot: The Negroes, Puerto Ricans, Jews, Italians, and Irish of New York City.* Cambridge, MA: M.I.T. Press, 1963.

Gleason, Philip. "American Identity and Americanization." In Thernstrom, Orlov, and Handlin, eds., *The Harvard Encyclopedia of American Ethnic Groups,* 31–58.

Gleason, Philip. "Americans All: World War II and the Shaping of American Identity." *Review of Politics* 43 (1981): 483–518.

Gleason, Philip. *Speaking of Diversity: Language and Ethnicity in Twentieth-Century America.* Baltimore: Johns Hopkins University Press, 1992.

Gordon, Milton. *Assimilation in American Life: The Role of Race, Religion, and National Origins.* New York: Oxford University Press, 1964.

Gottdiener, Mark. *The Theming of America: Dreams, Visions, and Commercial Spaces.* Boulder, CO: Westview Press, 1997.

Gray, Jack, and Donna Monson. "An Attraction Guest Study: Swiss Historical Village." Madison: University of Wisconsin Extension, Recreation Resources Center, 1982.

Greeley, Andrew. *Ethnicity in the United States: A Preliminary Reconnaissance.* New York: Wiley, 1974.

Greene, Victor. *American Immigrant Leaders, 1800–1910: Marginality and Identity.* Baltimore: Johns Hopkins University Press, 1987.

Greenwood, Davydd J. "Culture by the Pound: An Anthropological Perspective on Tourism as Cultural Commoditization." In *Hosts and Guests: The Antropology of Tourism,* edited by Valene Smith, 171–85. Philadelphia: University of Pennsylvania Press, 1989.

Grossenbacher, Paul. "Looking Back: From Burgdorf, Canton Bern, to New Glarus, Wisconsin. An Autobiographical Sketch of a Twentieth-Century Swiss Immigrant." *Swiss American Historical Review* 25 (1989): 5–49.

Hacker, Andrew. "Trans-National America." *New York Review of Books,* 22 November 1990, 19–24.

Hahn, Steve, and Jonathan Prude, eds. *The Countryside in the Age of Capitalist*

Transformation: Essays on the Social History of Rural America. Chapel Hill: University of North Carolina Press, 1985.

Halbwachs, Maurice. *The Social Frameworks of Memory.* Chicago: University of Chicago Press, 1992.

Hale, Frederick. *The Swiss in Wisconsin.* Madison: State Historical Society of Wisconsin, 1984.

Hamilton, Susan, et al. "Visitor Profile of the Wisconsin Dells' Area Attractions." Madison: University of Wisconsin Extension, Tourism Research and Resource Center, 1992.

Hand, Wayland D. "*Schweizer Schwingen:* Swiss Wrestling in California." *California Folklore Quarterly* 2 (1943): 77–84.

Handelman, Don. *Models and Mirrors: Towards an Anthropology of Public Events.* New York: Cambridge University Press, 1990.

Handler, Richard. *Nationalism and the Politics of Culture in Quebec.* Madison: University of Wisconsin Press, 1988.

Handler, Richard, and Eric Gable. *The New History in an Old Museum: Creating the Past at Colonial Williamsburg.* Durham: Duke University Press, 1997.

Handler, Richard, and Jocelyn Linnekin. "Tradition, Genuine or Spurious." *Journal of American Folklore* 97 (1984): 273–90.

Hansen, Marcus Lee. "The Problem of the Third Generation Immigrant." In Kivisto and Blanck, eds., *American Immigrants,* 191–203, [orig. pub. 1937].

Hareven, Tamara K. "The Search for Generational Memory: Tribal Rites in Industrial Society." *Daedalus* 107 (1978): 137–49.

Harney, Robert F. "E Pluribus Unum: Louis Adamic and the Meaning of Ethnic History." *Journal of Ethnic Studies* 14 (1986): 29–46.

Harvey, David. *The Condition of Postmodernity.* Oxford: Basil Blackwell, 1989.

Harvey, David. "From Space to Place and Back Again: Reflections on the Condition of Postmodernity." In *Mapping the Futures: Local Cultures, Global Change,* edited by Jon Bird et al., 3–29. London: Routledge, 1993.

Herold, J. Christopher. *The Swiss without Halos.* New York: Columbia University Press, 1948.

Hewison, Robert. *The Heritage Industry: Britain in an Age of Decline.* London: Methuen, 1987.

Higham, John. "Another American Dilemma: Integration vs. Pluralism." *The Center,* July/August 1974, 67–73.

Higham, John. "Current Trends in the Study of Ethnicity in the United States." *Journal of American Ethnic History* 3 (1982): 6–15.

Higham, John. "The Ethnic Historical Society in Changing Times." *Journal of American Ethnic History* 14 (1994): 30–44.

Higham, John. *Send These to Me: Jews and Other Immigrants in Urban America.* Baltimore: Johns Hopkins University Press, 1984.

Higham, John. *Strangers in the Land: Patterns of American Nativism, 1860–1925,* 2nd ed. New York: Atheneum, 1978.

Hill, George W. "The Use of the Culture-Area Concept in Social Research." *American Journal of Sociology* 47 (1941): 39–47.

Hill, George W., and Ronald A. Smith. "Man in the 'Cut-Over': A Study of Family-Farm Resources." Madison: Agricultural Experiment Station, University of Wisconsin, 1941.

Bibliography

An Historic Structure Inventory for Wisconsin, Including the Old World Wisconsin Research Project. Madison: State Historical Society of Wisconsin, 1969.

Hitchcock, Henry-Russell. *Architecture: Nineteenth and Twentieth Centuries,* 4th ed. Harmondsworth: Penguin Books, 1977.

Hobsbawm, Eric, and Terence Ranger, eds. *The Invention of Tradition.* Cambridge: Cambridge University Press, 1983.

Hoelscher, Steven, Jeffrey Zimmerman, and Timothy Bawden. "Milwaukee's German Renaissance Twice-Told: Inventing and Recycling Landscape in America's German Athens." In *Wisconsin Land and Life,* edited by Robert Ostergren and Thomas Vale, 376–409. Madison: University of Wisconsin Press, 1997.

Hoelscher, Steven, and Robert C. Ostergren. "Old European Homelands in the Middle West." *Journal of Cultural Geography* 13 (1993): 87–106.

Holcomb, Briaval. "Revisioning Place: De- and Re-constructing the Image of the Industrial City." In *Selling Places: The City as Cultural Capital, Past and Present,* edited by Gerry Kearns and Chris Philo, 133–44. Oxford: Pergamon Press, 1993.

Holdsworth, Deryck. "Revaluing the House." In *Place/Culture/Representation,* edited by James Duncan and David Ley, 95–109. London: Routledge, 1993.

Hollinger, David. *In the American Province: Studies in the History and Historiography of Ideas.* Bloomington: Indiana University Press, 1985.

Hollinger, David. "National Solidarity at the End of the Twentieth Century: Reflections on the United States and Liberal Nationalism." *Journal of American History* 84 (1997): 559–69.

Hollinger, David. *Postethnic America: Beyond Multiculturalism.* New York: Basic Books, 1995.

Holmes, Fred L. *Alluring Wisconsin.* Milwaukee: E. M. Hale, 1937.

Holmes, Fred L. *Badger Saints and Sinners.* Milwaukee: E. M. Hale, 1939.

Holmes, Fred L. *Old World Wisconsin: Around Europe in the Badger State.* Eau Claire, WI: E. M. Hale, 1944.

Holmes, Fred L. *Regulation of Railroads and Public Utilities in Wisconsin.* New York: D. Appleton and Co., 1915.

Holmes, Fred L. *Side Roads: Excursions into Wisconsin's Past.* Madison: State Historical Society of Wisconsin, 1949.

Hornung, Richard. "Unterrichtseinheit: Wilhelm Tell—Held oder Verbrecher?" *Anregung* 24 (1978): 275–82.

Horowitz, Helen Lefkowitz. *Culture and the City: Cultural Philanthropy in Chicago from the 1880s to 1917.* Chicago: University of Chicago Press, 1976.

Hout, Michael, and Joshua Goldstein. "How 4.5 Million Irish Immigrants Became 40 Million Irish Americans: Demographic and Subjective Aspects of the Ethnic Composition of White Americans." *American Sociological Review* 59 (1994): 64–82.

Howe, Irving. "The Limits of Ethnicity." *New Republic,* 25 June 1977, 17–19.

Hugger, Paul, ed. *Handbuch der Schweizerischen Volkskultur,* 3 vols. Zürich: Offizin, 1992.

Hull-House Maps and Papers: A Presentation of Nationalities and Wages in a Congested District of Chicago. New York: T. Y. Crowell, 1895.

Bibliography

Hummon, David M. "Tourist Worlds: Tourist Advertising, Ritual and American Culture." *Sociological Quarterly* 10 (1988): 179–202.

Hymes, Dell. "Folklore's Nature and the Sun's Myth." *Journal of American Folklore* 88 (1975): 345–69.

Ibarra, Robert A. "Ethnicity Genuine and Spurious: A Study of a Norwegian Community in Rural Wisconsin." Ph.D. diss., University of Wisconsin, 1976.

Ibson, John. "Virgin Land or Virgin Mary? Studying the Ethnicity of White Americans." *American Quarterly* 33 (1981): 284–308.

J.D.B. "An Unaltered Swiss Colony." *Nation*, 2 August 1883, 93–94.

Jackson, John Brinkerhoff. *Discovering the Vernacular Landscape.* New Haven: Yale University Press, 1984.

Jackson, J[ohn] B[rinkerhoff] "Other-Directed Houses." In *Landscapes: Selected Writings of J. B. Jackson,* edited by Ervin H. Zube, 55–72. Amherst: University of Massachusetts Press, 1970.

Jackson, John Brinkerhoff. *A Sense of Place, A Sense of Time.* New Haven: Yale University Press, 1994.

Jakle, John. *The American Small Town: Twentieth-Century Place Images.* Hamden, CT: Archon Books, 1982.

Jakle, John. *The Tourist: Travel in Twentieth-Century North America.* Lincoln: University of Nebraska Press, 1985.

Jencks, Charles. *The Language of Post-Modern Architecture,* 4th ed. New York: Rizzoli, 1984.

Johnson, Hildegard Binder. *Order Upon the Land: The U.S. Rectangular Land Survey and the Upper Mississippi Country.* New York: Oxford University Press, 1976.

Johnson, Nuala. "Where Geography and History Meet: Heritage Tourism and the Big House in Ireland." *Annals of the Association of American Geographers* 86 (1996): 551–66.

Kallen, Horace. *Culture and Democracy in the United States: Studies in the Group Psychology of the American People.* New York: Boni and Liveright, 1924.

Kamm, Henry. "The Swiss Debunk William Tell and All That." *New York Times,* 30 March 1994.

Kammen, Michael. *The Machine That Would Go of Itself: The Constitution in American Culture.* New York: Knopf, 1986.

Kammen, Michael. *Mystic Chords of Memory: the Transformation of Tradition in American Culture.* New York: Knopf, 1991.

Kamphoefner, Walter D. *The Westfalians: From Germany to Missouri.* Princeton: Princeton University Press, 1987.

Kapchan, Deborah. "Performance." *Journal of American Folklore* 108 (1995): 479–508.

Kazal, Russell A. "Revisiting Assimilation: The Rise, Fall, and Reappraisal of a Concept in American Ethnic History." *American Historical Review* 100 (1995): 437–71.

Kearns, Gerry, and Chris Philo, eds. *Selling Places: The City as Cultural Capital, Past and Present.* Oxford: Pergamon Press, 1993.

Kellner, Douglas. "Popular Culture and the Construction of Postmodern Identities." In *Modernity and Identity*, edited by Scott Lash and Jonathan Fried-man, 141–77. Oxford: Blackwell Publishers, 1992.

Keyes, Charles F. "Towards a New Formulation of the Concept of the Ethnic Group." *Ethnicity* 3 (1976): 202–13.

Kirshenblatt-Glimblett, Barbara. "Theorizing Heritage." *Ethnomusicology* 39 (1995): 367–80.

Kirshenblatt-Gimblett, Barbara. "Authenticity and Authority in the Represen-tation of Culture: The Poetics and Politics of Tourist Production." In *Kul-turkontakt, Kulturkonflikt: Zur Erfahrung des Fremden*, edited by Konrad Kos-tlin, Ina-Maria Greverus, and Heinz Schilling, 59–69. Frankfurt am Main: Institut für Kulturanthropologie und Europaische Ethnologie, 1988.

Kirshenblatt-Gimblett, Barbara. "Mistaken Dichotomies." *Journal of American Folklore* 101 (1988): 140–55.

Kirshenblatt-Gimblett, Barbara. "Objects of Ethnography." In *Exhibiting Cul-tures: The Poetics and Politics of Museum Display*, edited by Ivan Karp and Steven D. Lavine, 386–443. Washington, D.C.: Smithsonian Institution Press, 1991.

Kirshenblatt-Gimblett, Barbara. "Studying Immigrant and Ethnic Folklore." In *Handbook of American Folklore*, edited by Richard M. Dorson, 39–47. Bloomington: Indiana University Press, 1983.

Kirshenblatt-Gimblett, Barbara, and Edward M. Bruner. "Tourism." In Bau-man, ed., *Folklore*, 300–307.

Kivisto, Peter. "Overview: Thinking about Ethnicity." In *The Ethnic Enigma: The Salience of Ethnicity for European-Origin Groups*, edited by Peter Kivisto, 11–22. Philadelphia: The Balch Institute Press, 1989.

Kivisto, Peter, and Dag Blanck, eds. *American Immigrants and their Generations: Studies and Commentaries on the Hansen Thesis after Fifty Years*. Urbana: Uni-versity of Illinois Press, 1990.

Knowles, Anne Kelly. *Calvinists Incorporated: Welsh Immigrants on Ohio's Indus-trial Frontier*. Chicago: University of Chicago Press, 1997.

Kock, Claudia, and Kaspar Marti. "Geschichte(n) und Mythen: Vertraut and dennoch Fremd: Swiss-American Architecture in New Glarus." *Fridolin*, 28 September 1995.

Kolb, John H. "Dr. Galpin at Wisconsin." *Rural Sociology* 13 (1948): 130–45.

Kolb, John H. *Emerging Rural Communities: Group Relations in Rural Society, A Review of Wisconsin Research in Action*. Madison: University of Wisconsin Press, 1959.

Kolb, John H. "Neighborhood-Family Relations in Rural Society: A Review of Trends in Dane County, Wisconsin over a 35 Year Period." Madison: Agriculture Experiment Station, College of Agriculture, University of Wis-consin, 1957.

Kolb, John H. "Rural Primary Groups." Madison: Agriculture Experiment Sta-tion, College of Agriculture, University of Wisconsin, 1921.

Kolb, John H. "Trends in Country Neighborhoods." Madison: Agriculture Ex-periment Station, College of Agriculture, University of Wisconsin, 1933.

Kolb, J. H., and Edmund de S. Brunner. "Rural Life." In *Recent Social Trends in the United States*, 497–552. New York: McGraw-Hill, 1933.

Bibliography

Kolehmainen, John I., and George W. Hill. *Haven in the Woods: The Story of the Finns in Wisconsin.* Madison: State Historical Society of Wisconsin, 1951.

Kreis, Georg. *Helvetia: Im Wandel der Zeiten. Die Geschichte einer Nationalen Repräsentationsfigur.* Zürich: Verlag der Neue Zürcher Zeitung, 1991.

Kubly, Herbert. "An American Finds America." *Common Ground* 3 (1943): 49–56.

Kubly, Herbert. "Everyone We Call 'Du.'" *Common Ground* 3 (1943): 82–85.

Kubly, Herbert. "A Village of Wilhelm Tells." In *At Large*, 39–52. Garden City, NY: Doubleday, 1964.

Kubly, Herbert. "Where is Tell?" *Holiday*, July-August 1971, 49–51.

Kubly, Herbert. *Native's Return: An American of Swiss Descent Unmasks an Enigmatic Land and People.* New York: Stein and Day, 1981.

Kugelmass, Jack. "The Rites of the Tribe: American Jewish Tourism in Poland." In *Museums and Communities: The Politics of Public Culture*, edited by Ivan Karp, Christine Mullen Kreamer, and Steven D. Lavine, 382–427. Washington, D.C.: Smithsonian Institution Press, 1992.

Lal, Barbara Ballis. "Perspectives on Ethnicity: Old Wine in New Bottles." *Ethnic and Racial Studies* 6 (1983): 154–73.

Lampard, Eric E. *The Rise of the Dairy Industry in Wisconsin: A Study of Agricultural Change, 1820–1920.* Madison: State Historical Society of Wisconsin, 1963.

"Lands and Peoples: Swiss at New-Glarus." *Our Times: A Weekly Journal of Current Events*, 21 October 1905, 123–24.

Larson, Olaf. "The Contribution of Wisconsin Rural Sociology: The Early Years." Paper presented at the Rural Sociology Conference: The Wisconsin Contribution, Current Status and Future Directions, Madison, December 1981.

Lavenda, Robert H. "Family and Corporation: Two Styles of Celebration in Central Minnesota." In Manning, ed., *The Celebration of Society*, 51–66.

Lavenda, Robert H. "Festivals and the Creation of Public Culture: Whose Voice(s)?" In *Museums and Communities: The Politics of Public Culture*, edited by Ivan Karp, Christine Mullen Kreamer, and Steven D. Lavine, 76–104. Washington, D.C.: Smithsonian Institution Press, 1992.

Lears, T. J. Jackson. *No Place of Grace: Antimodernism and the Transformation of American Culture.* New York: Pantheon, 1981.

Leary, James P. *Yodeling in Dairyland: A History of Swiss Music in Wisconsin.* Mount Horeb, WI: Wisconsin Folk Museum, 1991.

Leon, Warren, and Roy Rosenzweig, eds. *History Museums in the United States: A Critical Appraisal.* Urbana: University of Illinois Press, 1989.

Levine, Lawrence. *Highbrow/Lowbrow: The Emergence of Cultural Hierarchy in America.* Cambridge, MA.: Harvard University Press, 1988.

Lewthwaite, Gordon R. "Midwestern Swiss Migrants and Foreign Cheese." *Yearbook of the Association of Pacific Coast Geographers* 34 (1972): 41–60.

Lieberson, Stanley. "Unhyphenated Whites in the United States." *Ethnic and Racial Studies* 8 (1985): 159–80.

Lieberson, Stanley, and Mary C. Waters. *From Many Strands: Ethnic and Racial Groups in Contemporary America.* New York: Russell Sage Foundation, 1988.

Lieberson, Stanley, and Mary C. Waters. "The Ethnic Responses of Whites:

What Causes Their Instability, Simplification, and Inconsistency?" *Social Forces* 72 (1993): 421–50.

Linton, Ralph. "Nativistic Movements." *American Anthropologist* 45 (1943): 230–40.

Lipsitz, George. *Time Passages: Collective Memory and American Popular Culture.* Minneapolis: University of Minnesota Press, 1990.

Lofgren, Orvar. "Anthropologizing America." *American Ethnologist* 16 (1989): 366–74.

Lowenthal, David. "Identity, Heritage, and History." In *Commemorations: The Politics of National Identity,* edited by John Gillis, 41–60. Princeton: Princeton University Press, 1994.

Lowenthal, David. *The Past is a Foreign County.* Cambridge: Cambridge University Press, 1985.

Lowenthal, David. "Pioneer Museums." In Leon and Rosenzweig, eds., *History Museums in the United States,* 115–27.

Lowenthal, David, *Possessed by the Past: The Heritage Crusade and the Spoils of History.* New York: The Free Press, 1996.

Luchsinger, John. "The History of a Great Industry." *Proceedings of the State Historical Society of Wisconsin* 19 (1899): 226–30.

Luchsinger, John. "The Planting of the Swiss Colony at New Glarus, Wis." *Wisconsin Historical Collections* 12 (1892): 335–82.

Luchsinger, John. "Report." In *Transactions of the Wisconsin State Agricultural Society.* Madison, 1882–1883.

Luchsinger, John. "The Swiss Colony of New Glarus." *Wisconsin Historical Collections* 8 (1879): 1–29.

Luebke, Frederick C. *Bonds of Loyalty: German Americans and World War I.* De Kalb, IL: Northern Illinois University Press, 1974.

Luod, Karl. "Schweizerischer als die Schweiz." In *Schweizer in Amerika: Karrieren und Misserfolge in der Neuen Welt,* 79–88. Olten, Switzerland: Walter Verlag AG, 1979.

Lutz, Catherine, and Jane L. Collins. *Reading National Geographic.* Chicago: University of Chicago Press, 1993.

MacCannell, Dean. "Reconstructed Ethnicity: Tourism and Cultural Identity in Third World Communities." In *Empty Meeting Grounds: The Tourist Papers,* 158–71. London: Routledge, 1992.

MacCannell, Dean. *The Tourist: A New Theory of the Leisure Class,* 2nd ed. New York: Schocken Books, 1989.

Manning, Frank, ed. *The Celebration of Society: Perspectives on Contemporary Cultural Performance.* Bowling Green, OH: Bowling Green University Popular Press, 1983.

March, Richard. "Statewide Folk Arts Fieldwork in Preparation for the Wisconsin Folklife Festival." Madison: Wisconsin Arts Board, 1994.

Marling, Karal Ann. *George Washington Slept Here: Colonial Revivals and American Culture, 1876–1986.* Cambridge, MA: Harvard University Press, 1988.

Marsh, John. "The Rocky and Selkirk Mountains and the Swiss Connection, 1885–1914." *Annals of Tourism Research* 12 (1985): 417–33.

Marti, Kaspar. "Von Kultur und Kulturen, von Kunst und Künsten." In *Art*

Glarus—Switzerland/Kunst Glarus—Schweiz, edited by Kaspar Marti, Madeleine Schuppli, and Susann Wintsch, 28–31. Glarus, Switzerland: Glarner Kunstverein, 1995.

Matthews, Fred. "Cultural Pluralism in Context: External History, Philosophical Premise and Theories of Ethnicity in Modern America." *Journal of Ethnic Studies* 12 (1984): 63–80.

Matthews, Fred. "Paradigm Changes in Interpretations of Ethnicity, 1930–80: From Process to Structure." In Kivisto and Blanck, eds., *American Immigrants,* 167–90.

Matthews, Fred. *The Quest for Community: Robert E. Park and the Chicago School of Sociology.* Montreal: McGill-Queens University Press, 1978.

May, Jon. "In Search of Authenticity off and *on* the Beaten Track." *Environment and Planning D: Society and Space* 14 (1996): 709–36.

McClymer, John. "The Federal Government and the Americanization Movement, 1915–1924." *Prologue: The Journal of the National Archives* 10 (1978): 22–41.

McCullough, Lawrence. "The Role of Language, Music, and Dance in the Revival of Irish Music in Chicago." *Ethnicity* 7 (1980): 436–44.

McGinn, Frank. "Ersatz Place." *Canadian Heritage,* February-March 1986, 25–29.

McKay, Ian. *The Quest of the Folk: Antimodernism and Cultural Selection in Twentieth-Century Nova Scotia.* Montreal and Kingston: McGill-Queen's University Press, 1994.

McKay, Ian. "Tartanism Triumphant: The Construction of Scottishness in Nova Scotia, 1933–1954." *Acadiensis* 21 (1992): 5–47.

McNutt, James. "Folk Festivals and the Semiotics of Tourism in Texas." *Kentucky Folklore Record* 32 (1986): 118–29.

McPhee, John. *La Place de la Concorde Suisse.* New York: Farrar, Straus, and Giroux, 1984.

McRae, Kenneth D. *Conflict and Comprise in Multilingual Societies: Switzerland.* Waterloo, ON: Wilfred Laurier University Press, 1983.

McRae, Kenneth D. *Switzerland: Example of Cultural Coexistence.* Toronto: Canadian Institute of International Affairs, 1964.

McWilliams, Carey. *Louis Adamic and Shadow-America.* Los Angeles: A. Whipple, 1935.

Meagher, Timothy J. "Why Should We Care for a Little Tro or Walk in the Mud? St. Patrick, Columbus Day Parades in Worcester, Massachusetts, 1845–1915." *New England Quarterly* 58 (1985): 5–10.

Meier, Heinz. *The Swiss American Historical Society.* Norfolk, VA: Donning, 1977.

Menolfi, Ernest, and Leo Schelbert. "The Wisconsin Swiss: A Portrait." *Clarion* 16 (1991): 57–63.

Merk, Frederick. *Economic History of Wisconsin During the Civil War Decade.* Madison: State Historical Society of Wisconsin, 1916.

Metraux, Guy Serge. "Social and Cultural Aspects of Swiss Immigration into the United States in the Nineteenth Century." Ph.D. diss., Yale University, 1949.

Mettler, Heinrich, and Heinz Lippuner. *Friedrich Schiller: Wilhelm Tell: Das Drama der Freiheit.* Paderborn, Germany: Fernand Schöningh, 1989.

Meyer-Arendt, Klaus J. "The Grand Isle, Louisiana Resort Cycle." *Annals of Tourism Research* 12 (1985): 449–65.

Meyrowitz, Joshua. *No Sense of Place: The Impact of Electronic Media on Social Behavior.* New York: Oxford University Press, 1985.

Millard, William E. "The Sale of Culture." Ph.D. diss., University of Minnesota, 1969.

Mitchell, Don. "Heritage, Landscape, and the Production of Community: Consensus History and its Alternatives in Johnstown, Pennsylvania." *Pennsylvania History* 59 (1992): 198–226.

Mitchell, Don. *The Lie of the Land: Migrant Workers and the California Landscape.* Minneapolis: University of Minnesota Press, 1996.

Montalto, Nicholas V. *A History of the Intercultural Educational Movement, 1924–1941.* New York: Garland, 1982.

Montgomery, Charles. "The Making of Spanish Heritage in the American Southwest, 1890–1940." Ph.D. diss., Cornell University, 1995.

Monthey, L. G. "Evaluation of Robert Schneider Land, with Respect to Developing a Swiss-Theme Village near New Glarus, Wis." Madison: University of Wisconsin-Extension, n.d.

Mooney-Melvin, Patricia. "Harnessing the Romance of the Past: Preservation, Tourism, and History." *The Public Historian* 13 (1991): 35–48.

Morawska, Ewa. "In Defense of the Assimilation Model." *Journal of American Ethnic History* 14 (1994): 76–87.

Morawska, Ewa. "The Sociology and Historiography of Immigration." In *Immigration Reconsidered: History, Sociology, and Politics,* edited by Virginia Yans-McLaughlin, 187–238. New York: Oxford University Press, 1990.

Moscardo, Gianna M., and Phillip L. Pearce. "Historic Theme Parks: An Australian Experience in Authenticity." *Annals of Tourism Research* 13 (1986): 467–79.

Mosse, George L. *Fallen Soldiers: Reshaping the Memory of the World Wars.* New York: Oxford University Press, 1990.

Mosse, George L. *The Nationalization of the Masses: Political Symbolism and Mass Movements in Germany from the Napoleonic Wars through the Third Reich.* New York: H. Fertig, 1975.

Myerhoff, Barbara. *Number Our Days.* New York: Simon and Schuster, 1980.

Myrdal, Gunnar. "The Case Against Romantic Ethnicity." *The Center,* July/August 1974, 26–30.

Nadel-Klein, Jane. "Reweaving the Fringe: Localism, Tradition, and Representation in British Ethnography." *American Ethnologist* 18 (1991): 500–517.

"Nation Watches U.W. Study of Political, Social, and Economic Folkway Institutions in Wisconsin." *University of Wisconsin Press Bulletin,* 4 September 1946.

Neff, Deborah, and Phillip B. Zarrilli. *Wilhelm Tell: In America's "Little Switzerland," New Glarus Wisconsin.* New Glarus, WI: Wilhelm Tell Guild, 1987.

Nicholas, William H. "Deep in the Heart of 'Swissconsin.'" *National Geographic,* June 1947, 781–800.

Nora, Pierre. "Between Memory and History: Les Lieux de Memoire." *Representations* 26 (1989): 7–25.

Norkunas, Martha. *The Politics of Public Memory: Tourism, History, and Ethnicity in Monterey, California.* Albany: State University of New York Press, 1993.

Novak, Michael. *The Rise of the Unmeltable Ethnics.* New York: Macmillan, 1972.

Oakes, Timothy. "Place and the Paradox of Modernity." *Annals of the Association of American Geographers* 87 (1997): 509–31.

Odell, Emery A. *Eighty Years of Swiss Cheese in Green County.* Monroe, WI: Monroe Evening Times, 1949.

Odell, Emery A. *Swiss Cheese Industry in Green County Wisconsin.* Monroe, WI: Monroe Evening Times, 1936.

Old World Wisconsin: An Outdoor, Ethnic Museum. Madison: Department of Landscape Architecture, University of Wisconsin, 1973.

Omni, Michael, and Howard Winant. *Racial Formation in the United States from the 1960s to the 1990s,* 2nd ed. New York: Routledge, 1994.

One Hundred Years of the Swiss Evangelical and Reformed Church. New Glarus, WI: n.p., 1950.

Orvell, Miles. *The Real Thing: Imitation and Authenticity in American Culture, 1880–1940.* Chapel Hill: University of North Carolina Press, 1989.

Out of Many . . . One: Old World Wisconsin. Madison: State Historical Society of Wisconsin, 1976.

Ostergren, Robert C. *A Community Transplanted: The Trans-Atlantic Experience of a Swedish Immigrant Settlement in the Upper Middle West, 1835–1915.* Madison: University of Wisconsin Press, 1988.

Padgett, Deborah. "Symbolic Ethnicity and Patterns of Ethnic Identity Assertion in American-Born Serbs." *Ethnic Groups* 3 (1980): 55–77.

Park, Robert E., and Herbert Miller. *Old World Traits Transplanted.* New York: Harper and Brothers, 1921.

Parsons, Talcott. "Some Theoretical Considerations on the Nature and Trends of Change of Ethnicity." In *Ethnicity: Theory and Experience,* edited by Nathan Glazer and Daniel P. Moynihan, 53–83. Cambridge: Harvard University Press, 1975.

Passage, Charles. *Friedrich von Schiller's William Tell.* New York: Frederick Unger, 1962.

Patterson, G. James. "A Critique of 'The New Ethnicity.'" *American Anthropologist* 81 (1979): 103–105.

Pederson, Jane Marie. *Between Memory and Reality: Family and Community in Rural Wisconsin, 1870–1970.* Madison: University of Wisconsin Press, 1992.

Persons, Stow. *Ethnic Studies at Chicago, 1905–45.* Urbana: University of Illinois Press, 1987.

Peter-Kubli, Susanne, ed. *Die Welt ist hier Weit.* Sonderdruck des Jahrbuchs des Historischen Vereins des Kantons Glarus, vol 75. Glarus, Switzerland: Historischen Vereins des Kantons Glarus, 1995.

Pirkova-Jakobson, Svatava. "Harvest Festivals among Czechs and Slovaks in America." *Journal of American Folklore* 69 (1956): 266–80.

Pitchford, Susan R. "Ethnic Tourism and Nationalism in Wales." *Annals of Tourism Research* 22 (1994): 35–52.

Portes, Alejandro, and Ruben G. Rumbaut. *Immigrant America: A Portrait.* Berkeley: University of California Press, 1990.

Posern-Zielinski, Aleksander. "Ethnicity, Ethnic Culture and Folk Tradition in the American Society." *Ethnologia Polona* 4 (1978): 105–25.

Raitz, Karl B. "Ethnic Maps of North America." *Geographical Review* 68 (1978): 335–50.

Redfoot, Donald. "Tourist Authenticity, Tourist Angst, and Modern Reality." *Qualitative Sociology* 7 (1984): 291–309.

Reitz, Jeffrey G. *The Survival of Ethnic Groups.* Toronto: McGraw-Hill, 1980.

Relph, Edward C. *The Modern Urban Landscape.* Baltimore: Johns Hopkins University Press, 1987.

Relph, Edward C. *Place and Placelessness.* London: Pion, 1976.

Rippley, LaVern J. "Ameliorized Americanization: The Effect of World War I on German Americans in the 1920s." In *America and the Germans: An Assessment of a 300-Year History,* edited by Frank Trommler and Joseph McVeigh, 217–31. Philadelphia: University of Pennsylvania Press, 1985.

Rippley, La Vern J. *The Immigrant Experience in Wisconsin.* Boston: Twayne, 1985.

Rockwell, Ethel. *A Century of Progress Cavalcade in Wisconsin: A Pageant Drama Based on Research in Wisconsin History Through the Century.* Madison: Wisconsin State Centennial Committee, 1948.

Rockwell, Ethel. "Historical Pageantry: A Treatise and a Bibliography." *State Historical Society of Wisconsin Bulletin of Information* 84 (1916): 5–19.

Rodriguez, Richard. "An American Writer." In Sollors, ed., *The Invention of Ethnicity,* 3–13.

Rodriguez, Sylvia. "Ethnic Reconstruction in Contemporary Taos." *Journal of the Southwest* 32 (1990): 541–55.

Roediger, David. *The Wages of Whiteness: Race and the Making of the American Working Class.* London: Verso, 1991.

Rosenbrook, Donald. "New Glarus, Wisconsin Comprehensive Planning Program." Platteville, WI: Southwestern Wisconsin Regional Planning Commission, 1977.

Rosenwaike, Ira. "Ancestry in the United States Census, 1980–1990." *Social Science Research* 22 (1993): 383–90.

Rosenzweig, Roy. *Eight Hours for What We Will: Workers and Leisure in an Industrial City, 1870–1920.* New York: Cambridge University Press, 1983.

Roucek, Joseph, Caroline F. Ware, and M. W. Royse. "Summary to the Discussion of Cultural Groups." In *The Cultural Approach to History,* edited by Caroline F. Ware, 86–92. New York: Columbia University Press, 1940.

Royce, Anya Peterson. *Ethnic Identity: Strategies of Diversity.* Bloomington: Indiana University Press, 1982.

Ryder, Frank G. "Schiller's *Tell* and the Cause of Freedom." *German Quarterly* 48 (1975): 487–504.

Sack, Robert David. *Human Territoriality: Its Theory and History.* Cambridge: Cambridge University Press, 1986.

Sack, Robert David. *Place, Modernity, and the Consumer's World: A Relational Framework for Geographical Analysis.* Baltimore: Johns Hopkins University Press, 1992.

Salis, J. R. von. "Ursprung, Gestalt und Wirkung des schweizerischen Mythos von Tell." In *Tell: Werden und Wandern eines Mythos,* edited by Lilly Stunzi, 9–88. Bern: Hallway Verlag, 1973.

Sanford, Mariellen R. "Tourism in Harlem: Between Negative Sighteeing and Gentrification." *Journal of American Culture* 10 (1987): 99–106.

Sarna, Jonathan. "From Immigrants to Ethnics: Toward a New Theory of 'Ethnicization.'" *Ethnicity* 5 (1978): 370–78.

Savage, Kirk. "The Politics of Memory: Black Emancipation and the Civil War Monument." In *Commemorations: The Politics of Public Memory,* edited by John Gillis, 127–49. Princeton: Princeton University Press, 1994.

Schafer, Joseph. "Editorial Comment: Peopling of the Middle West." *Wisconsin Magazine of History* 21 (1937): 94–101.

Schafer, Joseph. *Wisconsin Domesday Book, Town Studies.* Vol. 1. Madison: State Historical Society of Wisconsin, 1924.

Schafer, Joseph. "The Yankee and the Teuton in Wisconsin." *Wisconsin Magazine of History* 6–7 (1922–1923): 125–45, 261–79, 386–402.

Schechner, Richard. *Between Theater and Anthropology.* Philadelphia: University of Pennsylvania Press, 1985.

Schelbert, Leo. "Der Wilhelm-Tell Mythos in der Tradition der Vereinigten Staaten von Nordamerika." In *Tell: Werden und Wandern eines Mythos,* edited by Lilly Stunzi, 313–30. Bern: Hallway Verlag, 1973.

Schelbert, Leo, ed. *New Glarus 1845–1970: The Making of a Swiss American Town.* Glarus, Switzerland: Kommissionsverlag Tschudi, 1970.

Schiesser, Elda, and Linda Schiesser. *The Swiss Endure Year by Year: A Chronological History.* New Glarus, WI: n.p., 1994.

Schindler, John. "Historical Pageant Given at the Ninetieth Anniversary Celebration of the Settlement of New Glarus, by the Swiss Colonists." In *Program to the 90th Anniversary Pageant, Presented at New Glarus Wisconsin.* New Glarus, WI: n.p., 1935.

Schindler, John. "The Old Lead Trail." Monroe: Monroe Evening Times, 1934.

Schink, Donald. "Economic Impact of the Proposed Swiss Village, New Glarus." Madison: Recreation Resources Center, University of Wisconsin-Extension, 1975.

Schlapp, Hermann. "New Glarus: Amerikanische Impressionen eines Greenhorns." *Basler Nachrichten,* 14 August 1967, 3–5.

Schlesinger, Arthur, Jr. *The Disuniting of America: Reflections on a Multicultural Society.* New York: Norton, 1991.

Schmid, Carol L. *Conflict and Consensus in Switzerland.* Berkeley: University of California Press, 1981.

Schöpflin, George. "Nationalism and Ethnic Minorities in Post-Communist Europe." In *Europe's New Nationalism: States and Minorities in Conflict,* edited by Richard Caplan and John Feffer, 151–68. New York: Oxford University Press, 1996.

Schultz, April R. "'The Pride of the Race Had Been Touched': The 1925 Norse-American Immigration Centennial and Ethnic Identity." *Journal of American History* 77 (1991): 1265–95.

Schultz, April R. *Ethnicity on Parade: Inventing the Norwegian American through Celebration.* Amherst: University of Massachusetts Press, 1994.

Schultz, April R. "Searching for a Unified America." *American Quarterly* 45 (1993): 639–48.

Schwartz, Barry. "The Social Context of Commemoration: A Study of Collective Memory." *Social Forces* 61 (1982): 374–402.

Sharpe, Leslie. "Wilhelm Tell." In *Friedrich Schiller: Drama, Thought, and Politics,* 293–309. Cambridge: Cambridge University Press, 1991.

Shaw, Gareth. "Culture and Tourism: The Economics of Nostalgia." *World Futures* 33 (1992): 199–212.

Shils, Edward. "Primordial, Personal, Sacred and Civil Ties." *British Journal of Sociology* 8 (1957): 130–45.

Sinke, Suzanne. "Tulips are Blooming in Holland, Michigan: Analysis of a Dutch-American Festival." In *Immigration and Ethnicity: American Society— "Melting Pot" or "Salad Bowl"?,* edited by Michael D'Innocenzo and Joseph P. Sirefman, 3–14. Westport, CT: Greenwood, 1992.

Slocum, Walter. "Ethnic Stocks as Culture Types in Rural Wisconsin." Ph.D. diss., University of Wisconsin, 1940.

Slotkin, Richard. *Gunfighter Nation: The Myth of the Frontier in Twentieth-Century America.* New York: HarperCollins, 1992.

Smith, Anthony D. *The Ethnic Revival.* Cambridge: Cambridge University Press, 1981.

Smith, Rockwell. "Church Affiliation as Social Differentiator in Rural Wisconsin." Ph.D. diss., University of Wisconsin, 1942.

Sollors, Werner. *Beyond Ethnicity: Consent and Descent in American Culture.* New York: Oxford University Press, 1986.

Sollors, Werner, ed. *The Invention of Ethnicity.* New York: Oxford University Press, 1989.

Squire, Shelagh J. "The Cultural Values of Literary Tourism." *Annals of Tourism Research* 21 (1993): 103–20.

Steinberg, Stephen. *The Ethnic Myth: Race, Ethnicity, and Class in America.* New York: Atheneum, 1981.

Steiner, Michael, and Clarence Mondale. *Region and Regionalism in the United States: A Sourcebook for the Humanities and Social Sciences.* New York: Garland, 1988.

Stephenson, John B. "Escape to the Periphery: Commodifying Place in Rural Appalachia." *Appalachian Journal* 11 (1984): 187–200.

Stern, Stephen. "Ethnic Folklore and the Folklore of Ethnicity." *Western Folklore* 36 (1977): 7–32.

Stevens, John D. "Suppression of Expression in Wisconsin During World War I." Ph.D. diss., University of Wisconsin, 1967.

Stevens, John D. "When Sedition Laws were Enforced: Wisconsin in World War I." *Transactions of the Wisconsin Academy of Sciences, Arts, and Letters* 58 (1970): 39–60.

Stewart, Charles D. "Prussianizing Wisconsin." *Atlantic Monthly,* January 1919, 99–105.

Stoeltje, Beverly. "Festival." In Bauman, ed., *Folklore,* 261–71.

Stoeltje, Beverly J. "Festival in America." In *Handbook of American Folklore,* edited by Richard M. Dorson, 239–46. Bloomington: Indiana University Press, 1983.

Streiff-Frederickson, Marion R. "A History of the Swiss Historical Village Museum." In *Amerikas Little Switzerland Errinert Sich*, edited by Hans Rhyner, 273–82. Glarus, Switzerland: Schweizerischer Verein der Freunde von New Glarus, 1996.

Streissguth, Wilhelm. "New Glarus in 1850." *Wisconsin Magazine of History* 16 (1935 [1851]): 328–344.

Stüssi, Heinrich. "Auswanderung." In *Glarus und die Schweiz: Streiflichter auf wechselseitige Beziehungen*, edited by Jürg Davatz, 146–54. Glarus, Switzerland: Verlag Baeschlin, 1991.

"Swiss Defense of Freedom Symbolized in 'Wilhelm Tell.'" *Christian Science Monitor*, 3 May 1941.

Swiss American Historical Society. *Prominent Americans of Swiss Origin*. New York: James T. White, 1932.

Takaki, Ronald. *A Different Mirror: A History of Multicultural America*. Boston: Little, Brown, 1993.

Tetzlaff, Walter R., Brian J. Thorsen, and Gordon Mickelson. "Demand and Economic Analysis, Swiss Village, New Glarus." Madison: Recreation Resources Center, University of Wisconsin Extension, 1973.

Theiler, Mrs. Arthur J. "New Glarus, Wisconsin: Transplanted and Flourishing." *The Swiss Record: Yearbook of the Swiss-American Historical Society* 1 (1949): 18–24.

Thernstrom, Stephan. "Is America's Ethnic Revival a Fad Like Jogging?" *U.S. News and World Report*, 17 November 1980, 85.

Thernstrom, Stephan, Ann Orlov, and Oscar Handlin, eds. *The Harvard Encyclopedia of American Ethnic Groups*. Cambridge, MA: Harvard University Press, 1980.

Thomas, William I., and Florian Znaniecki. *The Polish Peasant in Europe and America*, 5 vols. Chicago: University of Chicago Press, 1918 (vols 1 and 2); Boston: Badger Press, 1919 and 1920 (vols. 3–5).

Tisdale, Sallie. "Never Let the Locals See Your Map: Why Most Travel Writers Should Stay Home." *Harper's*, September 1995, 66–74.

Tishler, William H. "Built From Tradition: Wisconsin's Rural Ethnic Folk Architecture." *Wisconsin Academy Review* 30 (1984): 14–18.

Tishler, William H. *Final Environmental Impact Statement for Old World Wisconsin, Bicentennial Park*. Madison: State Historical Society of Wisconsin, 1973.

Tishler, William H. *Old World Wisconsin: A Study Prepared for the State Historical Society of Wisconsin*. Madison: Department of Landscape Architecture, University of Wisconsin, 1972.

Trepanier, Cecyle. "The Cajunization of French Louisiana: Forging a Regional Identity." *The Geographical Journal* 157 (1991): 161–71.

Trewartha, Glen. "The Green County, Wis., Foreign Cheese Industry." *Economic Geography* 2 (1926): 292–308.

Trillin, Calvin. "U.S. Journal: New Glarus, Wis.: Swissness." *New Yorker*, 20 January 1975, 51–57.

Trilling, Lionel. *Sincerity and Authenticity*. Cambridge MA: Harvard University Press, 1972.

Troxler, Josef. *Wilhelm Tell*. Chapelle-sur-Moudon, Switzerland: Verlag Ketty and Alexandre, 1985.

311

Bibliography

Tschudy, J. J. "Additional Notes." *Wisconsin Historical Collections* 8 (1879): 30–35.

Tschudy, Millard. *New Glarus, Wisconsin: Mirror of Switzerland*, 7th ed. New Glarus, WI: n.p., 1995 [1965].

Tuan, Yi-Fu. *Cosmos and Hearth: A Cosmopolite's Viewpoint.* Minneapolis: University of Minnesota Press, 1996.

Tuan, Yi-Fu. *Segmented Worlds and Self: Group Life and Individual Consciousness.* Minneapolis: University of Minnesota Press, 1982.

Tuan, Yi-Fu. *Space and Place: The Perspective of Experience.* Minneapolis: University of Minnesota Press, 1977.

Turner, Frederick Jackson. "Dominant Forces in American Life." In *The Frontier in American History*, 222–42. Tucson: University of Arizona Press, 1992 [1897].

Turner, Frederick Jackson. "The Significance of the Frontier in American Life." In *The Turner Thesis: Concerning the Role of the Frontier in American History*, edited by George Rogers Taylor, 3–27. Lexington, MA: D.C. Heath, 1972 [1893].

Turner, Victor. *The Ritual Process: Structure and Anti-Structure.* Ithaca: Cornell University Press, 1969.

Turner, Victor, ed. *Celebrations: Studies in Festivity and Ritual.* Washington, D.C.: Smithsonian Institution Press, 1982.

Turner, Victor. "Are There Universals of Performance, Myth, Ritual, and Drama?" In *By Means of Performance: Intercultural Studies of Theatre and Ritual*, edited by Richard Schechner and Willa Appel, 8–18. New York: Cambridge University Press, 1990.

Urry, John. *Consuming Places.* London: Routledge, 1995.

Urry, John. *The Tourist Gaze: Leisure and Travel in Contemporary Societies.* London: Sage, 1990.

Utz, Peter. *Die ausgehöhlte Gasse: Stationen der Wirkungsgeschichte von Schillers "Wilhelm Tell."* Königsten: Forum Academicum, 1984.

van den Berghe, Pierre. *The Quest for the Other: Ethnic Tourism in San Cristóbal, Mexico.* Seattle: University of Washington Press, 1994.

van den Berghe, Pierre. "Tourism and the Ethnic Division of Labor." *Annals of Tourism Research* 19 (1992): 234–49.

van den Berghe, Pierre, and Charles Keyes. "Tourism and Re-Created Ethnicity." *Annals of Tourism Research* 11 (1984): 343–52.

Van Esterick, Penny. "Celebrating Ethnicity: Ethnic Flavor in an Urban Festival." *Ethnic Groups* 4 (1982): 207–28.

Vecoli, Rudolph. "Primo Maggio in the United States: An Invented Tradition of the Italian Anarchists." In *May Day Celebrations*, edited by Andrea Pannacione, 55–88. Venice: Marsilio Editori, 1988.

Vecoli, Rudolph. "Return to the Melting Pot: Ethnicity in the United States in the Eighties." *Journal of American Ethnic History* 5 (1985): 7–20.

Vetter, Rob. "New Glarus, Wisconsin Comprehensive Planning Program." Platteville, WI: Southwest Wisconsin Regional Planning Commission, 1995.

von Grueningen, John Paul. *The Swiss in the United States.* Madison: Swiss-American Historical Society, 1940.

Waldrop, Judith. "Happy Kwanzaa: African American Celebration of Culture and Heritage." *American Demographics*, December 1994, 4.

Wallace, Anthony F. C. "Revitalization Movements." *American Anthropologist* 58 (1956): 264–81.

Wallace, Mike. *Mickey Mouse History and Other Essays on American Memory*. Philadelphia: Temple University Press, 1996.

Wallar, Randall. *Settlement*, 3 vols, edited by Barbara Wyatt. Vol. 1, *Cultural Resources Management in Wisconsin*. Madison: State Historical Society of Wisconsin, 1986.

Walzer, Michael. "Pluralism: A Political Perspective." In *What it Means to Be an American: Essays on the American Experience*, 53–80. New York: Marsilio, 1992.

Ward, David. *Poverty, Ethnicity, and the American City, 1840–1925: Changing Conceptions of the Slum and the Ghetto*. Cambridge: Cambridge University Press, 1989.

Ware, Caroline F. "Cultural Groups in the United States." In *The Cultural Approach to History*, edited by Caroline F. Ware, 62–73. New York: Columbia University Press, 1940.

Waters, Mary C. "The Construction of a Symbolic Ethnicity: Suburban White Ethnics in the 1980s." In *Immigration and Ethnicity: American Society—"Melting Pot" or "Salad Bowl"?*, edited by Michael D'Innocenzo and Josef P. Sirefman, 75–90. Westport, CT: Greenwood, 1992.

Waters, Mary C. *Ethnic Options: Choosing Identities in America*. Berkeley: University of California Press, 1990.

Watson, G. Llewellyn, and Joseph P. Kopachevsky. "Interpretations of Tourism as Commodity." *Annals of Tourism Research* 21 (1994): 643–60.

Weiss, Richard. "Ethnicity and Reform: Minorities and the Ambience of the Depression Years." *Journal of American History* 66 (1979): 566–85.

Weiss, Richard. *Häuser und Landscaften der Schweiz*. Zürich: Eugen Rentsch Verlag, 1959.

Westergaard, Chris T. "My Days in the Schützen Game." *American Single Shot Rifle News* 41 (1987 [1951]): 1–5, 19.

Whisnant, David E. *All That is Native and Fine: The Politics of Culture in an American Region*. Chapel Hill: The University of North Carolina Press, 1983.

Whittet, Lawrence. "Frederick Lionel Holmes, 1883–1946." *Wisconsin Magazine of History* 30 (1946): 184–85.

Wileden, Arthur F. "Early History of the Department of Rural Sociology." Madison: Department of Rural Sociology, University of Wisconsin, 1979.

Williams, Raymond. *Keywords: A Vocabulary of Culture and Society*. New York: Oxford University Press, 1983.

Williams, Raymond. *Marxism and Literature*. New York: Oxford University Press, 1977.

Wilson, Chris. *The Myth of Santa Fe: Creating a Regional Tradition*. Albuquerque: University of New Mexico Press, 1997.

Wilson, William A. "Herder, Folklore, and Romantic Nationalism." In *Folk Groups and Folklore Genres: A Reader*, edited by Elliot Oring, 23–44. Logan, UT: Utah State University Press, 1989.

Winteler, Jakob. *Geschichte des Landes Glarus: von 1638 bis zur Gegenwart*, Band 2. Glarus, Switzerland: Kommissionsverlag E. Baeschlin, 1954.

Wirth, Louis. "The Problems of Minority Groups." In *The Science of Man in the*

World Crisis, edited by Ralph Linton, 347–72. New York: Columbia University Press, 1944.

Wisconsin State Division of Tourism. "Official Calendar of Events." Madison, 1993.

"Wisconsin's Changing Population." Bulletin of the University of Wisconsin, Serial no. 2642, Science Inquiry Publication 9, October 1942.

Wisconsin's Ethnic Settlement Trail, a Travel Brochure. Sheboyagn Falls, WI: WEST, 1995.

Wolfe, Alan. "The Return of the Melting Pot: What the New Immigrants can Learn from the Old." *New Republic,* 31 December 1990, 27–34.

Wood, Denis. *The Power of Maps.* New York: Guilford, 1992.

Wood, Joseph S. "Nothing Should Stand for Something that Never Existed." *Places* 2 (1985): 81–87.

Woods, Thomas A. "Filiopietism at Historic Sites and Contemporary Alternatives." Paper presented at the annual meeting of the American Association of State and Local History, Miami, 22 September 1992.

Wunderlin, Dominik. *Colorful Clothes, Natural Ice, Sap Sago, and Blue Gold: Exhibit for the 1995 New Glarus Sesquicentennial.* Translated by Catherine De Capitani-Dolf. Basel: Swiss Museum of Folk Arts and Traditions, 1995.

Yancey, William L., Eugene P. Erickson, and Richard N. Juliani. "Emergent Ethnicity: A Review and Reformulation." *American Sociological Review* 41 (1976): 391–402.

Young, James. *The Texture of Meaning: Holocaust Memorials and Meaning.* New Haven: Yale University Press, 1993.

Zarrilli, Phillip, and Deborah Neff. "Performance in 'America's Little Switzerland': New Glarus, Wisconsin." *The Drama Review* 26 (1982): 111–24.

Zelinsky, Wilbur. "America in Flux." In *The Cultural Geography of the United States,* 2nd ed., 143–85. Englewood Cliffs, NJ.: Prentice-Hall, 1992.

Zelinsky, Wilbur. *Nation into State: The Shifting Symbolic Foundations of American Nationalism.* Chapel Hill: University of North Carolina Press, 1988.

Zimmerman, Conrad. "Town of New Glarus." In *History of Green County, Wisconsin,* 1023–45. Springfield, IL: Union Publishing, 1884.

Zweifel, Jakob. "Die Entstehung der 'Hall of History.'" In *Amerikas Little Switzerland Erinnert Sich,* edited by Hans Rhyner, 281–88. Glarus, Switzerland: Schweizerischer Verein der Freunde von New Glarus, 1996.

Index

Abrahams, Roger, 23, 33
Adamic, Louis, 169–75
Addams, Jane, 150
African Americans, 20, 233–34, 240n.4
agriculture, 28–29, 56, 64–67, 165. *See also* farmers
Alluring Wisconsin (Holmes), 162
alpenhorns, 7, 43, 44, 47, 57, 91
Altdorf (Switzerland), 265n.30
American Revolution, 129–30, 132
"America's Little Switzerland." *See* New Glarus (Wisconsin)
antimodernism: ethnicity as surrogate for, 17–18, 30, 89, 185, 233; Holmes on, 167–68; New Glarus as symbol of, 64, 72, 133, 139–43; and non-European ethnic groups, 234; and progress, 55, 64–65, 89; and *Wilhelm Tell* performance, 123, 124, 139. *See also* authenticity
Appadurai, Arjun, 185
Appenzell (Swiss canton), 201
architecture (in New Glarus): of churches, 62, 87, 186; of Edwin Barlow's chalet, 117, 182, 189–91, 194, 195, 201, 208; of Hall of History, 205, 207–9, 224, 226, 230; non-Swiss, 86, 186–87, 196; of pioneers, 48, 59, 72, 191, 205, 208; postmodernism in, 201–2; "Swiss," 173, 183, 186, 188, 189–97, 219–20. *See also* ethnic place; landscape; "Swisscapes"; *specific buildings*
architecture (New Glarus): authenticity of, 19–20, 209, 214–16
Arn, Doris, 268n.86
Ashland (Wisconsin), 79

Asian Americans, 20, 234
assimilation, 240n.6; attempts at, 85, 116; as basis of democracy, 75; Frederick Jackson Turner on, 13, 16, 27, 28; as inevitable, 20; praise for, 83; reconsidering of, 123, 173, 176; vs. rooted cosmopolitanism, 159, 161, 167; as Wisconsin Loyalty Legion's objective, 80. *See also* cosmopolitanism
Atherton, Lewis, 250n.53, 251n.57
Augustana Historical Society, 99
Austria: as ancient Swiss enemy, 30, 43, 45, 112, 141; architecture in, 208; immigrants from, 222
Auswanderungsverein (emigration society), 27–28
authenticity: architectural, 19–20, 209, 214–16; contestation of, 213–20; emergent quality of, 199; and ethnic place, 19–20, 26, 94, 139, 199, 201–4, 230; and official culture, 185; as political economy of taste, 185; and power, 23, 185, 216–20, 231; as staged, 23, 143–44, 185; of theme parks, 214–15; tourism as quest for, 23, 134, 139–43, 173, 175, 185, 218–19; of *Wilhelm Tell's* cast, 120, 122, 125. *See also* heritage

Badger Saints and Sinners (Holmes), 162
Baltimore (Maryland), 53
"banality of difference," 168–69
Bank of New Glarus, 57
Barber, Benjamin, 52
Barlow, Edwin, 114–20, 138; chalet of, 117, 182, 189–91, 194–96, 201, 208; medal-

315

Index